Deep Reinforcement Learning

Aske Plaat

Deep Reinforcement Learning

 Springer

Aske Plaat
LIACS
Leiden University
Leiden, The Netherlands

ISBN 978-981-19-0637-4   ISBN 978-981-19-0638-1   (eBook)
https://doi.org/10.1007/978-981-19-0638-1

This Springer imprint is published by the registered company Springer Nature Singapore Pte Ltd.
The registered company address is: 152 Beach Road, #21-01/04 Gateway East, Singapore 189721, Singapore

Preface

Deep reinforcement learning has gathered much attention recently. Impressive results were achieved in activities as diverse as autonomous driving, game playing, molecular recombination, and robotics. In all these fields, computer programs have learned to solve difficult problems. They have learned to fly model helicopters and perform aerobatic manoeuvers such as loops and rolls. In some applications they have even become better than the best humans, such as in Atari, Go, poker, and StarCraft.

The way in which deep reinforcement learning explores complex environments reminds us how children learn, by playfully trying out things, getting feedback, and trying again. The computer seems to truly possess aspects of human learning; deep reinforcement learning touches the dream of artificial intelligence.

The successes in research have not gone unnoticed by educators, and universities have started to offer courses on the subject. The aim of this book is to provide a comprehensive overview of the field of deep reinforcement learning. The book is written for graduate students of artificial intelligence, and for researchers and practitioners who wish to better understand deep reinforcement learning methods and their challenges. We assume an undergraduate-level of understanding of computer science and artificial intelligence; the programming language of this book is Python.

We describe the foundations, the algorithms, and the applications of deep reinforcement learning. We cover the established model-free and model-based methods that form the basis of the field. Developments go quickly, and we also cover more advanced topics: deep multi-agent reinforcement learning, deep hierarchical reinforcement learning, and deep meta learning.

We hope that learning about deep reinforcement learning will give you as much joy as the many researchers experienced when they developed their algorithms, finally got them to work, and saw them learn!

Leiden, The Netherlands   Aske Plaat
January 2022

Acknowledgments

This book benefited from the help of many friends. First of all, I thank everyone at the Leiden Institute of Advanced Computer Science, for creating such a fun and vibrant environment to work in.

Many people contributed to this book. Some material is based on the book that we used in our previous reinforcement learning course and on lecture notes on policy-based methods written by Thomas Moerland. Thomas also provided invaluable critique on an earlier draft of the book. Furthermore, as this book was being prepared, we worked on survey articles on deep model-based reinforcement learning, deep meta-learning, and deep multi-agent reinforcement learning. I thank Mike Preuss, Walter Kosters, Mike Huisman, Jan van Rijn, Annie Wong, Anna Kononova, and Thomas Bäck, the co-authors on these articles.

I thank all members of the Leiden reinforcement learning community for their input and enthusiasm. I thank especially Thomas Moerland, Mike Preuss, Matthias Müller-Brockhausen, Mike Huisman, Hui Wang, and Zhao Yang, for their help with the course for which this book is written. I thank Wojtek Kowalczyk for insightful discussions on deep supervised learning, and Walter Kosters for his views on combinatorial search, as well as for his never-ending sense of humor.

A very special thank you goes to Thomas Bäck, for our many discussions on science, the universe, and everything (including, especially, evolution). Without you, this effort would not have been possible.

This book is a result of the graduate course on reinforcement learning that we teach at Leiden. I thank all students of this course, past, present, and future, for their wonderful enthusiasm, sharp questions, and many suggestions. This book was written for you and by you!

Finally, I thank Saskia, Isabel, Rosalin, Lily, and Dahlia, for being who they are, for giving feedback and letting me learn, and for their boundless love.

Contents

List of Tables

Chapter 1
Introduction

Deep reinforcement learning studies how we learn to solve complex problems, problems that require us to find a solution to a sequence of decisions in high-dimensional states. To make bread, we must use the right flour, add some salt, yeast and sugar, prepare the dough (not too dry and not too wet), pre-heat the oven to the right temperature, and bake the bread (but not too long); to win a ballroom dancing contest we must find the right partner, learn to dance, practice, and beat the competition; to win in chess we must study, practice, and make all the right moves.

1.1 What Is Deep Reinforcement Learning?

Deep reinforcement learning is the combination of deep learning and reinforcement learning.

The goal of deep reinforcement learning is to learn optimal actions that maximize our reward for all states that our environment can be in (the bakery, the dance hall, the chess board). We do this by interacting with complex, high-dimensional environments, trying out actions, and learning from the feedback.

The field of *deep* learning is about approximating functions in high-dimensional problems, problems that are so complex that tabular methods cannot find exact solutions anymore. Deep learning uses deep neural networks to find *approximations* for large, complex, high-dimensional environments, such as in image and speech recognition. The field has made impressive progress; computers can now recognize pedestrians in a sequence of images (to avoid running over them) and can understand sentences such as: "What is the weather going to be like tomorrow?"

The field of *reinforcement* learning is about learning from feedback; it learns by trial and error. Reinforcement learning does not need a pre-existing dataset to train on; it chooses its own actions and learns from the *feedback* that an environment provides. It stands to reason that in this process of trial and error, our agent will

Table 1.1 The constituents of deep reinforcement learning

	Low-dimensional states	High-dimensional states
Static dataset	Classic supervised learning	Deep supervised learning
Agent/environment interaction	Tabular reinforcement learning	*Deep reinforcement learning*

make mistakes (the fire extinguisher is essential to survive the process of learning to bake bread). The field of reinforcement learning is all about learning from success as well as from mistakes.

In recent years the two fields of *deep* and *reinforcement* learning have come together and have yielded new algorithms that are able to *approximate* high-dimensional problems by *feedback* on their actions. Deep learning has brought new methods and new successes, with advances in policy-based methods, model-based approaches, transfer learning, hierarchical reinforcement learning, and multi-agent learning.

The fields also exist separately, as deep supervised learning and tabular reinforcement learning (see Table 1.1). The aim of deep supervised learning is to generalize and approximate complex, high-dimensional, functions from pre-existing datasets, without interaction; Appendix B discusses deep supervised learning. The aim of tabular reinforcement learning is to learn by interaction in simpler, low-dimensional, environments such as Grid worlds; Chap. 2 discusses tabular reinforcement learning.

Let us have a closer look at the two fields.

1.1.1 Deep Learning

Classic machine learning algorithms learn a predictive model on data, using methods such as linear regression, decision trees, random forests, support vector machines, and artificial neural networks. The models aim to generalize, to make predictions. Mathematically speaking, machine learning aims to approximate a function from data.

In the past, when computers were slow, the neural networks that were used consisted of a few layers of fully connected neurons and did not perform exceptionally well on difficult problems. This changed with the advent of deep learning and faster computers. Deep neural networks now consist of many layers of neurons and use different types of connections.[1] Deep networks and deep learning have taken the accuracy of certain important machine learning tasks to a new level and have allowed machine learning to be applied to complex, high-dimensional, problems, such as recognizing cats and dogs in high-resolution (mega-pixel) images.

---

[1] Where *many* means *more than one* hidden layer in between the input and output layers.

Deep learning allows high-dimensional problems to be solved in real time; it has allowed machine learning to be applied to day-to-day tasks such as the face recognition and speech recognition that we use in our smartphones.

1.1.2 Reinforcement Learning

Let us look more deeply at reinforcement learning, to see what it means to learn from our own actions.

Reinforcement learning is a field in which an agent learns by interacting with an environment. In supervised learning we need pre-existing datasets of labeled examples to approximate a function; reinforcement learning only needs an environment that provides feedback signals for actions that the agent is trying out. This requirement is easier to fulfill, allowing reinforcement learning to be applicable to more situations than supervised learning.

Reinforcement learning agents generate, by their actions, their own on-the-fly data, through the environment's rewards. Agents can choose which actions to learn from; reinforcement learning is a form of active learning. In this sense, our agents are like children, that, through playing and exploring, teach themselves a certain task. This level of autonomy is one of the aspects that attracts researchers to the field. The reinforcement learning agent chooses which action to perform—which hypothesis to test—and adjusts its knowledge of what works, building up a policy of actions that are to be performed in the different states of the world that it has encountered. (This freedom is also what makes reinforcement learning hard, because when you are allowed to choose your own examples, it is all too easy to stay in your comfort zone, stuck in a positive reinforcement bubble, believing you are doing great, but learning very little of the world around you.)

1.1.3 Deep Reinforcement Learning

Deep reinforcement learning combines methods for learning high-dimensional problems with reinforcement learning, allowing *high-dimensional, interactive* learning. A major reason for the interest in deep reinforcement learning is that it works well on current computers and does so in seemingly different applications. For example, in Chap. 3 we will see how deep reinforcement learning can learn eye–hand coordination tasks to play 1980s video games, in Chap. 4 we see how a simulated robot cheetah learns to jump, and in Chap. 6 we see how it can teach itself to play complex games of strategy to the extent that world champions are beaten.

Let us have a closer look at the kinds of applications on which deep reinforcement learning does so well.

1.1.4 Applications

In its most basic form, reinforcement learning is a way to teach an agent to operate in the world. As a child learns to walk from actions and feedback, so do reinforcement learning agents learn from actions and feedback. Deep reinforcement learning can learn to solve large and complex decision problems—problems whose solution is not yet known, but for which an approximating trial-and-error mechanism exists that can learn a solution out of repeated interactions with the problem. This may sound a bit cryptical and convoluted, but approximation and trial and error are something that we do in real life all the time. Generalization and approximation allow us to infer patterns or rules from examples. Trial and error is a method by which humans learn how to deal with things that are unfamiliar to them ("What happens if I press this button? Oh. Oops." Or: "What happens if I do not put my leg before my other leg while moving forward? Oh. Ouch.").

1.1.4.1 Sequential Decision Problems

Learning to operate in the world is a high-level goal; we can be more specific. Reinforcement learning is about the agent's behavior. Reinforcement learning can find solutions for *sequential decision problems*, or optimal control problems, as they are known in engineering. There are many situations in the real world where, in order to reach a goal, a sequence of decisions must be made. Whether it is baking a cake, building a house, or playing a card game; a sequence of decisions has to be made. Reinforcement learning provides efficient ways to learn solutions to sequential decision problems.

Many real-world problems can be modeled as a sequence of decisions [33]. For example, in autonomous driving, an agent is faced with questions of speed control, finding drivable areas, and, most importantly, avoiding collisions. In healthcare, treatment plans contain many sequential decisions, and factoring the effects of delayed treatment can be studied. In customer centers, natural language processing can help improve chatbot dialogue, question answering, and even machine translation. In marketing and communication, recommender systems recommend news, personalize suggestions, deliver notifications to user, or otherwise optimize the product experience. In trading and finance, systems decide to hold, buy, or sell financial titles, in order to optimize future reward. In politics and governance, the effects of policies can be simulated as a sequence of decisions before they are implemented. In mathematics and entertainment, playing board games, card games, and strategy games consists of a sequence of decisions. In computational creativity, making a painting requires a sequence of esthetic decisions. In industrial robotics and engineering, the grasping of items and the manipulation of materials consist of a sequence of decisions. In chemical manufacturing, the optimization of production processes consists of many decision steps that influence the yield and quality of the product. Finally, in energy grids, the efficient and safe distribution of energy can be modeled as a sequential decision problem.

In all these situations, we must make a sequence of decisions. In all these situations, taking the wrong decision can be very costly.

The algorithmic research on sequential decision making has focused on two types of applications: (1) robotic problems and (2) games. Let us have a closer look at these two domains, starting with robotics.

1.1.4.2 Robotics

In principle, all actions that a robot should take can be pre-programmed step by step by a programmer in meticulous detail. In highly controlled environments, such as a welding robot in a car factory, this can conceivably work, although any small change or any new task requires reprogramming the robot.

It is surprisingly hard to manually program a robot to perform a complex task. Humans are not aware of their own operational knowledge, such as what "voltages" we put on which muscles when we pick up a cup. It is much easier to define a desired goal state, and let the system find the complicated solution by itself. Furthermore, in environments that are only slightly challenging, when the robot must be able to respond more flexibly to different conditions, an adaptive program is needed.

It will be no surprise that the application area of robotics is an important driver for machine learning research, and robotics researchers turned early on to finding methods by which the robots could teach themselves certain behavior.

The literature on robotics experiments is varied and rich. A robot can teach itself how to navigate a maze, how to perform manipulation tasks, and how to learn locomotion tasks.

Research into adaptive robotics has made quite some progress. For example, one of the recent achievements involves flipping pancakes [29] and flying an aerobatic model helicopter [1, 2]; see Figs. 1.1 and 1.2. Frequently, learning tasks are combined with computer vision, where a robot has to learn by visually interpreting the consequences of its own actions.

Fig. 1.1 Robot flipping pancakes [29]

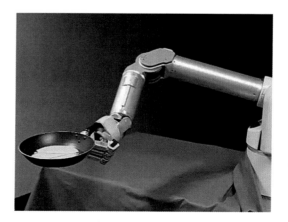

Fig. 1.2 Aerobatic model
helicopter [2]

Fig. 1.3 Chess

1.1.4.3   Games

Let us now turn to games. Puzzles and games have been used from the earliest days
to study aspects of intelligent behavior. Indeed, before computers were powerful
enough to execute chess programs, in the days of Shannon and Turing, paper designs
were made, in the hope that understanding chess would teach us something about
the nature of intelligence [38, 41].

Games allow researchers to limit the scope of their studies, to focus on intelligent
decision making in a limited environment, without having to master the full
complexity of the real world. In addition to board games such as chess and Go,
video games are being used extensively to test intelligent methods in computers.
Examples are Arcade-style games such as Pac-Man [32] and multiplayer strategy
games such as StarCraft [43]. See Figs. 1.3, 1.4, 1.5, and 1.6.

Fig. 1.4 Go

Fig. 1.5 Pac-Man [6]

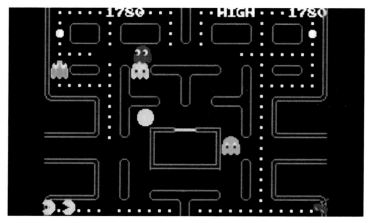

Fig. 1.6 StarCraft [43]

1.1.5   Four Related Fields

Reinforcement learning is a rich field that has existed in some form long before the artificial intelligence endeavor had started, as a part of biology, psychology, and education [9, 25, 40]. In artificial intelligence it has become one of the three main categories of machine learning, the other two being supervised and unsupervised learning [10]. This book is a book of algorithms that are inspired by topics from the natural and social sciences. Although the rest of the book will be about these algorithms, it is interesting to briefly discuss the links of deep reinforcement learning to human and animal learning. We will introduce the four scientific disciplines that have a profound influence on deep reinforcement learning.

1.1.5.1   Psychology

In psychology, reinforcement learning is also known as *learning by conditioning* or as *operant conditioning*. Figure 1.7 illustrates the folk psychological idea of how a dog can be conditioned. A natural reaction to food is that a dog salivates. By ringing a bell whenever the dog is given food, the dog learns to associate the sound with food, and after enough trials, the dog starts salivating as soon as it hears the bell, presumably in anticipation of the food, whether it is there or not.

The behavioral scientists Pavlov (1849–1936) and Skinner (1904–1990) are well known for their work on conditioning. Phrases such as *Pavlov-reaction* have entered

Fig. 1.7 Classical conditioning: (1) a dog salivates when seeing food, (2) but initially not when hearing a bell, (3) when the sound rings often enough together when food is served, the dog starts to associate the bell with food, and (4) also salivates when only the bell rings

Fig. 1.8 Who is conditioning whom?

our everyday language, and various jokes about conditioning exist (see, for example, Fig. 1.8). Psychological research into learning is one of the main influences on reinforcement learning as we know it in artificial intelligence.

1.1.5.2 Mathematics

Mathematical logic is another foundation of deep reinforcement learning. Discrete optimization and graph theory are of great importance for the formalization of reinforcement learning, as we will see in Sect. 2.2.2 on Markov decision processes. Mathematical formalizations have enabled the development of efficient planning and optimization algorithms that are at the core of current progress.

Planning and optimization are an important part of deep reinforcement learning. They are also related to the field of operations research, although there the emphasis is on (non-sequential) combinatorial optimization problems. In AI, planning and optimization are used as building blocks for creating learning systems for sequential, high-dimensional, problems that can include visual, textual, or auditory input.

The field of symbolic reasoning is based on logic, and it is one of the earliest success stories in artificial intelligence. Out of work in symbolic reasoning came heuristic search [34], expert systems, and theorem proving systems. Well-known systems are the STRIPS planner [17], the Mathematica computer algebra system [13], the logic programming language PROLOG [14], and also systems such as SPARQL for semantic (web) reasoning [3, 7].

Symbolic AI focuses on reasoning in discrete domains, such as decision trees, planning, and games of strategy, such as chess and checkers. Symbolic AI has driven success in methods to search the web, to power online social networks, and to power online commerce. These highly successful technologies are the basis of much of our modern society and economy. In 2011 the highest recognition in computer

Fig. 1.9 Turing-award winner Judea Pearl

science, the Turing award, was awarded to Judea Pearl for work in causal reasoning (Fig. 1.9).[2] Pearl later published an influential book to popularize the field [35].

Another area of mathematics that has played a large role in deep reinforcement learning is the field of continuous (numerical) optimization. Continuous methods are important, for example, in efficient gradient descent and backpropagation methods that are at the heart of current deep learning algorithms.

1.1.5.3 Engineering

In engineering, the field of reinforcement learning is better known as *optimal control*. The theory of optimal control of dynamical systems was developed by Richard Bellman and Lev Pontryagin [8]. Optimal control theory originally focused on dynamical systems, and the technology and methods relate to continuous optimization methods such as used in robotics (see Fig. 1.10 for an illustration of optimal control at work in docking two space vehicles). Optimal control theory is of central importance to many problems in engineering.

To this day reinforcement learning and optimal control use a different terminology and notation. States and actions are denoted as s and a in state-oriented reinforcement learning, where the engineering world of optimal control uses x and u. In this book the former notation is used.

---

[2] Joining a long list of AI researchers that have been honored earlier with a Turing award: Minsky, McCarthy, Newell, Simon, Feigenbaum, and Reddy.

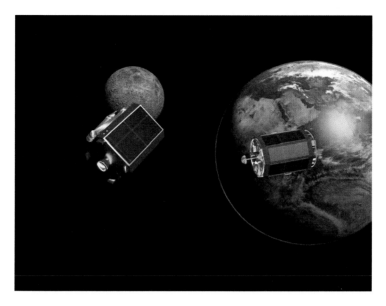

Fig. 1.10   Optimal control of dynamical systems at work

1.1.5.4   Biology

Biology has a profound influence on computer science. Many nature-inspired optimization algorithms have been developed in artificial intelligence. An important nature-inspired school of thought is connectionist AI.

Mathematical logic and engineering approach intelligence as a top-down deductive process; observable effects in the real world follow from the application of theories and the laws of nature, and intelligence follows deductively from theory. In contrast, connectionism approaches intelligence in a bottom-up fashion. Connectionist intelligence emerges out of many low-level interactions. Intelligence follows inductively from practice. Intelligence is embodied: the bees in bee hives, the ants in ant colonies, and the neurons in the brain all interact, and out of the connections and interactions arises behavior that we recognize as intelligent [11].

Examples of the connectionist approach to intelligence are nature-inspired algorithms such as ant colony optimization [15], swarm intelligence [11, 26], evolutionary algorithms [4, 18, 23], robotic intelligence [12], and, last but not least, artificial neural networks and deep learning [19, 21, 30].

It should be noted that both the symbolic and the connectionist school of AI have been very successful. After the enormous economic impact of search and symbolic AI (Google, Facebook, Amazon, Netflix), much of the interest in AI in the last two decades has been inspired by the success of the connectionist approach in computer language and vision. In 2018 the Turing award was awarded to three key researchers in deep learning: Bengio, Hinton, and LeCun (Fig. 1.11). Their most famous paper on deep learning may well be [30].

Fig. 1.11 Turing-award winners Geoffrey Hinton, Yann LeCun, and Yoshua Bengio

1.2 Three Machine Learning Paradigms

Now that we have introduced the general context and origins of deep reinforcement learning, let us switch gears and talk about machine learning. Let us see how deep reinforcement learning fits in the general picture of the field. At the same time, we will take the opportunity to introduce some notation and basic concepts.

In the next section we will then provide an outline of the book. But first it is time for machine learning. We start at the beginning, with function approximation.

Representing a Function
Functions are a central part in artificial intelligence. A function f transforms input x into output y according to some method, and we write $f(x) \rightarrow y$. In order to perform calculations with function f, the function must be represented as a computer program in some form in memory. We also write function

$$f : X \rightarrow Y,$$

where the domain X and range Y can be discrete or continuous; the dimensionality (the number of attributes in X) can be arbitrary.

Often, in the real world, the same input may yield a range of different outputs, and we would like our function to provide a conditional probability distribution, a function that maps

$$f : X \to p(Y).$$

Here the function maps the domain to a probability distribution p over the range. Representing a conditional probability allows us to model functions for which the input does not always give the same output. (Appendix A provides more mathematical background.)

Given Versus Learned Function

Sometimes the function that we are interested in is given, and we can represent the function by a specific algorithm that computes an analytical expression that is known exactly. This is, for example, the case for laws of physics, or when we make explicit assumptions for a particular system.

Example Newton's second law of motion states that for objects with constant mass

$$F = m \cdot a,$$

where F denotes the net force on the object, m denotes its mass, and a denotes its acceleration. In this case, the analytical expression defines the entire function, for every possible combination of the inputs.

However, for many functions in the real world, we do not have an analytical expression. Here, we enter the realm of machine learning, in particular of *supervised learning*. When we do not know an analytical expression for a function, our best approach is to collect data—examples of (x, y) pairs—and reverse engineer or *learn* the function from this data. See Fig. 1.12.

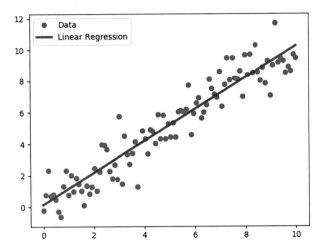

Fig. 1.12 Example of learning a function; data points are in blue, a possible learned linear function is the red line, which allows us to make predictions $\hat{y}$ for any new input x

Example A company wants to predict the chance that you buy a shampoo to color your hair, based on your age. They collect many data points of $x \in \mathbb{N}$, your age (a natural number), that map to $y \in \{0, 1\}$, a binary indicator whether you bought their shampoo. They then want to *learn* the mapping

$$\hat{y} = f(x),$$

where f is the desired function that tells the company who will buy the product and $\hat{y}$ is the predicted y (admittedly overly simplistic in this example).

Let us see which methods exist in machine learning to find function approximations.

Three Paradigms
There are three main paradigms for how the observations can be provided in machine learning: (1) supervised learning, (2) reinforcement learning, and (3) unsupervised learning.

1.2.1 Supervised Learning

The first and most basic paradigm for machine learning is supervised learning. In supervised learning, the data to learn the function $f(x)$ is provided to the learning algorithm in (x, y) example pairs. Here x is the input, and y the observed output value to be learned for that particular input value x. The y values can be thought of as supervising the learning process, and they teach the learning process the right answers for each input value x, hence the name *supervised* learning.

The data pairs to be learned from are organized in a dataset, which must be present in its entirety before the algorithm can start. During the learning process, an estimate of the real function that generated the data is created, $\hat{f}$. The x values of the pair are also called the *input*, and the y values are the *label* to be learned.

Two well-known problems in supervised learning are regression and classification. Regression predicts a continuous number, and classification a discrete category. The best known regression relation is the linear relation: the familiar straight line through a cloud of observation points that we all know from our introductory statistics course. Figure 1.12 shows such a linear relationship $\hat{y} = a \cdot x + b$. The linear function can be characterized with two parameters a and b. Of course, more complex functions are possible, such as quadratic regression, nonlinear regression, or regression with higher-order polynomials [16].

The supervisory signal is computed for each data item i as the difference between the current estimate and the given label, for example by $(\hat{f}(x_i) - y_i)^2$. Such an error function $(\hat{f}(x) - y)^2$ is also known as a loss function; it measures the quality of our prediction. The closer our prediction is to the true label, the lower the loss. There are many ways to compute this closeness, such as the mean squared error loss $\mathcal{L} = \frac{1}{N} \sum_1^N (\hat{f}(x_i) - y_i)^2$, which is used often for regression over N observations. This loss function can be used by a supervised learning algorithm to adjust model parameters a and b to fit the function $\hat{f}$ to the data. Some of the many possible learning algorithms are linear regression and support vector machines [10, 36].

In classification, a relation between an input value and a class label is learned. A well-studied classification problem is image recognition, where two-dimensional images are to be categorized. Table 1.2 shows a tiny dataset of labeled images of the proverbial cats and dogs. A popular loss function for classification is the cross-entropy loss $\mathcal{L} = - \sum_1^N y_i \log(\hat{f}(x_i))$, see also Sect. A.2.5.3. Again, such a loss function can be used to adjust the model parameters to fit the function to the data. The model can be small and linear, with few parameters, or it can be large, with many parameters, such as a neural network, which is often used for image classification.

In supervised learning a large dataset exists where all input items have an associated training label. Reinforcement learning is different, and it does not assume the pre-existence of a large labeled training set. Unsupervised learning does require a large dataset, but no user-supplied output labels; all it needs are the inputs.

Table 1.2 (Input/output) pairs for a supervised classification problem

| "Cat" | "Cat" | "Dog" | "Cat" | "Dog" | "Dog" |

Deep learning function approximation was first developed in a supervised setting. Although this book is about deep reinforcement learning, we will encounter supervised learning concepts frequently, whenever we discuss the deep learning aspect of deep reinforcement learning.

1.2.2 Unsupervised Learning

When there are no labels in the dataset, different learning algorithms must be used. Learning without labels is called *unsupervised learning*. In unsupervised learning an inherent metric of the data items is used, such as distance. A typical problem in unsupervised learning is to find patterns in the data, such as clusters or subgroups [42, 44].

Popular unsupervised learning algorithms are k-means algorithms and principal component analysis [24, 37]. Other popular unsupervised methods are dimensionality reduction techniques from visualization, such as t-SNE [31], minimum description length [20], and data compression [5]. A popular application of unsupervised learning are autoencoders, see Sect. B.2.6.3 [27, 28].

The relation between supervised and unsupervised learning is sometimes characterized as follows: supervised learning aims to learn the conditional probability distribution $p(x|y)$ of input data conditioned on a label y, whereas unsupervised learning aims to learn the *a priori* probability distribution $p(x)$ [22].

We will encounter unsupervised methods in this book in a few places, specifically, when autoencoders and dimension reduction are discussed, for example, in Chap. 5. At the end of this book explainable artificial intelligence is discussed, where interpretable models play an important role, in Chap. 10.

1.2.3 Reinforcement Learning

The last machine learning paradigm is, indeed, reinforcement learning. There are three differences between reinforcement learning and the previous paradigms. First, reinforcement learning learns by *interaction*; in contrast to supervised and unsupervised learning, in reinforcement learning data items come one by one. The dataset is produced dynamically, as it were. The objective in reinforcement learning is to find the policy: a function that gives us the best action in each state that the world can be in.

The approach of reinforcement learning is to learn the policy for the world by interacting with it. In reinforcement learning we recognize an *agent* that does the learning of the policy and an *environment* that provides feedback to the agent's actions (and that performs state changes, see Fig. 1.13). In reinforcement learning, the agent stands for the human, and the environment for the world. The goal of reinforcement learning is to find the actions for each state that maximize the long-

Fig. 1.13 Agent and environment

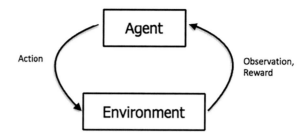

Table 1.3 Supervised vs. reinforcement learning

Concept	Supervised learning	Reinforcement learning
Inputs x	Full dataset of states	Partial (one state at a time)
Labels y	Full (correct action)	Partial (numeric action reward)

term accumulated expected reward. This optimal function of states to actions is called the *optimal policy*.

In reinforcement learning there is no teacher or supervisor, and there is no static dataset. There is, however, the environment that will tell us how good the state is in which we find ourselves. This brings us to the second difference: the *reward value*. Reinforcement learning gives us partial information, a number indicating the quality of the action that brought us to our state, where supervised learning gives full information: a label that provides the correct answer in that state (Table 1.3). In this sense, reinforcement learning is in between supervised learning, in which all data items have a label, and unsupervised learning, where no data has a label.

The third difference is that reinforcement learning is used to solve *sequential decision problems*. Supervised and unsupervised learning learn single-step relations between items; reinforcement learning learns a policy, which is the answer to a multi-step problem. Supervised learning can classify a set of images for you; unsupervised learning can tell you which items belong together; reinforcement learning can tell you the winning sequence of moves in a game of chess, or the action sequence that robot legs need to take in order to walk.

These three differences have consequences. Reinforcement learning provides the data to the learning algorithm step by step, action by action, whereas in supervised learning the data is provided all at once in one large dataset. The step-by-step approach is well suited to sequential decision problems. On the other hand, many deep learning methods were developed for supervised learning and may work differently when data items are generated one by one. Furthermore, since actions are selected using the policy function, and action rewards are used to update this same policy function, there is a possibility of circular feedback and local minima. Care must be taken to ensure convergence to global optima in our methods. Human learning also suffers from this problem, when a stubborn child refuses to explore outside of its comfort zone. This topic is discussed in Sect. 2.2.4.7.

Another difference is that in supervised learning the pupil learns from a finite-sized teacher (the dataset) and at some point may have learned all there is to learn.

The reinforcement learning paradigm allows a learning setup where the agent can continue to sample the environment indefinitely and will continue to become smarter as long as the environment remains challenging (which can be a long time, for example in games such as chess and Go).[3]

For these reasons there is great interest in reinforcement learning, although getting the methods to work is often harder than for supervised learning.

Most classical reinforcement learning use tabular methods that work for low-dimensional problems with small state spaces. Many real-world problems are complex and high-dimensional, with large state spaces. Due to steady improvements in learning algorithms, datasets, and compute power, deep learning methods have become quite powerful. Deep reinforcement learning methods have emerged that successfully combine step-by-step sampling in high-dimensional problems with large state spaces. We will discuss these methods in the subsequent chapters of this book.

1.3   Overview of the Book

The aim of this book is to present the latest insights in deep reinforcement learning in a single comprehensive volume, suitable for teaching a graduate level one-semester course.

In addition to covering state-of-the-art algorithms, we cover necessary background in classic reinforcement learning and deep learning. We also cover advanced, forward looking developments in self-play, and in multi-agent, hierarchical, and meta-learning.

1.3.1   Prerequisite Knowledge

In an effort to be comprehensive, we make modest assumptions about previous knowledge. We assume a bachelor level of computer science or artificial intelligence and an interest in artificial intelligence and machine learning. A good introductory textbook on artificial intelligence is Russell and Norvig: *Artificial Intelligence, A Modern Approach* [36].

Figure 1.14 shows an overview of the structure of the book. Deep reinforcement learning combines deep supervised learning and classical (tabular) reinforcement learning. The figure shows how the chapters are built on this dual foundation. For deep reinforcement learning, the field of deep supervised learning is of great importance. It is a large field, deep, and rich. Many students may have followed

---

[3] In fact, some argue that reward is enough for artificial general intelligence, see Silver, Singh, Precup, and Sutton [39].

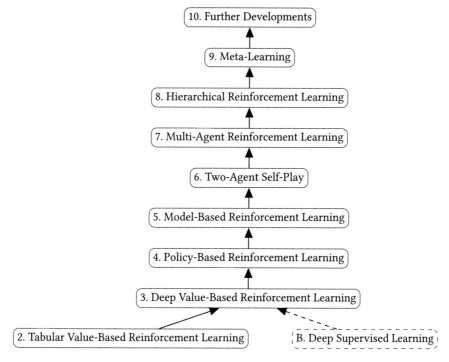

Fig. 1.14 Deep reinforcement learning is built on deep supervised learning and tabular reinforcement learning

a course on deep learning; if not, Appendix B provides you with the necessary background (dashed). Tabular reinforcement learning, on the other hand, may be new to you, and we start our story with this topic in Chap. 2.

We also assume undergraduate level familiarity with the Python programming language. Python has become the programming language of choice for machine learning research, and the host language of most machine learning packages. All example code in this book is in Python, and major machine learning environments such as scikit-learn, TensorFlow, Keras, and PyTorch work best from Python. See https://www.python.org for pointers on how to get started in Python. Use the latest stable version, unless the text mentions otherwise.

We assume an undergraduate level of familiarity with mathematics—a basic understanding of set theory, graph theory, probability theory, and information theory is necessary, although this is not a book of mathematics. Appendix A contains a summary to refresh your mathematical knowledge, and to provide an introduction to the notation that is used in the book.

1.3.1.1 Course

There is a lot of material in the chapters, both basic and advanced, with many pointers to the literature. One option is to teach a single course about all topics in the book. Another option is to go slower and deeper, to spend sufficient time on the basics, and create a course about Chaps. 2–5 to cover the basic topics (value-based, policy-based, and model-based learning), and to create a separate course about Chaps. 6–9 to cover the more advanced topics of multi-agent, hierarchical, and meta-learning.

1.3.1.2 Blogs and GitHub

The field of deep reinforcement learning is a highly active field, in which theory and practice go hand in hand. The culture of the field is open, and you will easily find many blog posts about interesting topics, some quite good. Theory drives experimentation, and experimental results drive theoretical insights. Many researchers publish their papers on arXiv and their algorithms, hyperparameter settings and environments on GitHub.

In this book we aim for the same atmosphere. Throughout the text we provide links to code, and we challenge you with hands-on sections to get your hands dirty to perform your own experiments. All links to web pages that we use have been stable for some time.

> **Website:** https://deep-reinforcement-learning.net is the companion website for this book. It contains updates, slides, and other course material that you are welcome to explore and use.

1.3.2 Structure of the Book

The field of deep reinforcement learning consists of two main areas: model-free reinforcement learning and model-based reinforcement learning. Both areas have two subareas. The chapters of this book are organized according to this structure:

- Model-free methods

  - Value-based methods: Chap. 2 (tabular) and 3 (deep)
  - Policy-based methods: Chap. 4

- Model-based methods

  - Learned model: Chap. 5
  - Given model: Chap. 6

Then, we have three chapters on more specialized topics.

* Multi-agent reinforcement learning: Chap. 7
* Hierarchical reinforcement learning: Chap. 8
* Transfer and Meta-learning: Chap. 9

Appendix B provides a necessary review of deep supervised learning.

The style of each chapter is to first provide the main idea of the chapter in an intuitive *example*, to then explain the kind of *problem* to be solved, and then to discuss algorithmic concepts that *agents* use, and the *environments* that have been solved in practice with these algorithms. The sections of the chapters are named accordingly: their names end in problem-agent-environment. At the end of each chapter we provide questions for quizzes to check your understanding of the concepts, and we provide exercises for larger programming assignments (some doable, some quite challenging). We also end each chapter with a summary and references to further reading.

Let us now look in more detail at what topics the chapters cover.

1.3.2.1 Chapters

After this introductory chapter, we continue with Chap. 2, in which we discuss in detail the basic concepts of tabular (non-deep) reinforcement learning. We start with Markov decision processes and discuss them at length. We will introduce tabular planning and learning and important concepts such as state, action, reward, value, and policy. We will encounter the first, tabular, value-based model-free learning algorithms (for an overview, see Table 2.1). Chapter 2 is the only non-deep chapter of the book. All other chapters cover deep methods.

Chapter 3 explains deep value-based reinforcement learning. The chapter covers the first deep algorithms that have been devised to find the optimal policy. We will still be working in the value-based, model-free, paradigm. At the end of the chapter we will analyze a player that teaches itself how to play 1980s Atari video games. Table 3.1 lists some of the many stable deep value-based model-free algorithms.

Value-based reinforcement learning works well with applications such as games, with discrete action spaces. The next chapter, Chap. 4, discusses a different approach: deep policy-based reinforcement learning (Table 4.1). In addition to discrete spaces, this approach is also suited for continuous actions spaces, such as robot arm movement and simulated articulated locomotion. We see how a simulated Half-Cheetah teaches itself to run.

The next chapter, Chap. 5, introduces deep model-based reinforcement learning with a learned model, a method that first builds up a transition model of the environment before it builds the policy. Model-based reinforcement learning holds the promise of higher sample efficiency and thus faster learning. New developments, such as latent models, are discussed. Applications are both in robotics and in games (Table 5.2).

The next chapter, Chap. 6, studies how a self-play system can be created for applications where the transition model is given by the problem description. This is the case in two-agent games, where the rules for moving in the game determine the transition function. We study how TD-Gammon and AlphaZero achieve *tabula rasa* learning: teaching themselves from zero knowledge to world champion level play through playing against a copy of itself (Table 6.2). In this chapter deep residual networks and Monte Carlo Tree Search result in curriculum learning.

Chapter 7 introduces recent developments in deep multi-agent and team learning. The chapter covers competition and collaboration, population-based methods, and playing in teams. Applications of these methods are found in games such as poker and StarCraft (Table 7.2).

Chapter 8 covers deep hierarchical reinforcement learning. Many tasks exhibit an inherent hierarchical structure, in which clear subgoals can be identified. The options framework is discussed, and methods that can identify subgoals, subpolicies, and meta policies. Different approaches for tabular and deep hierarchical methods are discussed (Table 8.1).

The final technical chapter, Chap. 9, covers deep meta-learning, or learning to learn. One of the major hurdles in machine learning is the long time it takes to learn to solve a new task. Meta-learning and transfer learning aim to speed up learning of new tasks by using information that has been learned previously for related tasks; algorithms are listed in Table 9.2. At the end of the chapter we will experiment with few-shot learning, where a task has to be learned without having seen more than a few training examples.

Chapter 10 concludes the book by reviewing what we have learned, and by looking ahead into what the future may bring.

Appendix A provides mathematical background information and notation. Appendix B provides a chapter-length overview of machine learning and deep supervised learning. If you wish to refresh your knowledge of deep learning, please go to this appendix before you read Chap. 3. Appendix C provides lists of useful software environments and software packages for deep reinforcement learning.

References

1. Pieter Abbeel, Adam Coates, and Andrew Y Ng. Autonomous helicopter aerobatics through apprenticeship learning. *The International Journal of Robotics Research*, 29(13):1608–1639, 2010. 5
2. Pieter Abbeel, Adam Coates, Morgan Quigley, and Andrew Y Ng. An application of reinforcement learning to aerobatic helicopter flight. In *Advances in Neural Information Processing Systems*, pages 1–8, 2007. 5, 6
3. Grigoris Antoniou and Frank Van Harmelen. *A Semantic Web Primer*. MIT Press Cambridge, MA, 2008. 9
4. Thomas Bäck and Hans-Paul Schwefel. An overview of evolutionary algorithms for parameter optimization. *Evolutionary Computation*, 1(1):1–23, 1993. 11
5. Andrew Barron, Jorma Rissanen, and Bin Yu. The minimum description length principle in coding and modeling. *IEEE Transactions on Information Theory*, 44(6):2743–2760, 1998. 16

6. Marc G Bellemare, Yavar Naddaf, Joel Veness, and Michael Bowling. The Arcade Learning Environment: An evaluation platform for general agents. *Journal of Artificial Intelligence Research*, 47:253–279, 2013. 7

7. Tim Berners-Lee, James Hendler, and Ora Lassila. The semantic web. *Scientific American*, 284(5):28–37, 2001. 9

8. Dimitri P Bertsekas, Dimitri P Bertsekas, Dimitri P Bertsekas, and Dimitri P Bertsekas. *Dynamic Programming and Optimal Control*, volume 1. Athena Scientific Belmont, MA, 1995. 10

9. Dimitri P Bertsekas and John Tsitsiklis. *Neuro-Dynamic Programming*. MIT Press Cambridge, 1996. 8

10. Christopher M Bishop. *Pattern Recognition and Machine Learning*. Information science and statistics. Springer Verlag, Heidelberg, 2006. 8, 15

11. Eric Bonabeau, Marco Dorigo, and Guy Theraulaz. *Swarm Intelligence: From Natural to Artificial Systems*. Oxford University Press, 1999. 11, 11

12. Rodney A Brooks. Intelligence without representation. *Artificial Intelligence*, 47(1–3):139–159, 1991. 11

13. Bruno Buchberger, George E Collins, Rüdiger Loos, and Rudolph Albrecht. Computer algebra symbolic and algebraic computation. *ACM SIGSAM Bulletin*, 16(4):5–5, 1982. 9

14. William F Clocksin and Christopher S Mellish. *Programming in Prolog: Using the ISO standard*. Springer Science & Business Media, 1981. 9

15. Marco Dorigo and Luca Maria Gambardella. Ant colony system: a cooperative learning approach to the traveling salesman problem. *IEEE Transactions on Evolutionary Computation*, 1(1):53–66, 1997. 11

16. Norman R Draper and Harry Smith. *Applied Regression Analysis*, volume 326. John Wiley & Sons, 1998. 15

17. Richard E Fikes and Nils J Nilsson. STRIPS: A new approach to the application of theorem proving to problem solving. *Artificial Intelligence*, 2(3-4):189–208, 1971. 9

18. David B Fogel. An introduction to simulated evolutionary optimization. *IEEE Transactions on Neural Networks*, 5(1):3–14, 1994. 11

19. Ian Goodfellow, Yoshua Bengio, and Aaron Courville. *Deep Learning*. MIT Press, Cambridge, 2016. 11

20. Peter D Grünwald. *The minimum description length principle*. MIT Press, 2007. 16

21. Simon Haykin. *Neural Networks: a Comprehensive Foundation*. Prentice Hall, 1994. 11

22. Geoffrey E Hinton and Terrence Joseph Sejnowski, editors. *Unsupervised Learning: Foundations of Neural Computation*. MIT Press, 1999. 16

23. John H Holland. Genetic algorithms. *Scientific American*, 267(1):66–73, 1992. 11

24. Ian T Jolliffe and Jorge Cadima. Principal component analysis: a review and recent developments. *Philosophical Transactions of the Royal Society A: Mathematical, Physical and Engineering Sciences*, 374(2065):20150202, 2016. 16

25. Leslie Pack Kaelbling, Michael L Littman, and Andrew W Moore. Reinforcement learning: A survey. *Journal of Artificial Intelligence Research*, 4:237–285, 1996. 8

26. James Kennedy. Swarm intelligence. In *Handbook of Nature-Inspired and Innovative Computing*, pages 187–219. Springer, 2006. 11

27. Diederik P Kingma and Max Welling. Auto-encoding variational Bayes. In *International Conference on Learning Representations*, 2014. 16

28. Diederik P Kingma and Max Welling. An introduction to variational autoencoders. *Found. Trends Mach. Learn.*, 12(4):307–392, 2019. 16

29. Petar Kormushev, Sylvain Calinon, and Darwin G Caldwell. Robot motor skill coordination with em-based reinforcement learning. In *2010 IEEE/RSJ International Conference on Intelligent Robots and Systems*, pages 3232–3237. IEEE, 2010. 5, 5

30. Yann LeCun, Yoshua Bengio, and Geoffrey Hinton. Deep learning. *Nature*, 521(7553):436, 2015. 11, 11

31. Laurens van der Maaten and Geoffrey Hinton. Visualizing data using t-SNE. *Journal of Machine Learning Research*, 9:2579–2605, Nov 2008. 16

32. Volodymyr Mnih, Koray Kavukcuoglu, David Silver, Andrei A. Rusu, Joel Veness, Marc G. Bellemare, Alex Graves, Martin A. Riedmiller, Andreas Fidjeland, Georg Ostrovski, Stig Petersen, Charles Beattie, Amir Sadik, Ioannis Antonoglou, Helen King, Dharshan Kumaran, Daan Wierstra, Shane Legg, and Demis Hassabis. Human-level control through deep reinforcement learning. *Nature*, 518(7540):529–533, 2015. 6

33. Derick Mwiti. Reinforcement learning applications. https://neptune.ai/blog/reinforcement-learning-applications. 4

34. Judea Pearl. *Heuristics: Intelligent Search Strategies for Computer Problem Solving*. Addison-Wesley, Reading, MA, 1984. 9

35. Judea Pearl and Dana Mackenzie. *The Book of Why: the New Science of Cause and Effect*. Basic Books, 2018. 10

36. Stuart J Russell and Peter Norvig. *Artificial intelligence: a modern approach*. Pearson Education Limited, Malaysia, 2016. 15, 18

37. Bernhard Schölkopf, Alexander Smola, and Klaus-Robert Müller. Kernel principal component analysis. In *International Conference on Artificial Neural Networks*, pages 583–588. Springer, 1997. 16

38. Claude E Shannon. Programming a computer for playing chess. In *Computer Chess Compendium*, pages 2–13. Springer, 1988. 6

39. David Silver, Satinder Singh, Doina Precup, and Richard S Sutton. Reward is enough. *Artificial Intelligence*, page 103535, 2021. 18

40. Richard S Sutton and Andrew G Barto. *Reinforcement learning, An Introduction, Second Edition*. MIT Press, 2018. 8

41. Alan M Turing. *Digital Computers Applied to Games*. Pitman & Sons, 1953. 6

42. Matthijs Van Leeuwen and Arno Knobbe. Diverse subgroup set discovery. *Data Mining and Knowledge Discovery*, 25(2):208–242, 2012. 16

43. Oriol Vinyals, Igor Babuschkin, Wojciech M. Czarnecki, Michaël Mathieu, Andrew Dudzik, Junyoung Chung, David H. Choi, Richard Powell, Timo Ewalds, Petko Georgiev, Junhyuk Oh, Dan Horgan, Manuel Kroiss, Ivo Danihelka, Aja Huang, Laurent Sifre, Trevor Cai, John P. Agapiou, Max Jaderberg, Alexander Sasha Vezhnevets, Rémi Leblond, Tobias Pohlen, Valentin Dalibard, David Budden, Yury Sulsky, James Molloy, Tom Le Paine, Çaglar Gülçehre, Ziyu Wang, Tobias Pfaff, Yuhai Wu, Roman Ring, Dani Yogatama, Dario Wünsch, Katrina McKinney, Oliver Smith, Tom Schaul, Timothy P. Lillicrap, Koray Kavukcuoglu, Demis Hassabis, Chris Apps, and David Silver. Grandmaster level in StarCraft II using multi-agent reinforcement learning. *Nature*, 575(7782):350–354, 2019. 6, 7

44. Jilles Vreeken, Matthijs Van Leeuwen, and Arno Siebes. KRIMP: mining itemsets that compress. *Data Mining and Knowledge Discovery*, 23(1):169–214, 2011. 16

Chapter 2
Tabular Value-Based Reinforcement Learning

This chapter will introduce the classic, tabular, field of reinforcement learning, to build a foundation for the next chapters. First, we will introduce the concepts of agent and environment. Next come Markov decision processes, the formalism that is used to reason mathematically about reinforcement learning. We discuss at some length the elements of reinforcement learning: states, actions, values, and policies.

We learn about transition functions, and solution methods that are based on dynamic programming using the transition model. There are many situations where agents do not have access to the transition model, and state and reward information must be acquired from the environment. Fortunately, methods exist to find the optimal policy without a model, by querying the environment. These methods, appropriately named model-free methods, will be introduced in this chapter. Value-based model-free methods are the most basic learning approach of reinforcement learning. They work well in problems with deterministic environments and discrete action spaces, such as mazes and games. Model-free learning makes few demands on the environment, building up the policy function $\pi(s) \rightarrow a$ by sampling the environment.

After we have discussed these concepts, it is time to apply them and to understand the kinds of sequential decision problems that we can solve. We will look at Gym, a collection of reinforcement learning environments. We will also look at simple Grid world puzzles and see how to navigate those.

This is a non-deep chapter: in this chapter functions are exact, and states are stored in tables, an approach that works as long as problems are small enough to fit in memory. The next chapter shows how function approximation with neural networks works when there are more states than fit in memory.

The chapter is concluded with exercises, a summary, and pointers to further reading.

Core Concepts

- Agent, environment
- MDP: state, action, reward, value, policy
- Planning and learning
- Exploration and exploitation
- Gym, baselines

Core Problem

- Learn a policy from interaction with the environment
- Value iteration (Listing 2.1)
- Temporal difference learning (Sect. 2.2.4.5)
- Q-learning (Listing 2.5)

Finding a Supermarket

Imagine that you have just moved to a new city, you are hungry, and you want to buy some groceries. There is a somewhat unrealistic catch: you do not have a map of the city and you forgot to charge your smartphone. It is a sunny day, you put on your hiking shoes, and after some random exploration, you have found a way to a supermarket and have bought your groceries. You have carefully noted your route in a notebook, and you retrace your steps, finding your way back to your new home.

What will you do the next time that you need groceries? One option is to follow exactly the same route, exploiting your current knowledge. This option is guaranteed to bring you to the store, at no additional cost for exploring possible alternative routes. Or you could be adventurous, and explore, trying to find a new route that may actually be quicker than the old route. Clearly, there is a trade-off: you should not spend so much time exploring that you can not recoup the gains of a potential shorter route before you move elsewhere.

Reinforcement learning is a natural way of learning the optimal route as we go, by trial and error, from the effects of the actions that we take in our environment.

This little story contained many of the elements of a reinforcement learning problem, and how to solve it. There is an *agent* (you), an *environment* (the city); there are *states* (your location at different points in time), *actions* (assuming a Manhattan-style grid, moving a block left, right, forward, or back); there are *trajectories* (the routes to the supermarket that you tried); there is a *policy* (that tells which action you will take at a particular location); there is a concept of *cost/reward* (the length of your current path); we see *exploration* of new routes, *exploitation* of old routes, a trade-off between them, and your notebook in which you have been sketching a map of the city (your local *transition model*).

By the end of this chapter you will have learned which role all these topics play in reinforcement learning.

2.1 Sequential Decision Problems

Reinforcement learning is used to solve sequential decision problems [3, 20]. Before we dive into the algorithms, let us have a closer look at these problems, to better understand the challenges that the agents must solve.

In a sequential decision problem the agent has to make a sequence of decisions in order to solve a problem. Solving implies to find the sequence with the highest (expected cumulative future) reward. The solver is called the *agent*, and the problem is called *environment* (or sometimes the *world*).

We will now discuss basic examples of sequential decision problems.

2.1.1 Grid Worlds

Some of the first environments that we encounter in reinforcement learning are *Grid worlds* (Fig. 2.1). These environments consist of a rectangular grid of squares, with a start square, and a goal square. The aim is for the agent to find the sequence of actions that it must take (up, down, left, right) to arrive at the goal square. In fancy versions, a "loss" square is added, that scores minus points, or a "wall" square, that is impenetrable for the agent. By exploring the grid, taking different actions, and recording the reward (whether it reached the goal square), the agent can find a route—and when it has a route, it can try to improve that route, to find a shorter route to the goal.

Grid world is a simple environment that is well suited for manually playing around with reinforcement learning algorithms, to build up intuition of what the algorithms do. In this chapter we will model reinforcement learning problems formally and encounter algorithms that find optimal routes in Grid world.

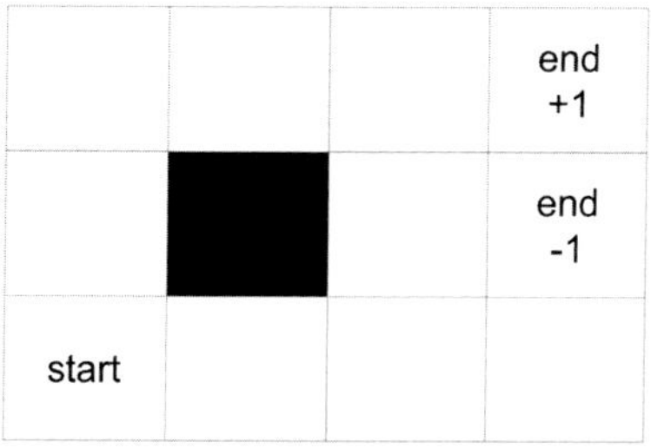

Fig. 2.1 Grid World with a goal, an "un-goal," and a wall

2.1.2 Mazes and Box Puzzles

After Grid world problems, there are more complicated problems, with extensive wall structures to make navigation more difficult (see Fig. 2.2). Trajectory planning algorithms play a central role in robotics [21, 34]; there is a long tradition of using 2D and 3D mazes for path-finding problems in reinforcement learning. The Taxi domain was introduced by Dietterich [15], and box-pushing problems such as Sokoban have also been used frequently [16, 28, 39, 60], see Fig. 2.3. The challenge in Sokoban is that boxes can only be pushed, not pulled. Actions can have the effect of creating an inadvertent dead-end for into the future, making Sokoban a difficult puzzle to play. The action space of these puzzles and mazes is discrete.

Small versions of the mazes can be solved exactly by planning, and larger instances are only suitable for approximate planning or learning methods. Solving these planning problems exactly is NP-hard or PSPACE-hard [14, 24]; as a consequence, the computational time required to solve problem instances exactly grows exponentially with the problem size and becomes quickly infeasible for all but the smallest problems.

Let us see how we can model agents to act in these types of environments.

Fig. 2.2 The Longleat Hedge Maze in Wiltshire, England

Fig. 2.3 Sokoban Puzzle [13]

2.2 Tabular Value-Based Agents

Reinforcement learning finds the best policy to operate in the environment by interacting with it. The reinforcement learning paradigm consists of an agent (you, the learner) and an environment (the world, which is in a certain state and gives you feedback on your actions).

2.2.1 Agent and Environment

In Fig. 2.4 the agent and environment are shown, together with action a_t, next state s_{t+1}, and its reward r_{t+1}. Let us have a closer look at the figure.

The environment is in a certain state s_t at time t. Then, the agent performs action a_t, resulting in a transition in the environment from state s_t to s_{t+1} at the next time step, also denoted as $s \rightarrow s'$. Along with this new state comes a reward value r_{t+1} (which may be a positive or a negative value). The goal of reinforcement learning is

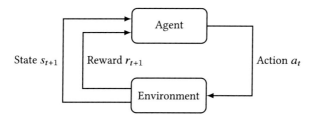

Fig. 2.4 Agent and environment [51]

to find the sequence of actions that gives the best reward. More formally, the goal is to find the optimal policy function π^* that gives in each state the best action to take in that state. By trying different actions, and accumulating the rewards, the agent can find the best action for each state. In this way, with the reinforcing reward values, the optimal policy is learned from repeated interaction with the environment, and the problem is "solved."

In reinforcement learning, the environment gives us only a number as an indication of the quality of an action that we performed, and we are left to derive the correct action policy from that, as we can see in Fig. 2.4. On the other hand, reinforcement learning allows us to generate as many action–reward pairs as we need, without a large hand-labeled dataset, and we can choose ourselves which actions to try.

2.2.2 Markov Decision Process

Sequential decision problems can be modelled as Markov decision processes (MDPs) [35]. Markov decision problems have the Markov property: the next state depends only on the current state and the actions available in it (no historical memory of previous states or information from elsewhere influences the next state) [27]. The no-memory property is important because it makes reasoning about future states possible using only the information present in the current state. If previous histories would influence the current state, and these would all have to be taken into account, then reasoning about the current state would be much harder or even infeasible.

Markov processes are named after Russian mathematician Andrey Markov (1856–1922) who is best known for his work on these stochastic processes. See [3, 20] for an introduction into MDPs. The MDP formalism is the mathematical basis under reinforcement learning, and we will introduce the relevant elements in this chapter. We follow Moerland [37] and François-Lavet et al. [20] for some of the notation and examples in this section.

Formalism
We define a Markov decision process for reinforcement learning as a 5-tuple (S, A, T_a, R_a, γ):

- S is a finite set of legal *states* of the environment; the initial state is denoted as s_0.
- A is a finite set of *actions* (if the set of actions differs per state, then A_s is the finite set of actions in state s).
- $T_a(s, s') = \Pr(s_{t+1} = s' | s_t = s, a_t = a)$ is the probability that action a in state s at time t will *transition* to state s' at time $t + 1$ in the environment.
- $R_a(s, s')$ is the *reward* received after action a transitions state s to state s'.
- $\gamma \in [0, 1]$ is the *discount factor* representing the difference between the future and present rewards.

2.2.2.1 State S

Let us have a deeper look at the Markov tuple S, A, T_a, R_a, γ, to see their role in the reinforcement learning paradigm, and how, together, they can model and describe reward-based learning processes.

At the basis of every Markov decision process is a description of the state s_t of the system at a certain time t.

State Representation

The state s contains the information to uniquely represent the configuration of the environment.

Often there is a straightforward way to uniquely represent the state in a computer memory. For the supermarket example, each identifying location is a state (such as: I am at the corner of 8th Av and 27nd St). For chess, this can be the location of all pieces on the board (plus information for the 50 move repetition rule, castling rights, and en-passant state). For robotics this can be the orientation of all joints of the robot, and the location of the limbs of the robot. For Atari, the state comprises the values of all screen pixels.

Using its current behavior policy, the agent chooses an action a, which is performed in the environment. How the environment reacts to the action is defined by the transition model $T_a(s, s')$ that is internal to the environment, which the agent does not know. The environment returns the new state s', as well as a reward value r' for the new state.

Deterministic and Stochastic Environment

In discrete deterministic environments the transition function defines a one-step transition, as each action (from a certain old state) deterministically leads to a single new state. This is the case in Grid worlds, Sokoban, and in games such as chess and checkers, where a move action deterministically leads to one new board position.

An example of a non-deterministic situation is a robot movement in an environment. In a certain state, a robot arm is holding a bottle. An agent action can be turning the bottle in a certain orientation (presumably to pour a drink in a cup). The next state may be a full cup, or it may be a mess, if the bottle was not poured in the correct orientation, or location, or if something happened in the environment such as someone bumping the table. The outcome of the action is unknown beforehand by the agent and depends on elements in the environment that are not known to the agent.

2.2.2.2 Action A

Now that we have looked at the state, it is time to look at the second item that defines an MDP, the *action*.

Irreversible Environment Action

When the agent is in state s, it chooses an action A to perform, based on its current behavior policy $\pi(a|s)$ (policies are explained soon). The agent communicates the selected action a to the environment (Fig. 2.4). For the supermarket example, an example of an action could be walking along a block in a certain direction (such as: *East*). For Sokoban, an action can be pushing a box to a new location in the warehouse. Note that in different states the possible actions may differ. For the supermarket example, walking East may not be possible at each street corner, and in Sokoban, pushing a box in a certain direction will only be possible in states where this direction is not blocked by a wall.

An action changes the state of the environment irreversibly. In the reinforcement learning paradigm, there is no *undo* operator for the environment (nor is there in the real world). When the environment has performed a state transition, it is final. The new state is communicated to the agent, together with a reward value. The actions that the agent performs in the environment are also known as its *behavior*, just as the actions of a human in the world constitute the human's behavior.

Discrete or Continuous Action Space

The actions are discrete in some applications and continuous in others. For example, the actions in board games, and choosing a direction in a navigation task in a grid, are discrete.

In contrast, arm and joint movements of robots, and bet sizes in certain games, are continuous (or span a very large range of values). Applying algorithms to continuous or very large action spaces either requires discretization of the continuous space (into buckets) or the development of a different kind of algorithm. As we will see in Chaps. 3 and 4, value-based methods work well for discrete action spaces, and policy-based methods work well for both action spaces.

For the supermarket example we can actually choose between modeling our actions discrete or continuous. From every state, we can move any number of steps, small or large, integer or fractional, in any direction. We can even walk a curvy path. So, strictly speaking, the action space is continuous. However, if, as in some cities, the streets are organized in a rectangular Manhattan-pattern, then it makes sense to discretize the continuous space, and to only consider discrete actions that take us to the next street corner. Then, our action space has become discrete, by using extra knowledge of the problem structure.[1]

---

[1] If we assume that supermarkets are large, block-sized, items that typically can be found on street corners, then we can discretize the action space. Note that we may miss small sub-block-sized supermarkets because of this simplification. Another, better, simplification would be to discretize the action space into walking distances of the size of the smallest supermarket that we expect to ever encounter.

2.2.2.3 Transition T_a

After having discussed state and action, it is time to look at the transition function $T_a(s, s')$. The transition function T_a determines how the state changes after an action has been selected. In model-free reinforcement learning the transition function is implicit to the solution algorithm: the environment has access to the transition function and uses it to compute the next state s', but the agent has *not*. (In Chap. 5 we will discuss model-based reinforcement learning. There the agent has its own transition function, an approximation of the environment's transition function, which is learned from the environment feedback.)

Graph View of the State Space

We have discussed states, actions, and transitions. The dynamics of the MDP are modelled by transition function $T_a(\cdot)$ and reward function $R_a(\cdot)$. The imaginary space of all possible states is called the *state space*. The state space is typically large. The two functions define a two-step transition from state s to s', via action a: $s \rightarrow a \rightarrow s'$.

To help our understanding of the transitions between states, we can use a graphical depiction, as in Fig. 2.5.

In the figure, states and actions are depicted as nodes (vertices), and transitions are links (edges) between the nodes. States are drawn as open circles, and actions as smaller black circles. In a certain state s, the agent can choose which action a to perform, which is then acted out in the environment. The environment returns the new state s' and the reward r'.

Figure 2.5 shows a transition graph of the elements of the MDP tuple s, a, t_a, r_a as well as s', and policy π, and how the value can be calculated. The root node at the top is state s, where policy π allows the agent to choose between three actions a, that, following distribution Pr, each can transition to two possible states s', with their reward r'. In the figure, a single transition is shown. Please use your imagination to picture the other transitions as the graph extends down.

In the left panel of the figure the environment can choose which new state it returns in response to the action (stochastic environment); in the middle panel there is only one state for each action (deterministic environment); the tree can then be simplified, showing only the states, as in the right panel.

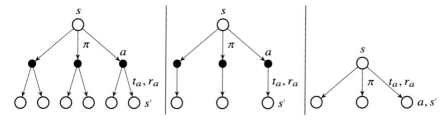

Fig. 2.5 Backup diagrams for MDP transitions: stochastic (left) and deterministic (middle and right) [51]

To calculate the value of the root of the tree, a backup procedure can be followed. Such a procedure calculates the value of a parent from the values of the children, recursively, in a bottom-up fashion, summing or maxing their values from the leaves to the root of the tree. This calculation uses discrete time steps, indicated by subscripts to the state and action, as in $s_t, s_{t+1}, s_{t+2}, \ldots$. For brevity, s_{t+1} is sometimes written as s'. The figure shows a single transition step; an episode in reinforcement learning typically consists of a sequence of many time steps.

Trial and Error, Down and Up

A graph such as the one in the center and right panel of Fig. 2.5, where child nodes have only one parent node and without cycles, is known as a *tree*. In computer science the root of a tree is at the top, and branches grow downward to the leaves.

As actions are performed and states and rewards are returned backup the tree, a learning process is taking place in the agent. We can use Fig. 2.5 to better understand the learning process that is unfolding.

The rewards of actions are learned by the agent by interacting with the environment, performing the actions. In the tree of Fig. 2.5, an action selection moves *downward*, toward the leaves. At the deeper states, we find the rewards, which we propagate to the parent states *upward*. Reward learning is learning by backpropagation: in Fig. 2.5 the reward information flows upward in the diagram from the leaves to the root. Action selection moves down, and reward learning flows up.

Reinforcement learning is learning by trial and error. *Trial* is selecting an action down (using the behavior policy) to perform in the environment. *Error* is moving up the tree, receiving a feedback reward from the environment, and reporting that back up the tree to the state to update the current behavior policy. The downward selection policy chooses which actions to explore, and the upward propagation of the error signal performs the learning of the policy.

Figures such as the one in Fig. 2.5 are useful for seeing how values are calculated. The basic notions are trial, and error, or down, and up.

2.2.2.4 Reward R_a

The reward function R_a is of central importance in reinforcement learning. It indicates the measure of quality of that state, such as *solved*, or *distance*. Rewards are associated with single states, indicating their quality. However, we are most often interested in the quality of a full decision making sequence from root to leaves (this sequence of decisions would be one possible answer to our sequential decision problem).

The reward of such a full sequence is called the *return*, sometimes denoted confusingly as R, just as the reward. The expected cumulative discounted future reward of a state is called the *value* function $V^\pi(s)$. The value function $V^\pi(s)$ is the expected cumulative reward of s where actions are chosen according to policy π. The value function plays a central role in reinforcement learning algorithms; in a few moments, we will look deeper into return and value.

2.2.2.5 Discount Factor γ

We distinguish between two types of tasks: (1) continuous time, long running, tasks and (2) episodic tasks—tasks that end. In continuous and long running tasks it makes sense to discount rewards from far in the future in order to more strongly value current information at the present time. To achieve this, a discount factor γ is used in our MDP that reduces the impact of far away rewards. Many continuous tasks use discounting, $\gamma \neq 1$.

However, in this book we will often discuss episodic problems, where γ is irrelevant. Both the supermarket example and the game of chess are episodic, and discounting does not make sense in these problems, $\gamma = 1$.

2.2.2.6 Policy π

Of central importance in reinforcement learning is the policy function π. The policy function π answers the question how the different actions a at state s should be chosen. Actions are anchored in states. The central question of MDP optimization is how to choose our actions. The policy π is a *conditional probability distribution* that for each possible state specifies the probability of each possible action. The function π is a mapping from the state space to a probability distribution over the action space:

$$\pi : S \to p(A),$$

where $p(A)$ can be a discrete or continuous probability distribution. For a particular probability (density) from this distribution we write

$$\pi(a|s).$$

Example For a discrete state space and discrete action space, we may store an explicit policy as a table, e.g.:

s	$\pi(a{=}up|s)$	$\pi(a{=}down|s)$	$\pi(a{=}left|s)$	$\pi(a{=}right|s)$
1	0.2	0.8	0.0	0.0
2	0.0	0.0	0.0	1.0
3	0.7	0.0	0.3	0.0
etc.	.	.	.	.

A special case of a policy is a *deterministic policy*, denoted by

$$\pi(s)$$

where

$$\pi : S \rightarrow A.$$

A deterministic policy selects a single action in every state. Of course the deterministic action may differ between states, as in the example below:

Example An example of a deterministic discrete policy is

s	$\pi(a=$up$|s)$	$\pi(a=$down$|s)$	$\pi(a=$left$|s)$	$\pi(a=$right$|s)$
1	0.0	1.0	0.0	0.0
2	0.0	0.0	0.0	1.0
3	1.0	0.0	0.0	0.0
etc.	.	.	.	.

We would write $\pi(s = 1) = $ down, $\pi(s = 2) = $ right, etc.

2.2.3   MDP Objective

Finding the optimal policy function is the goal of the reinforcement learning problem, and the remainder of this book will discuss many different algorithms to achieve this goal under different circumstances. Let us have a closer look at the objective of reinforcement learning. Before we can do so, we will look at traces, their return, and value functions.

2.2.3.1   Trace τ

As we start interacting with the MDP, at each time step t, we observe s_t, take an action a_t, and then observe the next state $s_{t+1} \sim T_{a_t}(s)$ and reward $r_t = R_{a_t}(s_t, s_{t+1})$. Repeating this process leads to a sequence of *trace* in the environment, which we denote by τ_t^n:

$$\tau_t^n = \{s_t, a_t, r_t, s_{t+1}, .., a_{t+n}, r_{t+n}, s_{t+n+1}\}.$$

Here, n denotes the length of the τ. In practice, we often assume $n = \infty$, which means that we run the trace until the domain terminates. In those cases, we will

Fig. 2.6 Single transition
step versus full 3-step
trace/episode/trajectory

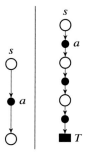

simply write $\tau_t = \tau_t^\infty$. Traces are one of the basic building blocks of reinforcement learning algorithms. They are a single full rollout of a sequence from the sequential decision problem. They are also called trajectory, episode, or simply sequence (Fig. 2.6 shows a single transition step, and an example of a three-step trace).

> **Example** A short trace with three actions could look like:
>
> $\tau_0^2 = \{s_0=1, a_0=\text{up}, r_0=-1, s_1=2, a_1=\text{up}, r_1=-1, s_2=3, a_2=\text{left}, r_2=20, s_3=5\}.$

Since both the policy and the transition dynamics can be stochastic, we will not always get the same trace from the start state. Instead, we will get a *distribution* over traces. The distribution of traces from the start state (distribution) is denoted by $p(\tau_0)$. The probability of each possible trace from the start is actually given by the product of the probability of each specific transition in the trace:

$$p(\tau_0) = p_0(s_0) \cdot \pi(a_0|s_0) \cdot T_{a_0}(s_0, s_1) \cdot \pi(a_1|s_1) \ldots$$

$$= p_0(s_0) \cdot \prod_{t=0}^{\infty} \pi(a_t|s_t) \cdot T_{a_t}(s_t, s_{t+1}). \tag{2.1}$$

Policy-based reinforcement learning depends heavily on traces, and we will discuss traces more deeply in Chap. 4. Value-based reinforcement learning (this chapter) uses single transition steps.

2.2.3.2 Return R

We have not yet formally defined what we actually want to achieve in the sequential decision making task—which is, informally, the best policy. The sum of the future reward of a trace is known as the *return*. The return of trace τ_t is

$$R(\tau_t) = r_t + \gamma \cdot r_{t+1} + \gamma^2 \cdot r_{t+2} + \ldots$$

$$= r_t + \sum_{i=1}^{\infty} \gamma^i r_{t+i}, \qquad (2.2)$$

where $\gamma \in [0, 1]$ is the discount factor. Two extreme cases are:

- $\gamma = 0$: A myopic agent, which only considers the immediate reward, $R(\tau_t) = r_t$
- $\gamma = 1$: A far-sighted agent, which treats all future rewards as equal, $R(\tau_t) = r_t + r_{t+1} + r_{t+2} + \ldots$

Note that if we would use an *infinite-horizon* return (Eq. 2.2) and $\gamma = 1.0$, then the cumulative reward may become unbounded. Therefore, in continuous problems, we use a discount factor close to 1.0, such as $\gamma = 0.99$.

> **Example** For the previous trace example we assume $\gamma = 0.9$. The return (cumulative reward) is equal to
>
> $$R(\tau_0^2) = -1 + 0.9 \cdot -1 + 0.9^2 \cdot 20 = 16.2 - 1.9 = 14.3.$$

2.2.3.3 State Value V

The real measure of optimality that we are interested in is not the return of just one trace. The environment can be stochastic and so can our policy, and for a given policy, we do not always get the same trace. Therefore, we are actually interested in the *expected* cumulative reward that a certain policy achieves. The expected cumulative discounted future reward of a state is better known as the *value* of that state.

We define the state value $V^{\pi}(s)$ as the return we expect to achieve when an agent starts in state s and then follows policy π, as

$$V^{\pi}(s) = \mathbb{E}_{\tau_t \sim p(\tau_t)} \Big[\sum_{i=0}^{\infty} \gamma^i \cdot r_{t+i} | s_t = s \Big]. \qquad (2.3)$$

> **Example** Imagine that we have a policy π, which from state s can result in two traces. The first trace has a cumulative reward of 20 and occurs in 60% of the times. The other trace has a cumulative reward of 10 and occurs 40% of the times. What is the value of state s?
>
> $$V^\pi(s) = 0.6 \cdot 20 + 0.4 \cdot 10 = 16.$$
>
> The average return (cumulative reward) that we expect to get from state s under this policy is 16.

Every policy π has one unique associated value function $V^\pi(s)$. We often omit π to simplify notation, simply writing $V(s)$, knowing a state value is always conditioned on a certain policy.

The state value is defined for every possible state $s \in S$. $V(s)$ maps every state to a real number (the expected return):

$$V : S \to \mathbb{R}.$$

> **Example** In a discrete state space, the value function can be represented as a table of size $|S|$.
>
> | s | $V^\pi(s)$ |
> | --- | --- |
> | 1 | 2.0 |
> | 2 | 4.0 |
> | 3 | 1.0 |
> | etc. | . |

Finally, the state value of a terminal state is by definition zero:

$$s = \text{terminal} \quad \Rightarrow \quad V(s) := 0.$$

2.2.3.4 State–Action Value Q

In addition to state values $V^\pi(s)$, we also define state–action value $Q^\pi(s, a)$.[2] The only difference is that we now condition on a state *and action*. We estimate the average return we expect to achieve when taking action a in state s and follow

---

[2] The reason for the choice for letter Q is lost in the mists of time. Perhaps it is meant to indicate quality.

policy π afterward:

$$Q^\pi(s, a) = \mathbb{E}_{\tau_t \sim p(\tau_t)}\Big[\sum_{i=0}^{\infty} \gamma^i \cdot r_{t+i} | s_t = s, a_t = a\Big]. \tag{2.4}$$

Every policy π has only one unique associated state–action value function $Q^\pi(s, a)$. We often omit π to simplify notation. Again, the state–action value is a function

$$Q : S \times A \to \mathbb{R},$$

which maps every state–action pair to a real number.

Example For a discrete state and action space, $Q(s, a)$ can be represented as a table of size $|S| \times |A|$. Each table entry stores a $Q(s, a)$ estimate for the specific s, a combination:

	a=up	a=down	a=left	a=right
s=1	4.0	3.0	7.0	1.0
s=2	2.0	-4.0	0.3	1.0
s=3	3.5	0.8	3.6	6.2
etc.	.	.	.	.

The state–action value of a terminal state is by definition zero:

$$s = \text{terminal} \quad \Rightarrow \quad Q(s, a) := 0, \quad \forall a.$$

2.2.3.5 Reinforcement Learning Objective

We now have the ingredients to formally state the objective $J(\cdot)$ of reinforcement learning. The objective is to achieve the highest possible average return from the start state:

$$J(\pi) = V^\pi(s_0) = \mathbb{E}_{\tau_0 \sim p(\tau_0 | \pi)}\Big[R(\tau_0)\Big] \tag{2.5}$$

for $p(\tau_0)$ given in Eq. 2.1. There is one optimal value function, which achieves higher or equal value than all other value functions. We search for a policy that achieves this optimal value function, which we call the optimal policy $\pi^\star$:

$$\pi^\star(a|s) = \arg\max_{\pi} V^\pi(s_0). \tag{2.6}$$

This function $\pi^\star$ is the optimal policy, and it uses the arg max function to select the policy with the optimal value. The goal in reinforcement learning is to find this optimal policy for start state s_0.

A potential benefit of state–action values Q over state values V is that state–action values directly tell what every action is worth. This may be useful for action selection, since, for discrete action spaces,

$$a^\star = \arg\max_{a \in A} Q^\star(s, a),$$

the Q function directly identifies the best action. Equivalently, the optimal policy can be obtained directly from the optimal Q function:

$$\pi^\star(s) = \arg\max_{a \in A} Q^\star(s, a).$$

We will now turn to construct algorithms to compute the value function and the policy function.

2.2.3.6 Bellman Equation

To calculate the value function, let us look again at the tree in Fig. 2.5 on page 33, and imagine that it is many times larger, with subtrees that extend to fully cover the state space. Our task is to compute the value of the root, based on the reward values at the real leaves, using the transition function T_a. One way to calculate the value $V(s)$ is to traverse this full state space tree, computing the value of a parent node by taking the reward value and the sum of the children, discounting this value by γ.

This intuitive approach was first formalized by Richard Bellman in 1957. Bellman showed that discrete optimization problems can be described as a recursive backward induction problem [7]. He introduced the term dynamic programming to recursively traverse the states and actions. The so-called *Bellman equation* shows the relationship between the value function in state s and the future child state s', when we follow the transition function.

The discrete Bellman equation of the value of state s after following policy π is[3]

$$V^\pi(s) = \sum_{a \in A} \pi(a|s) \left[\sum_{s' \in S} T_a(s, s') \left[R_a(s, s') + \gamma \cdot V^\pi(s') \right] \right], \tag{2.7}$$

where π is the probability of action a in state s, T is the stochastic transition function, R is the reward function, and γ is the discount rate. Note the recursion on the value function and that for the Bellman equation, the transition and reward functions must be known for all states by the agent.

---

[3] State–action value and continuous Bellman equations can be found in Appendix A.4.

Together, the transition and reward models are referred to as the *dynamics model* of the environment. The dynamics model is often not known by the agent, and model-free methods have been developed to compute the value function and policy function without them.

The recursive Bellman equation is the basis of algorithms to compute the value function, and other relevant functions to solve reinforcement learning problems. In the next section we will study these solution methods.

2.2.4   MDP Solution Methods

The Bellman equation is a recursive equation: it shows how to calculate the value of a state, out of the values of applying the function specification again on the successor states. Figure 2.7 shows a recursive picture, of a picture in a picture, in a picture, etc. In algorithmic form, dynamic programming calls its own code on states that are closer and closer to the leaves, until the leaves are reached, and the recursion cannot go further.

Dynamic programming uses the principle of divide and conquer: it begins with a start state whose value is to be determined by searching a large subtree, which it does by going down into the recursion, finding the value of sub-states that are closer

Fig. 2.7 Recursion: Droste effect

```
1  def value_iteration():
2      initialize(V)
3      while not convergence(V):
4          for s in range(S):
5              for a in range(A):
6                  Q[s,a] = ∑_{s'∈S} T_a(s, s')(R_a(s, s') + γ V[s'])
7              V[s] = max_a(Q[s,a])
8      return V
```

Listing 2.1 Value iteration pseudocode

to terminals. At terminals the reward values are known, and these are then used in the construction of the parent values, as it goes up, back out of the recursion, and ultimately arrives at the root value itself.

A simple dynamic programming method to iteratively traverse the state space to calculate Bellman's equation is *value iteration* (VI). Pseudocode for a basic version of VI is shown in Listing 2.1, based on [2]. Value iteration converges to the optimal value function by iteratively improving the estimate of $V(s)$. The value function $V(s)$ is first initialized to random values. Value iteration repeatedly updates $Q(s, a)$ and $V(s)$ values, looping over the states and their actions, until convergence occurs (when the values of $V(s)$ stop changing much).

Value iteration works with a finite set of actions. It has been proven to converge to the optimal values, but, as we can see in the pseudocode in Listing 2.1, it does so quite inefficiently by essentially repeatedly enumerating the entire state space in a triply nested loop, traversing the state space many times. Soon we will see more efficient methods.

2.2.4.1 Hands On: Value Iteration in Gym

We have discussed in detail how to model a reinforcement learning problem with an MDP. We have talked in depth and at length about states, actions, and policies. It is now time for some hands-on work, to experiment with the theoretical concepts. We will start with the environment.

2.2.4.2 OpenAI Gym

OpenAI has created the Gym suite of *environments* for Python, which has become the de facto standard in the field [11]. The Gym suite can be found at OpenAI[4] and on GitHub.[5] Gym works on Linux, macOS, and Windows. An active community

---

[4] https://gym.openai.com.

[5] https://github.com/openai/gym.

```
1  import gym
2
3  env = gym.make('CartPole-v0')
4  env.reset()
5  for _ in range(1000):
6      env.render()
7      env.step(env.action_space.sample()) # take a random action
8  env.close()
```

Listing 2.2 Running the Gym CartPole environment from Gym

exists, and new environments are created continuously and uploaded to the Gym website. Many interesting environments are available for experimentation, to create your own agent algorithm for, and test it.

If you browse Gym on GitHub, you will see different sets of environments, from easy to advanced. There are the classics, such as Cartpole and Mountain car. There are also small text environments. Taxi is there, and the Arcade Learning Environment [6], which was used in the paper that introduced DQN [36], as we will discuss at length in the next chapter. MuJoCo[6] is also available, an environment for experimentation with simulated robotics [54], or you can use PyBullet.[7]

You should now install Gym. Go to the Gym page on https://gym.openai.com and read the documentation. Make sure Python is installed on your system (does typing Python at the command prompt work?) and that your Python version is up to date (version 3.10 at the time of this writing). Then type

```
pip install gym
```

to install Gym with the Python package manager. Soon, you will also be needing deep learning suites, such as TensorFlow or PyTorch. It is recommended to install Gym in the same virtual environment as your upcoming PyTorch and TensorFlow installation, so that you can use both at the same time (see Sect. B.3.3.1). You may have to install or update other packages, such as numpy, scipy, and pyglet, to get Gym to work, depending on your system installation.

You can check if the installation works by trying if the CartPole environment works, see Listing 2.2. A window should appear on your screen in which a Cartpole is making random movements (your window system should support OpenGL, and you may need a version of pyglet newer than version 1.5.11 on some operating systems).

---

[6] http://www.mujoco.org.

[7] https://pybullet.org/wordpress/.

2.2.4.3 Taxi Example with Value Iteration

The Taxi example (Fig. 2.8) is an environment where taxis move up, down, left, and right, and pick up and drop off passengers. Let us see how we can use value iteration to solve the Taxi problem. The Gym documentation describes the Taxi world as follows. There are four designated locations in the Grid world indicated by R(ed), B(lue), G(reen), and Y(ellow). When the episode starts, the taxi starts off at a random square and the passenger is at a random location. The taxi drives to the passenger's location, picks up the passenger, drives to the passenger's destination (another one of the four specified locations), and then drops off the passenger. Once the passenger is dropped off, the episode ends.

The Taxi problem has 500 discrete states: there are 25 taxi positions, five possible locations of the passenger (including the case when the passenger is in the taxi), and 4 destination locations ($25 \times 5 \times 4$).

The environment returns a new result tuple at each step. There are six discrete deterministic actions for the Taxi driver:

0: Move south
1: Move north
2: Move east
3: Move west
4: Pick up passenger
5: Drop off passenger

There is a reward of -1 for each action and an additional reward of $+20$ for delivering the passenger, and a reward of -10 for executing actions *pickup* and *dropoff* illegally.

Fig. 2.8 Taxi world [30]

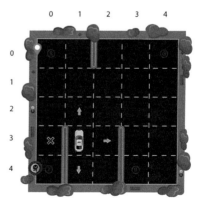

The Taxi environment has a simple transition function, which is used by the agent in the value iteration code.[8] Listing 2.3 shows an implementation of value iteration that uses the Taxi environment to find a solution. This code is written by Mikhail Trofimov and illustrates clearly how value iteration first creates the value function for the states, and then that a policy is formed by finding the best action in each state, in the build-greedy-policy function.[9]

To get a feeling for how the algorithms work, please use the value iteration code with the Gym Taxi environment, see to Listing 2.3. Run the code, and play around with some of the hyperparameters to familiarize yourself a bit with Gym and with planning by value iteration. Try to visualize for yourself what the algorithm is doing. This will prepare you for the more complex algorithms that we will look into next.

2.2.4.4 Model-Free Learning

The value iteration algorithm can compute the policy function. It uses the transition model in its computation. Frequently, we are in a situation when the exact transition probabilities are not known to the agent, and we need other methods to compute the policy function. For this situation, model-free algorithms have been developed, where the agent can compute the policy without knowing any of the transition probabilities itself.

The development of these model-free methods is a major milestone of reinforcement learning, and we will spend some time to understand how they work. We will start with value-based model-free algorithms. We will see how, when the agent does not know the transition function, an optimal policy can be learned by sampling rewards from the environment. Table 2.1 lists value iteration in conjunction with the value-based model-free algorithms that we cover in this chapter. (Policy-based model-free algorithms will be covered in Chap. 4.)

First we will discuss how the principle of *temporal difference* uses sampling and bootstrapping to construct the value function. We will see how the value function can be used to find the best actions, to form the policy. Second, we will discuss which mechanisms for action selection exist, where we will encounter the *exploration/exploitation* trade-off. Third, we will discuss how to learn from the rewards of the selected actions. We will encounter *on-policy* learning and *off-policy* learning. We will discuss two simple algorithms: SARSA and Q-learning. Let us now start by having a closer look at sampling actions with temporal difference learning.

---

[8] Note that the code uses the environment to compute the next state, so that we do not have to implement a version of the transition function for the agent.

[9] https://gist.github.com/geffy/b2d16d01cbca1ae9e13f11f678fa96fd#file-taxi-vi-py.

```
1   import gym
2   import numpy as np
3
4   def iterate_value_function(v_inp, gamma, env):
5       ret = np.zeros(env.nS)
6       for sid in range(env.nS):
7           temp_v = np.zeros(env.nA)
8           for action in range(env.nA):
9               for (prob, dst_state, reward, is_final) in env.P[sid
                    ][action]:
10                  temp_v[action] += prob*(reward + gamma*v_inp[
                        dst_state]*(not is_final))
11          ret[sid] = max(temp_v)
12      return ret
13
14  def build_greedy_policy(v_inp, gamma, env):
15      new_policy = np.zeros(env.nS)
16      for state_id in range(env.nS):
17          profits = np.zeros(env.nA)
18          for action in range(env.nA):
19              for (prob, dst_state, reward, is_final) in env.P[
                    state_id][action]:
20                  profits[action] += prob*(reward + gamma*v[
                        dst_state])
21          new_policy[state_id] = np.argmax(profits)
22      return new_policy
23
24
25  env = gym.make('Taxi-v3')
26  gamma = 0.9
27  cum_reward = 0
28  n_rounds = 500
29  env.reset()
30  for t_rounds in range(n_rounds):
31      # init env and value function
32      observation = env.reset()
33      v = np.zeros(env.nS)
34
35      # solve MDP
36      for _ in range(100):
37          v_old = v.copy()
38          v = iterate_value_function(v, gamma, env)
39          if np.all(v == v_old):
40              break
41      policy = build_greedy_policy(v, gamma, env).astype(np.int)
42
43      # apply policy
44      for t in range(1000):
45          action = policy[observation]
46          observation, reward, done, info = env.step(action)
47          cum_reward += reward
48          if done:
49              break
50      if t_rounds % 50 == 0 and t_rounds > 0:
51          print(cum_reward * 1.0 / (t_rounds + 1))
52  env.close()
```

Listing 2.3 Value iteration for Gym Taxi

Table 2.1 Tabular value-based approaches

Name	Approach	Ref
Value iteration	Model-based enumeration	[2, 7]
SARSA	On-policy temporal difference model-free	[44]
Q-learning	Off-policy temporal difference model-free	[56]

2.2.4.5 Temporal Difference Learning

In the previous section the value function was calculated recursively by using the value function of successor states, following Bellman's equation (Eq. 2.7).

Bootstrapping is a process of subsequent refinement by which old estimates of a value are refined with new updates. It means literally: pull yourself up by your boot straps. Bootstrapping solves the problem of computing a final value when we only know how to compute step-by-step intermediate values. Bellman's recursive computation is a form of bootstrapping. In model-free learning, the role of the transition function is replaced by an iterative sequence of environment samples.

A bootstrapping method that can be used to process the samples, and to refine them to approximate the final value, is *temporal difference learning*. Temporal difference learning, TD for short, was introduced by Sutton [50] in 1988. The temporal difference in the name refers to the difference in values between two time steps, which are used to calculate the value at the new time step.

Temporal difference learning works by updating the current estimate of the state value $V(s)$ (the bootstrap value) with an error value based on the estimate of the next state that it has gotten through sampling the environment:

$$V(s) \leftarrow V(s) + \alpha[r' + \gamma V(s') - V(s)]. \qquad (2.8)$$

Here s is the current state, s' the new state, and r' the reward of the new state. Note the introduction of α, the learning rate, which controls how fast the algorithm learns (bootstraps). It is an important parameter; setting the value too high can be detrimental since the last value then dominates the bootstrap process too much. Finding the optimal value will require experimentation. The γ parameter is the discount rate. The last term $-V(s)$ subtracts the value of the current state, to compute the temporal difference. Another way to write this update rule is

$$V(s) \leftarrow \alpha[r' + \gamma V(s')] + (1 - \alpha)V(s)$$

as the difference between the new temporal difference target and the old value. Note the absence of transition model T in the formula; temporal difference is a model-free update formula.

The introduction of the temporal difference method has allowed model-free methods to be used successfully in various reinforcement learning settings. Most notably, it was the basis of the program TD-Gammon that beats human world champions in the game of Backgammon in the early 1990s [52].

2.2.4.6 Find Policy by Value-Based Learning

The goal of reinforcement learning is to construct the policy with the highest cumulative reward. Thus, we must find the best action a in each state s. In the value-based approach we know the value functions $V(s)$ or $Q(s, a)$. How can that help us to find action a? In a discrete action space, there is at least one discrete action with the highest value. Thus, if we have the optimal state value $V^\star$, then the optimal policy can be found by finding the action with that value. This relationship is given by

$$\pi^\star = \max_\pi V^\pi(s) = \max_{a,\pi} Q^\pi(s, a),$$

and the arg max function finds the best action for us

$$a^\star = \arg\max_{a \in A} Q^\star(s, a).$$

In this way the optimal policy sequence of best actions $\pi^\star(s)$ can be recovered from the value functions, hence the name *value-based method* [58].

A full reinforcement learning algorithm consists of a rule for the selection (downward) and a rule for the learning (upward) step. Now that we know how to calculate the value function (the up motion in the tree diagram), let us see how we can select the action in our model-free algorithm (the down motion in the tree diagram).

2.2.4.7 Exploration

Since there is no local transition function, model-free methods perform their state changes directly in the environment. This may be an expensive operation, for example, when a real-world robot arm has to perform a movement. The sampling policy should choose promising actions to reduce the number of samples as much as possible, and not waste any actions. What behavior policy should we use? It is tempting to favor at each state the actions with the highest Q-value, since then we would be following what is currently thought to be the best policy.

This approach is called the *greedy* approach. It appears attractive but is short-sighted and risks settling for local maxima. Following the trodden path based on only a few early samples risks missing a potential better path. Indeed, the greedy approach is high variance, using values based on few samples, resulting in a high uncertainty. We run the risk of circular reinforcement, if we update the same behavior policy that we use to choose our samples from. In addition to exploiting known good actions, a certain amount of exploration of unknown actions is necessary. Smart sampling strategies use a mix of the current behavior policy (*exploitation*) and randomness (*exploration*) to select which action to perform in the environment.

2.2.4.8 Bandit Theory

The exploration/exploitation trade-off, the question of how to get the most reliable information at the least cost, has been studied extensively in the literature [26, 57]. The field has the colorful name of multi-armed bandit theory [4, 22, 33, 43]. A *bandit* in this context refers to a casino slot machine, with not one arm, but many arms, each with a different and unknown payout probability. Each trial costs a coin. The multi-armed bandit problem is then to find a strategy that finds the arm with the highest payout at the least cost.

A multi-armed bandit is a single-state single-decision reinforcement learning problem, a one-step non-sequential decision making problem, with the arms representing the possible actions. This simplified model of stochastic decision making allows the in-depth study of exploration/exploitation strategies.

Single-step exploration/exploitation questions arise for example in clinical trials, where new drugs are tested on test subjects (real people). The bandit is the trial, and the arms are the choice how many of the test subjects are given the real experimental drug, and how many are given the placebo. This is a serious setting, since the cost may be measured in the quality of human lives.

In a conventional fixed randomized controlled trial (supervised setup) the sizes of the groups that get the experimental drugs and the control group would be fixed, and the confidence interval and the duration of the test would also be fixed. In an adaptive trial (bandit setup) the sizes would adapt during the trial depending on the outcomes, with more people getting the drug if it appears to work, and fewer if it does not.

Let us have a look at Fig. 2.9. Assume that the learning process is a clinical trial in which three new compounds are tested for their medical effect on test subjects.

Fig. 2.9 Adaptive trial [1]

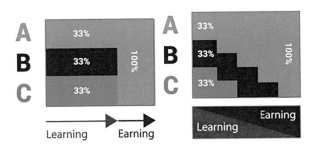

In the fixed trial (left panel) all test subjects receive the medicine of their group to the end of the test period, after which the data set is complete and we can determine which of the compounds has the best effect. At that point we know which group has had the best medicine, and which two thirds of the subjects did not, with possibly harmful effect. Clearly, this is not a satisfactory situation. It would be better if we could gradually adjust the proportion of the subjects that receive the medicine that currently looks best, as our confidence in our test results increases as the trial progresses. Indeed, this is what reinforcement learning does (Fig. 2.9, right panel). It uses a mix of exploration and exploitation, adapting the treatment, giving more subjects the promising medicine, while achieving the same confidence as the static trial at the end [32, 33].

2.2.4.9 ϵ-Greedy Exploration

A popular pragmatic approach is to use a fixed ratio of exploration versus exploitation. This approach is known as the ϵ-greedy approach, which is to mostly try the (greedy) action that currently has the highest policy value except to explore an ϵ fraction of times a randomly selected other action. If $\epsilon = 0.1$, then 90% of the times the currently best action is taken, and 10% of the times a random other action.

The algorithmic choice between greedily exploiting known information and exploring unknown actions to gain new information is called the exploration/exploitation trade-off. It is a central concept in reinforcement learning; it determines how much confidence we have in our outcome, and how quickly the confidence can be increased and the variance reduced. A second approach is to use an adaptive ϵ-ratio that changes over time or over other statistics of the learning process.

Other popular approaches to add exploration are to add Dirichlet noise [31] or to use Thompson sampling [46, 53].

2.2.4.10 Off-Policy Learning

In addition to the exploration/exploitation selection question, another main theme in the design of reinforcement learning algorithms is how the behavior policy learns. Should it update strictly on-policy—only from its recent actions—or off-policy, learning from all available information?

Reinforcement learning is concerned with learning a policy from actions and rewards. The agent selects an action and learns from the reward that it gets back from the environment.

Normally, the learning takes place by consistently backing up the value of the selected action back to the same behavior policy function that was used to select the action: the learning is *on-policy*. There is, however, an alternative. If the learning takes place by backing up values of another action, not the one selected by the behavior policy, then this is known as *off-policy learning*. This makes sense when the behavior policy explores by selecting a non-optimal action (it does not perform the greedy exploitation action; this usually results in an inferior reward value). On-policy learning would then have to back up the value of the non-optimal exploration action (since otherwise it would not learn on-policy). Off-policy learning, however, is free to back up the value of the best action instead, and not the inferior one selected by the exploration policy, not polluting the policy with a known inferior choice. Thus, in the case of exploration, off-policy learning can be more efficient, by not stubbornly backing up the value of the action selected by the behavior policy, but the value of an older, better, action instead. (In the case of an exploiting step by the policy, on-policy learning and off-policy learning are the same.)

Learning methods face a trade-off: they try to learn the best target action from current behavior that is known so far (learning exploitation), or they choose new (but sometimes non-optimal) behavior. Do we penalize a node when exploration found a worse value (as expected) or do we keep the optimistic value? Thus, off-policy learning uses *two* policies: first, the behavior policy that is used for actual action selection behavior (sometimes exploring), and second, the target policy that is updated by backing up values. The first policy performs *downward* selection to expand states, and the second policy performs *upward* value propagation to update the target policy. In on-policy learning, these policies are the same, and in off-policy learning, the behavior policy and the target policy are separate.

A well-known tabular on-policy algorithm is SARSA.[10] An even more well-known tabular off-policy algorithm is Q-learning. Both SARSA and Q-learning converge to the optimal action value function, although convergence in off-policy can be slower, since older, non-current, values are used.

2.2.4.11 On-Policy SARSA

In on-policy learning a single policy function is used for (downward) action selection and (upward) value backup toward the learning target. SARSA is an on-policy algorithm [44]. On-policy learning updates values directly on the single policy. The same policy function is used for exploration behavior and for the target policy.

The SARSA update formula is

$$Q(s_t, a_t) \leftarrow Q(s_t, a_t) + \alpha[r_{t+1} + \gamma Q(s_{t+1}, a_{t+1}) - Q(s_t, a_t)]. \tag{2.9}$$

Going back to temporal difference (Eq. 2.8), we see that the SARSA formula looks very much like TD, although now we deal with state–action values, and temporal difference dealt with state values.

On-policy learning selects an action, evaluates it in the environment, and moves on to better actions, guided by the behavior policy (which is not specified in the

---

[10] The name of the SARSA algorithm is a play on the MDP symbols as they occur in the action value update formula: s, a, r, s, a.

```
1   def qlearn(environment, alpha=0.001, gamma=0.9, epsilon=0.05):
2       Q[TERMINAL,_] = 0 # policy
3       for episode in range(max_episodes):
4           s = s0
5           while s not TERMINAL: # perform steps of one full episode
6               a = epsilongreedy(Q[s], epsilon)
7               (r, sp) = environment(s, a)
8               Q[s,a] = Q[s,a] + alpha*(r+gamma*max(Q[sp])-Q[s,a])
9               s = sp
10      return Q
```

Listing 2.4 Q-learning pseudocode [51, 56]

formula but might be ϵ-greedy). On-policy learning begins with a behavior policy, samples the state space with this policy, and improves the policy by backing up values of the selected actions. Note that the term $Q(s_{t+1}, a_{t+1})$ can also be written as $Q(s_{t+1}, \pi(s_{t+1}))$, highlighting the difference with off-policy learning. SARSA updates its Q-values using the Q-value of the next state s and the current policy's action.

The primary advantage of on-policy learning is that it directly optimizes the target of interest and converges quickly by learning with the direct behavior values. The biggest drawback is sample inefficiency, since the target policy is updated with suboptimal explorative rewards.

2.2.4.12 Off-Policy Q-Learning

Off-policy learning is more complicated; it uses separate behavior and target policies: one for exploratory downward selection behavior, and one to update as the current target backup policy. Learning (backing up) is from data *off* the downward behavior policy, and the whole method is therefore called off-policy learning.

The most well-known off-policy algorithm is Q-learning [56]. It gathers information from explored moves, and it evaluates states as if a greedy policy was used, even when the actual behavior performed an exploration step.

The Q-learning update formula is

$$Q(s_t, a_t) \leftarrow Q(s_t, a_t) + \alpha[r_{t+1} + \gamma \max_a Q(s_{t+1}, a) - Q(s_t, a_t)]. \qquad (2.10)$$

The difference from on-policy learning is that the $\gamma Q(s_{t+1}, a_{t+1})$ term from Eq. 2.9 has been replaced by $\gamma \max_a Q(s_{t+1}, a)$. The learning is from backup values of the best action, not the one that was actually evaluated. Listing 2.4 shows the full pseudocode for Q-learning.

The reason that Q-learning is off-policy is that it updates its Q-values using the Q-value of the next state s *and the greedy action* (not necessarily the behavior policy's action—it is learning off the behavior policy). In this sense, off-policy learning collects all available information and uses it simultaneously to construct the best target policy.

On-policy targets follow the behavioral policy, and convergence is typically more stable (low variance). Off-policy targets can learn the optimal policy/value (low bias), but they can be unstable due to the max operation, especially in combination with function approximation, as we will see in the next chapter.

2.2.4.13   Sparse Rewards and Reward Shaping

Before we conclude this section, we should discuss sparsity. For some environments a reward exists for each state. For the supermarket example a reward can be calculated for each state that the agent has walked to. The reward is actually the opposite of the cost expended in walking. Environments in which a reward exists in each state are said to have a dense reward structure.

For other environments rewards may exist for only some of the states. For example, in chess, rewards only exist at terminal board positions where there is a win or a draw. In all other states the return depends on the future states and must be calculated by the agent by propagating reward values from future states up toward the root state s_0. Such an environment is said to have a sparse reward structure.

Finding a good policy is more complicated when the reward structure is sparse. A graph of the landscape of such a sparse reward function would show a flat landscape with a few sharp mountain peaks. Many of the algorithms that we will see in future chapters use the reward gradient to find good returns. Finding the optimum in a flat landscape where the gradient is zero is hard. In some applications it is possible to change the reward function to have a shape more amenable to gradient-based optimization algorithms such as we use in deep learning. Reward shaping can make all the difference when no solution can be found with a naive reward function. It is a way of incorporating heuristic knowledge into the MDP. A large literature on reward shaping and heuristic information exists [40]. The use of heuristics on board games such as chess and checkers can also be regarded as reward shaping.

2.2.4.14   Hands on: Q-Learning on Taxi

To get a feeling for how these algorithms work in practice, let us see how Q-learning solves the Taxi problem.

In Sect. 2.2.4.3 we discussed how value iteration can be used for the Taxi problem, provided that the agent has access to the transition model. We will now see how we solve this problem if we do not have the transition model. Q-learning samples actions and records the reward values in a Q-table, converging to the state–action value function. When in all states the best values of the best actions are known, then these can be used to sequence the optimal policy.

Let us see how a value-based model-free algorithm solves a simple 5×5 Taxi problem. Refer to Fig. 2.8 on page 45 for an illustration of Taxi world.

```
1   # Q learning for OpenAI Gym Taxi environment
2   import gym
3   import numpy as np
4   import random
5   #Environment Setup
6   env = gym.make("Taxi-v2")
7   env.reset()
8   env.render()
9   # Q[state,action] table implementation
10  Q = np.zeros([env.observation_space.n, env.action_space.n])
11  gamma = 0.7    # discount factor
12  alpha = 0.2    # learning rate
13  epsilon = 0.1 # epsilon greedy
14  for episode in range(1000):
15      done = False
16      total_reward = 0
17      state = env.reset()
18      while not done:
19          if random.uniform(0, 1) < epsilon:
20              action = env.action_space.sample() # Explore state
                    space
21          else:
22              action = np.argmax(Q[state]) # Exploit learned values
23          next_state, reward, done, info = env.step(action) #
                invoke Gym
24          next_max = np.max(Q[next_state])
25          old_value = Q[state,action]
26
27          new_value = old_value + alpha * (reward + gamma *
                next_max - old_value)
28
29          Q[state,action] = new_value
30          total_reward += reward
31          state = next_state
32      if episode % 100 == 0:
33          print("Episode_{}_Total_Reward:_{}".format(episode,
                total_reward))
```

Listing 2.5 Q-learning Taxi example, after [30]

Please recall that in Taxi world, the taxi can be in one of 25 locations, and there are $25 \times (4 + 1) \times 4 = 500$ different states that the environment can be in.

We follow the reward model as it is used in the Gym Taxi environment. Recall that our goal is to find a policy (actions in each state) that leads to the highest cumulative reward. Q-learning learns the best policy through guided sampling. The agent records the rewards that it gets from actions that it performs in the environment. The Q-values are the expected rewards of the actions in the states. The agent uses the Q-values to guide which actions it will sample. Q-values $Q(s, a)$ are stored in an array that is indexed by state and action. The Q-values guide the exploration, and higher values indicate better actions.

```
1   total_epochs, total_penalties = 0, 0
2   ep = 100
3   for _ in range(ep):
4       state = env.reset()
5       epochs, penalties, reward = 0, 0, 0
6       done = False
7       while not done:
8           action = np.argmax(Q[state])
9           state, reward, done, info = env.step(action)
10          if reward == -10:
11              penalties += 1
12          epochs += 1
13      total_penalties += penalties
14      total_epochs += epochs
15  print(f"Results_after_{ep}_episodes:")
16  print(f"Average_timesteps_per_episode:_{total_epochs_/_ep}")
17  print(f"Average_penalties_per_episode:_{total_penalties_/_ep}")
```

Listing 2.6 Evaluate the optimal Taxi result, after [30]

Listing 2.5 shows the full Q-learning algorithm, in Python, after [30]. It uses an ϵ-greedy behavior policy: mostly the best action is followed, but in a certain fraction a random action is chosen, for exploration. Recall that the Q-values are updated according to the Q-learning formula:

$$Q(s_t, a_t) \leftarrow Q(s_t, a_t) + \alpha[r_{t+1} + \gamma \max_a Q(s_{t+1}, a) - Q(s_t, a_t)],$$

where $0 \leq \gamma \leq 1$ is the discount factor and $0 < \alpha \leq 1$ is the learning rate. Note that Q-learning uses bootstrapping, and the initial Q-values are set to a random value (their value will disappear slowly due to the learning rate).

Q-learning learns the best action in the current state by looking at the reward for the current state–action combination, plus the maximum rewards for the next state. Eventually the best policy is found in this way, and the taxi will consider the route consisting of a sequence of the best rewards.

To summarize informally:

1. Initialize the Q-table to random values.
2. Select a state s.
3. For all possible actions from s, select the one with the highest Q-value and travel to this state, which becomes the new s, or, with ϵ-greedy, explore.
4. Update the values in the Q-array using the equation.
5. Repeat until the goal is reached; when the goal state is reached, repeat the process until the Q-values stop changing (much), then stop.

Listing 2.5 shows Q-learning code for finding the policy in Taxi world.

The optimal policy can be found by sequencing together the actions with the highest Q-value in each state. Listing 2.6 shows the code for this. The number of illegal pickups/drop-offs is shown as penalty.

This example shows how the optimal policy can be found by the introduction of a Q-table that records the quality of irreversible actions in each state, and uses that table to converge the rewards to the value function. In this way the optimal policy can be found model-free.

2.2.4.15 Tuning Your Learning Rate

Go ahead, implement and run this code, and play around to become familiar with the algorithm. Q-learning is an excellent algorithm to learn the essence of how reinforcement learning works. Try out different values for hyperparameters, such as the exploration parameter ϵ, the discount factor γ, and the learning rate α. To be successful in this field, it helps to have a feeling for these hyperparameters. A choice close to 1 for the discount parameter is usually a good start, and a choice close to 0 for the learning rate is a good start. You may feel a tendency to do the opposite, to choose the learning rate as high as possible (close to 1) to learn as quickly as possible. Please go ahead and see which works best in Q-learning (you can have a look at [17]). In many deep learning environments a high learning rate is a recipe for disaster, your algorithm may not converge at all, and Q-values can become unbounded. Play around with tabular Q-learning, and approach your deep learning slowly, with gentle steps!

The Taxi example is small, and you will get results quickly. It is well suited to build up useful intuition. In later chapters, we will do experiments with deep learning that take longer to converge, and acquiring intuition for tuning hyperparameter values will be more expensive.

Conclusion

We have now seen how a value function can be learned by an agent without having the transition function, by sampling the environment. Model-free methods use actions that are irreversible for the agent. The agent samples states and rewards from the environment, using a behavior policy with the current best action, and following an exploration/exploitation trade-off. The backup rule for learning is based on bootstrapping and can follow the rewards of the actions on-policy, including the value of the occasional explorative action, or off-policy, always using the value of the best action. We have seen two model-free tabular algorithms, SARSA and Q-learning, where the value function is assumed to be stored in an exact table structure.

In the next chapter we will move to network-based algorithms for high-dimensional state spaces, based on function approximation with a deep neural network.

2.3   Classic Gym Environments

Now that we have discussed at length the tabular agent algorithms, it is time to
have a look at the environments, the other part of the reinforcement learning model.
Without them, progress cannot be measured, and results cannot be compared in a
meaningful way. In a real sense, environments define the kind of intelligence that
our artificial methods can be trained to perform.

In this chapter we will start with a few smaller environments that are suited for the
tabular algorithms that we have discussed. Two environments that have been around
since the early days of reinforcement learning are Mountain car and Cartpole (see
Fig. 2.10).

2.3.1   Mountain Car and Cartpole

Mountain car is a physics puzzle in which a car on a one-dimensional road is in
a valley between two mountains. The goal for the car is to drive up the mountain
and reach the flag on the right. The car's engine can go forward and backward. The
problem is that the car's engine is not strong enough to climb the mountain by itself
in a single pass [38], but it can do so with the help of gravity: by repeatedly going
back and forth the car can build up momentum. The challenge for the reinforcement
learning agent is to apply alternating backward and forward forces at the right
moment.

Cartpole is a pole-balancing problem. A pole is attached by a joint to a movable
cart, which can be pushed forward or backward. The pendulum starts upright and
must be kept upright by applying either a force of $+1$ or -1 to the cart. The puzzle
ends when the pole falls over, or when the cart runs too far left or right [5]. Again
the challenge is to apply the right force at the right moment, solely by feedback of
the pole being upright or too far down.

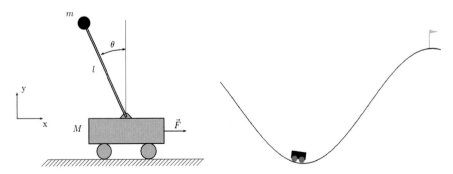

Fig. 2.10 Cartpole and Mountain car

2.3.2   Path Planning and Board Games

Navigation tasks and board games provide environments for reinforcement learning that are simple to understand. They are well suited to reason about new agent algorithms. Navigation problems, and the heuristic search trees built for board games, can be of moderate size and are then suited for determining the best action by dynamic programming methods, such as tabular Q-learning, A*, branch and bound, and alpha–beta [45]. These are straightforward search methods that do not attempt to generalize to new, unseen, states. They find the best action in a space of states, all of which are present at training time—the optimization methods do not perform generalization from training to test time.

2.3.2.1   Path Planning

Path planning (Fig 2.1) is a classic problem that is related to robotics [21, 34]. Popular versions are mazes, as we have seen earlier (Fig. 2.2). The Taxi domain (Fig. 2.8) was originally introduced in the context of hierarchical problem solving [15]. Box-pushing problems such as Sokoban are frequently used as well [16, 28, 39, 60], see Fig. 2.3. The action space of these puzzles and mazes is discrete. Basic path and motion planning can enumerate possible solutions [14, 24].

Small versions of mazes can be solved exactly by enumeration, and larger instances are only suitable for approximation methods. Mazes can be used to test algorithms for path-finding problems and are frequently used to do so. Navigation tasks and box-pushing games such as Sokoban can feature rooms or subgoals that may then be used to test algorithms for hierarchically structured problems [18, 19, 23, 42] (Chap. 8). The problems can be made more difficult by enlarging the grid and by inserting more obstacles.

2.3.2.2   Board Games

Board games are a classic group of benchmarks for planning and learning since the earliest days of artificial intelligence. Two-person zero-sum perfect information board games such as tic tac toe, chess, checkers, Go, and shogi have been used to test algorithms since the 1950s. The action space of these games is discrete. Notable achievements were in checkers, chess, and Go, where human world champions were defeated in 1994, 1997, and 2016, respectively [12, 47, 49].

The board games are typically used "as is" and are not changed for different experiments (in contrast to mazes that are often adapted in size or complexity for specific purposes of the experiment). Board games are used for the difficulty of the challenge. The ultimate goals are to beat human grandmasters or even the world champion. Board games have been traditional mainstays of artificial intelligence, mostly associated with the search-based symbolic reasoning approach to artificial intelligence [45]. In contrast, the benchmarks in the next chapter are associated with connectionist artificial intelligence.

Summary and Further Reading

This has been a long chapter, to provide a solid basis for the rest of the book. We will summarize the chapter and provide references for further reading.

Summary

Reinforcement learning can learn behavior that achieves high rewards, using feedback from the environment. Reinforcement learning has no supervisory labels, and it can learn beyond a teacher, as long as there is an environment that provides feedback.

Reinforcement learning problems are modeled as a Markov decision problem, consisting of a 5-tuple (S, A, T_a, R_a, γ) for states, actions, transition, reward, and discount factor. The agent performs an action, and the environment returns the new state and the reward value to be associated with the new state.

Games and robotics are two important fields of application. Fields of application can be episodic (they end—such as a game of chess) or continuous (they do not end—a robot remains in the world). In continuous problems it often makes sense to discount behavior that is far from the present, and episodic problems typically do not bother with a discount factor—a win is a win.

Environments can be deterministic (many board games are deterministic—boards do not move) or stochastic (many robotic worlds are stochastic—the world around a robot moves). The action space can be discrete (a piece either moves to a square or it does not) or continuous (typical robot joints move continuously over an angle).

The goal in reinforcement learning is to find the optimal policy that gives for all states the best actions, maximizing the cumulative future reward. The policy function is used in two different ways. In a discrete environment the policy function $a = \pi(s)$ returns for each state the best action in that state. (Alternatively, the value function returns the value of each action in each state, out of which the argmax function can be used to find the actions with the highest value.)

The optimal policy can be found by finding the maximal value of a state. The value function $V(s)$ returns the expected reward for a state. When the transition function $T_a(s, s')$ is present, the agent can use Bellman's equation, or a dynamic programming method to recursively traverse the behavior space. Value iteration is one such dynamic programming method. Value iteration traverses all actions of all states, backing up reward values, until the value function stops changing. The state–action value $Q(s, a)$ determines the value of an action of a state.

Bellman's equation calculates the value of a state by calculating the value of successor states. Accessing successor states (by following the action and transition) is also called expanding a successor state. In a tree diagram successor states are called child nodes, and expanding is a downward action. Backpropagating the reward values to the parent node is a movement upward in the tree.

Methods where the agent makes use of the transition model are called model-based methods. When the agent does not use the transition model, they are model-free methods. In many situations the learning agent does not have access to the transition model of the environment, and planning methods cannot be used by the agent. Value-based model-free methods can find an optimal policy by using only irreversible actions, sampling the environment to find the value of the actions.

A major determinant in model-free reinforcement learning is the exploration/exploitation trade-off, or how much of the information that has been learned from the environment is used in choosing actions to sample. We discussed the advantages of exploiting the latest knowledge in settings where environment actions are very costly, such as clinical trials. A well-known exploration/exploitation method is ϵ-greedy, where the greedy (best) action is followed from the behavior policy, except in ϵ times, when random exploration is performed. Always following the policy's best action runs the risk of getting stuck in a cycle. Exploring random nodes allows breaking free of such cycles.

So far we have discussed the action selection operation. How should we process the rewards that are found at nodes? Here we introduced another fundamental element of reinforcement learning: bootstrapping, or finding a value by refining a previous value. Temporal difference learning uses the principle of bootstrapping to find the value of a state by adding appropriately discounted future reward values to the state value function.

We have now discussed both up and down motions and can construct full model-free algorithms. The best known algorithm may well be Q-learning, which learns the action value function of each action in each state through off-policy temporal difference learning. Off-policy algorithms improve the best policy function always with the value of the best action, even if the behavior action was less. Off-policy learning thus has a behavior policy that is different from the policy that is optimized.

In the next chapters we will look at value-based and policy-based model-free methods for large, complex problems, which make use of function approximation (deep learning).

Further Reading

There is a rich literature on tabular reinforcement learning. A standard work for tabular value-based reinforcement learning is Sutton and Barto [51]. Two condensed introductions to reinforcement learning are [3, 20]. Another major work on reinforcement learning is Bertsekas and Tsitsiklis [9]. Kaelbling has written an important survey article on the field [29]. The early works of Richard Bellman on dynamic programming and planning algorithms are [7, 8]. For a recent treatment of games and reinforcement learning, with a focus on heuristic search methods and the methods behind AlphaZero, see [41].

The methods of this chapter are based on bootstrapping [7] and temporal difference learning [50]. The on-policy algorithm SARSA [44] and the off-policy algorithm Q-Learning [56] are among the best known exact, tabular, value-based model-free algorithms.

Mazes and Sokoban grids are sometimes procedurally generated [25, 48, 55]. The goal for the algorithms is typically to find a solution for a grid of a certain difficulty class, to find a shortest path solution, or, in transfer learning, to learn to solve a class of grids by training on a different class of grids [59].

For general reference, one of the major textbooks on artificial intelligence is written by Russell and Norvig [45]. A more specific textbook on machine learning is by Bishop [10].

Exercises

We will end with questions on key concepts, with programming exercises to build up more experience.

Questions

The questions below are meant to refresh your memory and should be answered with yes, no, or short answers of one or two sentences.

1. In reinforcement learning the agent can choose which training examples are generated. Why is this beneficial? What is a potential problem?
2. What is Grid world?
3. Which five elements does an MDP have to model reinforcement learning problems?
4. In a tree diagram, is successor selection of behavior up or down?
5. In a tree diagram, is learning values through backpropagation up or down?
6. What is τ?
7. What is $\pi(s)$?
8. What is $V(s)$?
9. What is $Q(s, a)$?
10. What is dynamic programming?
11. What is recursion?
12. Do you know a dynamic programming method to determine the value of a state?
13. Is an action in an environment reversible for the agent?
14. Mention two typical application areas of reinforcement learning.

15. Is the action space of games typically discrete or continuous?
16. Is the action space of robots typically discrete or continuous?
17. Is the environment of games typically deterministic or stochastic?
18. Is the environment of robots typically deterministic or stochastic?
19. What is the goal of reinforcement learning?
20. Which of the five MDP elements is not used in episodic problems?
21. Which model or function is meant when we say "model-free" or "model-based"?
22. What type of action space and what type of environment are suited for value-based methods?
23. Why are value-based methods used for games and not for robotics?
24. Name two basic Gym environments.

Exercises

There is an even better way to learn about deep reinforcement learning than reading about it, and that is to perform experiments yourself, to see the learning processes unfold before your own eyes. The following exercises are meant as starting points for your own discoveries in the world of deep reinforcement learning.

Consider using Gym to implement these exercises. Section 2.2.4.2 explains how to install Gym.

1. *Q-learning* Implement Q-learning for Taxi, including the procedure to derive the best policy for the Q-table. Go to Sect. 2.2.4.14 and implement it. Print the Q-table, to see the values on the squares. You could print a live policy as the search progresses. Try different values for ϵ, the exploration rate. Does it learn faster? Does it keep finding the optimal solution? Try different values for α, the learning rate. Is it faster?
2. *SARSA* Implement SARSA, and the code is in Listing 2.7. Compare your results to Q-learning, can you see how SARSA chooses different paths? Try different ϵ and α.
3. *Problem size* How large can problems be before converging starts taking too long?
4. *Cartpole* Run Cartpole with the greedy policy computed by value iteration. Can you make it work? Is value iteration a suitable algorithm for Cartpole? If not, why do you think it is not?

```
1   # SARSA for OpenAI Gym Taxi environment
2   import gym
3   import numpy as np
4   import random
5   #Environment Setup
6   env = gym.make("Taxi-v2")
7   env.reset()
8   env.render()
9   # Q[state,action] table implementation
10  Q = np.zeros([env.observation_space.n, env.action_space.n])
11  gamma = 0.7    # discount factor
12  alpha = 0.2    # learning rate
13  epsilon = 0.1 # epsilon greedy
14  for episode in range(1000):
15      done = False
16      total_reward = 0
17      current_state = env.reset()
18      if random.uniform(0, 1) < epsilon:
19          current_action = env.action_space.sample() # Explore
                state space
20      else:
21          current_action = np.argmax(Q[current_state]) # Exploit
                learned values
22      while not done:
23          next_state, reward, done, info = env.step(current_action)
                # invoke Gym
24          if random.uniform(0, 1) < epsilon:
25              next_action = env.action_space.sample() # Explore
                    state space
26          else:
27              next_action = np.argmax(Q[next_state]) # Exploit
                    learned values
28          sarsa_value = Q[next_state,next_action]
29          old_value = Q[current_state,current_action]
30
31          new_value = old_value + alpha * (reward + gamma *
                sarsa_value - old_value)
32
33          Q[current_state,current_action] = new_value
34          total_reward += reward
35          current_state = next_state
36          current_action = next_action
37      if episode % 100 == 0:
38          print("Episode_{}_Total_Reward:_{}".format(episode,
                total_reward))
```

Listing 2.7 SARSA Taxi example, after [30]

References

1. Abhishek. Multi-arm bandits: a potential alternative to a/b tests https://medium.com/brillio-data-science/multi-arm-bandits-a-potential-alternative-to-a-b-tests-a647d9bf2a7e, 2019. 50
2. Ethem Alpaydin. *Introduction to Machine Learning*. MIT Press, 2009. 43, 48
3. Kai Arulkumaran, Marc Peter Deisenroth, Miles Brundage, and Anil Anthony Bharath. Deep reinforcement learning: A brief survey. *IEEE Signal Processing Magazine*, 34(6):26–38, 2017. 27, 30, 61
4. Peter Auer. Using confidence bounds for exploitation-exploration trade-offs. *Journal of Machine Learning Research*, 3(Nov):397–422, 2002. 50
5. Andrew G Barto, Richard S Sutton, and Charles W Anderson. Neuronlike adaptive elements that can solve difficult learning control problems. *IEEE Transactions on Systems, Man, and Cybernetics*, (5):834–846, 1983. 58
6. Marc G Bellemare, Yavar Naddaf, Joel Veness, and Michael Bowling. The Arcade Learning Environment: An evaluation platform for general agents. *Journal of Artificial Intelligence Research*, 47:253–279, 2013. 44
7. Richard Bellman. *Dynamic Programming*. Courier Corporation, 1957, 2013. 41, 48, 62, 62
8. Richard Bellman. On the application of dynamic programing to the determination of optimal play in chess and checkers. *Proceedings of the National Academy of Sciences*, 53(2):244–247, 1965. 62
9. Dimitri P Bertsekas and John Tsitsiklis. *Neuro-Dynamic Programming*. MIT Press Cambridge, 1996. 61
10. Christopher M Bishop. *Pattern Recognition and Machine Learning*. Information science and statistics. Springer Verlag, Heidelberg, 2006. 62
11. Greg Brockman, Vicki Cheung, Ludwig Pettersson, Jonas Schneider, John Schulman, Jie Tang, and Wojciech Zaremba. OpenAI Gym. *arXiv preprint arXiv:1606.01540*, 2016. 43
12. Murray Campbell, A Joseph Hoane Jr, and Feng-Hsiung Hsu. Deep Blue. *Artificial Intelligence*, 134(1-2):57–83, 2002. 59
13. Yang Chao. Share and play new Sokoban levels. http://Sokoban.org, 2013. 29
14. Joseph Culberson. Sokoban is PSPACE-complete. Technical report, University of Alberta, 1997. 28, 59
15. Thomas G Dietterich. Hierarchical reinforcement learning with the MAXQ value function decomposition. *Journal of Artificial Intelligence Research*, 13:227–303, 2000. 28, 59
16. Dorit Dor and Uri Zwick. Sokoban and other motion planning problems. *Computational Geometry*, 13(4):215–228, 1999. 28, 59
17. Eyal Even-Dar, Yishay Mansour, and Peter Bartlett. Learning rates for Q-learning. *Journal of machine learning Research*, 5(1), 2003. 57
18. Gregory Farquhar, Tim Rocktäschel, Maximilian Igl, and SA Whiteson. TreeQN and ATreeC: Differentiable tree planning for deep reinforcement learning. In *International Conference on Learning Representations*, 2018. 59
19. Dieqiao Feng, Carla P Gomes, and Bart Selman. Solving hard AI planning instances using curriculum-driven deep reinforcement learning. *arXiv preprint arXiv:2006.02689*, 2020. 59
20. Vincent François-Lavet, Peter Henderson, Riashat Islam, Marc G Bellemare, and Joelle Pineau. An introduction to deep reinforcement learning. *Foundations and Trends in Machine Learning*, 11(3-4):219–354, 2018. 27, 30, 30, 61
21. Alessandro Gasparetto, Paolo Boscariol, Albano Lanzutti, and Renato Vidoni. Path planning and trajectory planning algorithms: A general overview. In *Motion and Operation Planning of Robotic Systems*, pages 3–27. Springer, 2015. 28, 59
22. John C Gittins. Bandit processes and dynamic allocation indices. *Journal of the Royal Statistical Society: Series B (Methodological)*, 41(2):148–164, 1979. 50
23. Arthur Guez, Mehdi Mirza, Karol Gregor, Rishabh Kabra, Sébastien Racanière, Theophane Weber, David Raposo, Adam Santoro, Laurent Orseau, Tom Eccles, Greg Wayne, David Silver, and Timothy P. Lillicrap. An investigation of model-free planning. In *International Conference on Machine Learning*, pages 2464–2473, 2019. 59

24. Robert A Hearn and Erik D Demaine. *Games, Puzzles, and Computation*. CRC Press, 2009. 28, 59
25. Mark Hendrikx, Sebastiaan Meijer, Joeri Van Der Velden, and Alexandru Iosup. Procedural content generation for games: A survey. *ACM Transactions on Multimedia Computing, Communications, and Applications*, 9(1):1–22, 2013. 62
26. John Holland. Adaptation in natural and artificial systems: an introductory analysis with application to biology. *Control and Artificial Intelligence*, 1975. 50
27. Ronald A Howard. *Dynamic programming and Markov processes*. New York: John Wiley, 1964. 30
28. Andreas Junghanns and Jonathan Schaeffer. Sokoban: Enhancing general single-agent search methods using domain knowledge. *Artificial Intelligence*, 129(1-2):219–251, 2001. 28, 59
29. Leslie Pack Kaelbling, Michael L Littman, and Andrew W Moore. Reinforcement learning: A survey. *Journal of Artificial Intelligence Research*, 4:237–285, 1996. 61
30. Satwik Kansal and Brendan Martin. Learn data science webpage., 2018. 45, 55, 56, 56, 64
31. Samuel Kotz, Narayanaswamy Balakrishnan, and Norman L Johnson. *Continuous Multivariate Distributions, Volume 1: Models and Applications*. John Wiley & Sons, 2004. 51
32. Tze Leung Lai. Adaptive treatment allocation and the multi-armed bandit problem. *The Annals of Statistics*, pages 1091–1114, 1987. 51
33. Tze Leung Lai and Herbert Robbins. Asymptotically efficient adaptive allocation rules. *Advances in Applied Mathematics*, 6(1):4–22, 1985. 50, 51
34. Jean-Claude Latombe. *Robot Motion Planning*, volume 124. Springer Science & Business Media, 2012. 28, 59
35. Michael L Littman. Markov games as a framework for multi-agent reinforcement learning. In *Machine Learning Proceedings 1994*, pages 157–163. Elsevier, 1994. 30
36. Volodymyr Mnih, Koray Kavukcuoglu, David Silver, Alex Graves, Ioannis Antonoglou, Daan Wierstra, and Martin Riedmiller. Playing Atari with deep reinforcement learning. *arXiv preprint arXiv:1312.5602*, 2013. 44
37. Thomas Moerland. Continuous Markov decision process and policy search. Lecture notes for the course reinforcement learning, Leiden University, 2021. 30
38. Andrew William Moore. Efficient memory-based learning for robot control. Technical Report UCAM-CL-TR-209, University of Cambridge, UK, https://www.cl.cam.ac.uk/techreports/UCAM-CL-TR-209.pdf, 1990. 58
39. Yoshio Murase, Hitoshi Matsubara, and Yuzuru Hiraga. Automatic making of Sokoban problems. In *Pacific Rim International Conference on Artificial Intelligence*, pages 592–600. Springer, 1996. 28, 59
40. Andrew Y Ng, Daishi Harada, and Stuart Russell. Policy invariance under reward transformations: Theory and application to reward shaping. In *International Conference on Machine Learning*, volume 99, pages 278–287, 1999. 54
41. Aske Plaat. *Learning to Play: Reinforcement Learning and Games*. Springer Verlag, Heidelberg, https://learningtoplay.net, 2020. 62
42. Sébastien Racanière, Theophane Weber, David P. Reichert, Lars Buesing, Arthur Guez, Danilo Jimenez Rezende, Adrià Puigdomènech Badia, Oriol Vinyals, Nicolas Heess, Yujia Li, Razvan Pascanu, Peter W. Battaglia, Demis Hassabis, David Silver, and Daan Wierstra. Imagination-augmented agents for deep reinforcement learning. In *Advances in Neural Information Processing Systems*, pages 5690–5701, 2017. 59
43. Herbert Robbins. Some aspects of the sequential design of experiments. *Bulletin of the American Mathematical Society*, 58(5):527–535, 1952. 50
44. Gavin A Rummery and Mahesan Niranjan. On-line Q-learning using connectionist systems. Technical report, University of Cambridge, Department of Engineering Cambridge, UK, 1994. 48, 52, 62
45. Stuart J Russell and Peter Norvig. *Artificial intelligence: a modern approach*. Pearson Education Limited, Malaysia, 2016. 59, 60, 62
46. Daniel Russo, Benjamin Van Roy, Abbas Kazerouni, Ian Osband, and Zheng Wen. A tutorial on Thompson sampling. *Found. Trends Mach. Learn.*, 11(1):1–96, 2018. 51

47. Jonathan Schaeffer, Robert Lake, Paul Lu, and Martin Bryant. CHINOOK, the world man-machine checkers champion. *AI Magazine*, 17(1):21, 1996. 59
48. Noor Shaker, Julian Togelius, and Mark J Nelson. *Procedural Content Generation in Games*. Springer, 2016. 62
49. David Silver, Aja Huang, Chris J. Maddison, Arthur Guez, Laurent Sifre, George van den Driessche, Julian Schrittwieser, Ioannis Antonoglou, Veda Panneershelvam, Marc Lanctot, Sander Dieleman, Dominik Grewe, John Nham, Nal Kalchbrenner, Ilya Sutskever, Timothy Lillicrap, Madeleine Leach, Koray Kavukcuoglu, Thore Graepel, and Demis Hassabis. Mastering the game of Go with deep neural networks and tree search. *Nature*, 529(7587):484, 2016. 59
50. Richard S Sutton. Learning to predict by the methods of temporal differences. *Machine Learning*, 3(1):9–44, 1988. 48, 62
51. Richard S Sutton and Andrew G Barto. *Reinforcement learning, an Introduction, Second Edition*. MIT Press, 2018. 29, 33, 53, 61
52. Gerald Tesauro. TD-Gammon: A self-teaching backgammon program. In *Applications of Neural Networks*, pages 267–285. Springer, 1995. 49
53. William R Thompson. On the likelihood that one unknown probability exceeds another in view of the evidence of two samples. *Biometrika*, 25(3/4):285–294, 1933. 51
54. Emanuel Todorov, Tom Erez, and Yuval Tassa. MuJoCo: A physics engine for model-based control. In *IEEE/RSJ International Conference on Intelligent Robots and Systems (IROS)*, pages 5026–5033, 2012. 44
55. Julian Togelius, Alex J Champandard, Pier Luca Lanzi, Michael Mateas, Ana Paiva, Mike Preuss, and Kenneth O Stanley. Procedural content generation: Goals, challenges and actionable steps. In *Artificial and Computational Intelligence in Games*. Schloss Dagstuhl-Leibniz-Zentrum für Informatik, 2013. 62
56. Christopher JCH Watkins. *Learning from Delayed Rewards*. PhD thesis, King's College, Cambridge, 1989. 48, 53, 53, 62
57. Ian H Witten. The apparent conflict between estimation and control—a survey of the two-armed bandit problem. *Journal of the Franklin Institute*, 301(1-2):161–189, 1976. 50
58. Annie Wong, Thomas Bäck, Anna V. Kononova, and Aske Plaat. Deep multi-agent reinforcement learning: Challenges and directions. *Artificial Intelligence Review*, 2022. 49
59. Zhao Yang, Mike Preuss, and Aske Plaat. Transfer learning and curriculum learning in Sokoban. *arXiv preprint arXiv:2105.11702*, 2021. 62
60. Neng-Fa Zhou and Agostino Dovier. A tabled Prolog program for solving Sokoban. *Fundamenta Informaticae*, 124(4):561–575, 2013. 28, 59

Chapter 3
Deep Value-Based Reinforcement Learning

The previous chapter introduced the field of classic reinforcement learning. We learned about agents and environments, and about states, actions, values, and policy functions. We also saw our first planning and learning algorithms: value iteration, SARSA, and Q-learning. The methods in the previous chapter were exact, tabular, methods, which work for problems of moderate size that fit in memory.

In this chapter we move to high-dimensional problems with large state spaces that no longer fit in memory. We will go beyond tabular methods and use methods to *approximate* the value function and to *generalize* beyond trained behavior. We will do so with *deep learning*.

The methods in this chapter are deep, model-free, value-based methods, related to Q-learning. We will start by having a closer look at the new, larger, environments that our agents must now be able to solve (or rather, approximate). Next, we will look at deep reinforcement learning algorithms. In reinforcement learning the current behavior policy determines which action is selected next, a process that can be self-reinforcing. There is no ground truth, as in supervised learning. The targets of loss functions are no longer static, or even stable. In deep reinforcement learning convergence to V and Q values is based on a bootstrapping process, and the first challenge is to find training methods that converge to stable function values. Furthermore, since with neural networks the states are approximated based on their features, convergence proofs can no longer count on identifying states individually. For many years it was assumed that deep reinforcement learning is inherently unstable due to a so-called *deadly triad* of bootstrapping, function approximation, and off-policy learning.

However, surprisingly, solutions have been found for many of the challenges. By combining a number of approaches (such as the replay buffer and increased exploration) the Deep Q-Networks algorithm (DQN) was able to achieve stable learning in a high-dimensional environment. The success of DQN spawned a large research effort to improve training further. We will discuss some of these new methods.

A. Plaat, *Deep Reinforcement Learning*,
https://doi.org/10.1007/978-981-19-0638-1_3

The chapter is concluded with exercises, a summary, and pointers to further reading.

> **Deep Learning:** Deep reinforcement learning builds on deep supervised learning, and this chapter and the rest of the book assume a basic understanding of deep learning. When your knowledge of parameterized neural networks and function approximation is rusty, this is the time to go to Appendix B and take an in-depth refresher. The appendix also reviews essential concepts such as training, testing, accuracy, overfitting, and the bias–variance trade-off. When in doubt, try to answer the questions on page 385.

Core Concepts

- Stable convergence
- Replay buffer

Core Problem

- Achieve stable deep reinforcement learning in large problems.

Core Algorithm

- Deep Q-network (Listing 3.6)

End-to-End Learning

Before the advent of deep learning, traditional reinforcement learning had been used mostly on smaller problems such as puzzles, or the supermarket example. Their state space fits in the memories of our computers. Reward shaping, in the form of domain-specific heuristics, can be used to shoehorn the problem into a computer, for example, in chess and checkers [10, 32, 67]. Impressive results are achieved, but at the cost of extensive problem-specific reward shaping and heuristic engineering [61]. Deep learning changed this situation, and reinforcement learning is now used on high-dimensional problems that are too large to fit into memory.

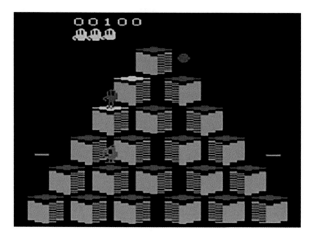

Fig. 3.1 Example game from the arcade learning environment [6]

In the field of supervised learning, a yearly competition had created years of steady progress in which the accuracy of image classification had steadily improved. Progress was driven by the availability of ImageNet, a large database of labeled images [15, 20], by increases in computation power through GPUs, and by steady improvement of machine learning algorithms, especially in deep convolutional neural networks. In 2012, a paper by Krizhevsky, Sutskever and Hinton presented a method that outperformed other algorithms by a large margin, and approached the performance of human image recognition [42]. The paper introduced the AlexNet architecture (after the first name of the first author), and 2012 is often regarded as the year of the breakthrough of deep learning. (See Appendix B.3.1.1 for details.) This breakthrough raised the question whether something similar in deep reinforcement learning could be achieved.

We did not have to wait long, only a year later, in 2013, at the deep learning workshop of one of the main AI conferences, a paper was presented on an algorithm that could play 1980s Atari video games just by training on the pixel input of the video screen (Fig. 3.1). The algorithm used a combination of deep learning and Q-learning and was named Deep Q-Network, or DQN [53, 54]. An illuminating video of how it learned to play the game Breakout is here.[1] This was a breakthrough for reinforcement learning. Many researchers at the workshop could relate to this achievement, perhaps because they had spent hours playing Space Invaders, Pac-Man, and Pong themselves when they were younger. Two years after the presentation at the deep learning workshop a longer article appeared in the journal Nature in which a refined and expanded version of DQN was presented (see Fig. 3.2 for the journal cover).

---

[1] https://www.youtube.com/watch?v=TmPfTpjtdgg.

Fig. 3.2 Atari experiments
on the cover of Nature

Why was this such a momentous achievement? Besides the fact that the problem that was solved was easily understood, true eye–hand coordination of this complexity had not been achieved by a computer before; furthermore, the end-to-end learning from pixel to joystick implied artificial behavior that was close to how humans play games. DQN essentially launched the field of deep reinforcement learning. For the first time the power of deep learning had been successfully combined with behavior learning, for an imaginative problem.

A major technical challenge that was overcome by DQN was the instability of the deep reinforcement learning process. In fact, there were convincing theoretical analyses at the time that this instability was fundamental, and it was generally assumed that it would be next to impossible to overcome [3, 25, 72, 77], since the target of the loss function depended on the convergence of the reinforcement learning process itself. By the end of this chapter we will have covered the problems of convergence and stability in reinforcement learning. We will have seen how DQN addresses these problems, and we will also have discussed some of the many further solutions that were devised after DQN.

But let us first have a look at the kind of new, high-dimensional, environments that were the cause of these developments.

3.1 Large, High-Dimensional, Problems

In the previous chapter, Grid worlds and mazes were introduced as basic sequential decision making problems in which exact, tabular, reinforcement learning methods work well. These are problems of moderate complexity. The complexity of a problem is related to the number of unique states that a problem has, or how large the state space is. Tabular methods work for small problems, where the entire state space fits in memory. This is for example the case with linear regression, which has only one variable x and two parameters a and b, or the Taxi problem, which has a state space of size 500. In this chapter we will be more ambitious and introduce various games, most notably Atari arcade games. The state space of a single frame of Atari video input is 210×160 pixels of 256 RGB color values $= 256^{33600}$.

There is a qualitative difference between small (500) and large (256^{33600}) problems.[2] For small problems the policy can be learned by loading all states of a problem in memory. States are identified individually, and each has its own best action that we can try to find. Large problems, in contrast, do not fit in memory, the policy cannot be memorized, and states are grouped together based on their features (see Sect. B.1.3, where we discuss feature learning). A parameterized network maps states to actions and values; states are no longer individually identifiable in a lookup table.

When deep learning methods were introduced in reinforcement learning, larger problems than before could be solved. Let us have a look at those problems.

3.1.1 Atari Arcade Games

Learning actions directly from high-dimensional sound and vision inputs is one of the long-standing challenges of artificial intelligence. To stimulate this research, in 2012 a test-bed was created designed to provide challenging reinforcement learning tasks. It was called the Arcade Learning Environment, or ALE [6], and it was based on a simulator for 1980s Atari 2600 video games. Figure 3.3 shows a picture of a distinctly retro Atari 2600 gaming console.

Among other things ALE contains an emulator of the Atari 2600 console. ALE presents agents with a high-dimensional[3] visual input (210×160 RGB video at 60 Hz, or 60 images per second) of tasks that were designed to be interesting and challenging for human players (Fig. 3.1 shows an example of such a game, and Fig. 3.4 shows a few more). The game cartridge ROM holds 2–4 kB of game code, while the console random-access memory is small, just 128 bytes (really, just 128 bytes, although the video memory is larger, of course). The actions can be selected

---

[2] See Sect. B.1.2, where we discuss the curse of dimensionality.

[3] That is, high dimensional for machine learning. 210×160 pixels is not exactly high-definition video quality.

Fig. 3.3 Atari 2600 console

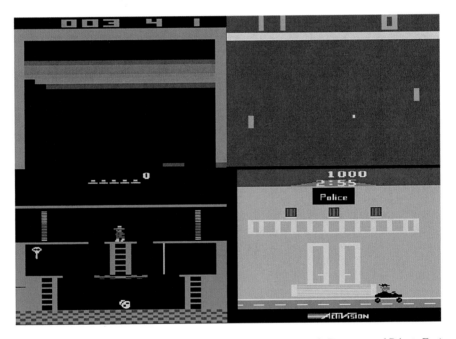

Fig. 3.4 Screenshots of 4 Atari Games (Breakout, Pong, Montezuma's Revenge, and Private Eye)

via a joystick (9 directions), which has a fire button (fire on/off), giving 18 actions in total.

The Atari games provide challenging eye–hand coordination and reasoning tasks that are both familiar and challenging to humans, providing a good test-bed for learning sequential decision making.

Atari games, with high-resolution video input at high frame rates, are an entirely different kind of challenge than Grid worlds or board games. Atari is a step closer to a human environment in which visual inputs should quickly be followed by correct actions. Indeed, the Atari benchmark called for very different agent algorithms, prompting the move from tabular algorithms to algorithms based on function approximation and deep learning. ALE has become a standard benchmark in deep reinforcement learning research.

3.1.2 Real-Time Strategy and Video Games

Real-time strategy games provide an even greater challenge than simulated 1980s Atari consoles. Games such as StarCraft (Fig. 1.6) [60] and Capture the Flag [37] have very large state spaces. These are games with large maps, many players, many pieces, and many types of actions. The state space of StarCraft is estimated at 10^{1685} [60], more than 1500 orders of magnitude larger than Go (10^{170}) [58, 76] and more than 1635 orders of magnitude large than chess (10^{47}) [33]. Most real-time strategy games are multiplayer, non-zero-sum, imperfect information games that also feature high-dimensional pixel input, reasoning, and team collaboration. The action space is stochastic and is a mix of discrete and continuous actions.

Despite the challenging nature, impressive achievements have been reported recently in three games where human performance was matched or even exceeded [7, 37, 80], see also Chap. 7.

Let us have a look at the methods that can solve these very different types of problems.

3.2 Deep Value-Based Agents

We will now turn to agent algorithms for solving large sequential decision problems. The main challenge of this section is to create an agent algorithm that can learn a good policy by interacting with the world—with a large problem, not a toy problem. From now on, our agents will be deep learning agents.

The questions that we are faced with are the following. How can we use deep learning for high-dimensional and large sequential decision making environments? How can tabular value and policy functions V, Q, and π be transformed into θ parameterized functions V_θ, Q_θ, and π_θ?

3.2.1 Generalization of Large Problems with Deep Learning

Recall from Appendix B that deep supervised learning uses a static dataset to approximate a function and that the labels are static targets in an optimization process where the loss function is minimized.

Deep reinforcement learning is based on the observation that bootstrapping is also a kind of minimization process in which an error (or difference) is minimized. In reinforcement learning this bootstrapping process converges on the true state value and state–action value functions. However, the Q-learning bootstrapping process lacks static ground truths; our data items are generated dynamically, and our loss function targets move. The movement of the loss function targets is influenced by the same policy function that the convergence process is trying to learn.

It has taken quite some effort to find deep learning algorithms that converge to stable functions on these moving targets. Let us try to understand in more detail how the supervised methods have to be adapted in order to work in reinforcement learning. We do this by comparing three algorithmic structures: supervised minimization, tabular Q-learning, and deep Q-learning.

3.2.1.1 Minimizing Supervised Target Loss

Listing 3.1 shows pseudocode for a typical supervised deep learning training algorithm, consisting of an input dataset, a forward pass that calculates the network output, a loss computation, and a backward pass. See Appendix B or [23] for more details.

We see that the code consists of a double loop: the outer loop controls the training epochs. Epochs consist of forward approximation of the target value using the parameters, computation of the gradient, and backward adjusting of the parameters with the gradient. In each epoch the inner loop serves all examples of the static dataset to the forward computation of the output value, the loss and the gradient computation, so that the parameters can be adjusted in the backward pass.

The dataset is static, and all that the inner loop does is delivering the samples to the backpropagation algorithm. Note that each sample is independent of the other, and samples are chosen with equal probability. After an image of a white horse is sampled, the probability that the next image is of a black grouse or a blue moon is equally (un)likely.

```
1  def train_sl(data, net, alpha=0.001):      # train classifier
2      for epoch in range(max_epochs):        # an epoch is one pass
3          sum_sq = 0                         # reset to zero for each pass
4          for (image, label) in data:
5              output = net.forward_pass(image) # predict
6              sum_sq += (output - label)**2 # compute error
7          grad = net.gradient(sum_sq)        # derivative of error
8          net.backward_pass(grad, alpha)     # adjust weights
9      return net
```

Listing 3.1 Network training pseudocode for supervised learning

```
1  def qlearn(environment, alpha=0.001, gamma=0.9, epsilon=0.05):
2      Q[TERMINAL,_] = 0 # policy
3      for episode in range(max_episodes):
4          s = s0
5          while s not TERMINAL: # perform steps of one full episode
6              a = epsilongreedy(Q[s], epsilon)
7              (r, sp) = environment(s, a)
8              Q[s,a] = Q[s,a] + alpha*(r+gamma*max(Q[sp])-Q[s,a])
9              s = sp
10     return Q
```

Listing 3.2 Q-learning pseudocode [72, 82]

3.2.1.2 Bootstrapping Q-Values

Let us now look at Q-learning. Reinforcement learning chooses the training examples differently. For convergence of algorithms such as Q-learning, the selection rule must guarantee that eventually all states will be sampled by the environment [82]. For large problems, this is not the case; this condition for convergence to the value function does not hold.

Listing 3.2 shows the short version of the bootstrapping tabular Q-learning pseudocode from the previous chapter. As in the previous deep learning algorithm, the algorithm consists of a double loop. The outer loop controls the Q-value convergence episodes, and each episode consists of a single trace of (time) steps from the start state to a terminal state. The Q-values are stored in a Python-array indexed by s and a, since Q is the state–action value. Convergence of the Q-values is assumed to have occurred when enough episodes have been sampled. The Q-formula shows how the Q-values are built up by bootstrapping on previous values, and how Q-learning is learning off-policy, taking the max value of an action.

A difference with the supervised learning is that in Q-learning subsequent samples are not independent. The next action is determined by the current policy and will most likely be the best action of the state (ϵ-greedy). Furthermore, the next state will be correlated to the previous state in the trajectory. After a state of the ball in the upper left corner of the field has been sampled, the next sample will with very high probability also be of a state where the ball is close to the upper left corner of the field. Training can be stuck in local minima, hence the need for exploration.

```
1  def train_qlearn(environment, Qnet, alpha=0.001, gamma=0.0,
        epsilon=0.05
2      s = s0                # initialize start state
3      for epoch in range(max_epochs): # an epoch is one pass
4          sum_sq = 0  # reset to zero for each pass
5          while s not TERMINAL: # perform steps of one full episode
6              a = epsilongreedy(Qnet(s,a)) # net: Q[s,a]-values
7              (r, sp) = environment(a)
8              output = Qnet.forward_pass(s, a)
9              target = r + gamma * max(Qnet(sp))
10             sum_sq += (target - output)**2
11             s = sp
12         grad = Qnet.gradient(sum_sq)
13         Qnet.backward_pass(grad, alpha)
14     return Qnet     # Q-values
```

Listing 3.3 Network training pseudocode for reinforcement learning

3.2.1.3   Deep Reinforcement Learning Target-Error

The two algorithms—deep learning and Q-learning—look similar in structure. Both consist of a double loop in which a target is optimized, and we can wonder if bootstrapping can be combined with loss function minimization. This is indeed the case, as Mnih et al. [53] showed in 2013. Our third listing, Listing 3.3, shows a naive deep learning version of Q-learning [53, 55], based on the double loop that now bootstraps Q-values by minimizing a loss function through adjusting the θ parameters.

Indeed, a Q-network can be trained with a gradient by minimizing a sequence of loss functions. The loss function for this bootstrap process is quite literally based on the Q-learning update formula. The loss function is the squared difference between the new Q-value $Q_{\theta_t}(s, a)$ from the forward pass and the old update target $r + \gamma \max_{a'} Q_{\theta_{t-1}}(s', a')$.[4]

An important observation is that the update targets depend on the previous network weights θ_{t-1} (the optimization targets move during optimization); this is in contrast with the targets used in a supervised learning process, which are fixed before learning begins [53]. In other words, the loss function of deep Q-learning minimizes a moving target, a target that depends on the network being optimized.

---

[4] Deep Q-learning is a fixed-point iteration [52]. The gradient of this loss function is $\nabla_{\theta_i} \mathcal{L}_i(\theta_i) = \mathbb{E}_{s,a\sim\rho(\cdot);s'\sim\mathcal{E}} \left[\left(r + \gamma \max_{a'} Q_{\theta_{i-1}}(s', a') - Q_{\theta_i}(s, a) \right) \nabla_{\theta_i} Q_{\theta_i}(s, a) \right]$, where ρ is the behavior distribution and $\mathcal{E}$ the Atari emulator. Further details are in [53].

3.2.2   Three Challenges

Let us have a closer look at the challenges that deep reinforcement learning faces. There are three problems with our naive deep Q-learner. First, convergence to the optimal Q-function depends on full coverage of the state space, yet the state space is too large to sample fully. Second, there is a strong correlation between subsequent training samples, with a real risk of local optima. Third, the loss function of gradient descent literally has a moving target, and bootstrapping may diverge. Let us have a look at these three problems in more detail.

3.2.2.1   Coverage

Proofs that algorithms such as Q-learning converge to the optimal policy depend on the assumption of full state space coverage; all state–action pairs must be sampled. Otherwise, the algorithms will not converge to an optimal action value for each state. Clearly, in large state spaces where not all states are sampled, this situation does not hold, and there is no guarantee of convergence.

3.2.2.2   Correlation

In reinforcement learning a sequence of states is generated in an agent/environment loop. The states differ only by a single action, one move or one stone, all other features of the states remain unchanged, and thus, the values of subsequent samples are correlated, which may result in a biased training. The training may cover only a part of the state space, especially when greedy action selection increases the tendency to select a small set of actions and states. The bias can result in the so-called specialization trap (when there is too much exploitation and too little exploration).

Correlation between subsequent states contributes to the low coverage that we discussed before, reducing convergence toward the optimal Q-function, increasing the probability of local optima and feedback loops. This happens, for example, when a chess program has been trained on a particular opening, and the opponent plays a different one. When test examples are different from training examples, then generalization will be bad. This problem is related to out-of-distribution training, see, for example, [48].

3.2.2.3   Convergence

When we naively apply our deep supervised methods to reinforcement learning, we encounter the problem that in a bootstrap process, the optimization target is part of the bootstrap process itself. Deep supervised learning uses a static dataset to approximate a function, and loss function targets are therefore stable. However, deep reinforcement learning uses as bootstrap target the Q-update from the previous time step, which changes during the optimization.

The loss is the squared difference between the Q-value $Q_{\theta_t}(s, a)$ and the old update target $r + \gamma \max_{a'} Q_{\theta_{t-1}}(s', a')$. Since both depend on parameters θ that are optimized, the risk of overshooting the target is real, and the optimization process can easily become unstable. It has taken quite some effort to find algorithms that can tolerate these moving targets.

3.2.2.4 Deadly Triad

Multiple works [3, 25, 77] showed that a combination of off-policy reinforcement learning with nonlinear function approximation (such as deep neural networks) could cause Q-values to diverge. Sutton and Barto [72] further analyze three elements for divergent training: function approximation, bootstrapping, and off-policy learning. Together, they are called *deadly triad*.

Function approximation may attribute values to states inaccurately. In contrast to exact tabular methods that are designed to identify individual states exactly, neural networks are designed to individual *features* of states. These features can be shared by different states, and values attributed to those features are shared also by other states. Function approximation may thus cause mis-identification of states and reward values and Q-values that are not assigned correctly. If the accuracy of the approximation of the true function values is good enough, then states may be identified well enough to reduce or prevent divergent training processes and loops [54].

Bootstrapping of values builds up new values on the basis of older values. This occurs in Q-learning and temporal difference learning where the current value depends on the previous value. Bootstrapping increases the efficiency of the training because values do not have to be calculated from the start. However, errors or biases in initial values may persist and spill over to other states as values are propagated incorrectly due to function approximation. Bootstrapping and function approximation can thus increase divergence.

Off-policy learning uses a behavior policy that is different from the target policy that we are optimizing for (Sect. 2.2.4.10). When the behavior policy is improved, the off-policy values may not improve. Off-policy learning converges generally less well than on-policy learning as it converges independently from the behavior policy. With function approximation convergence may be even slower, due to values being assigned to incorrect states.

3.2.3 Stable Deep Value-Based Learning

These considerations discouraged further research in deep reinforcement learning for many years. Instead, research focused for some time on linear function approximators, which have better convergence guarantees. Nevertheless, work on convergent deep reinforcement learning continued [8, 29, 50, 66], and algorithms such as neural fitted Q-learning were developed, which showed some promise [44, 47, 64]. After the further results of DQN [53] showed convincingly that convergence and stable learning could be achieved in a non-trivial problem, even more experimental studies were performed to find out under which circumstances convergence can be achieved and the deadly triad can be overcome. Further convergence and diversity-enhancing techniques were developed, some of which we will cover in Sect. 3.2.4.

Although the theory provides reasons why function approximation may preclude stable reinforcement learning, there were, in fact, indications that stable training is possible. Starting at the end of the 1980s, Tesauro had written a program that played very strong Backgammon based on a neural network. The program was called Neurogammon and used supervised learning from grandmaster games [73]. In order to improve the strength of the program, he switched to temporal difference reinforcement learning from self-play games [75]. TD-Gammon [74] learned by playing against itself, achieving stable learning in a shallow network. TD-Gammon's training used a temporal difference algorithm similar to Q-learning, approximating the value function with a network with one hidden layer, using raw board input enhanced with hand-crafted heuristic features [74]. Perhaps some form of stable reinforcement learning was possible, at least in a shallow network?

TD-Gammon's success prompted attempts with TD learning in checkers [11] and Go [13, 71]. Unfortunately, the success could not be replicated in these games, and it was believed for some time that Backgammon was a special case, well suited for reinforcement learning and self-play [63, 69].

However, as there came further reports of successful applications of deep neural networks in a reinforcement learning setting [29, 66], more work followed. The results in Atari [54] and later in Go [70] as well as further work [78] have now provided clear evidence that both stable training and generalizing deep reinforcement learning are indeed possible, and have improved our understanding of the circumstances that influence stability and convergence.

Let us have a closer look at the methods that are used to achieve stable deep reinforcement learning.

3.2.3.1 Decorrelating States

As mentioned in the introduction of this chapter, in 2013 Mnih et al. [53, 54] published their work on *end-to-end* reinforcement learning in Atari games.

The original focus of DQN is on breaking correlations between subsequent states and also on slowing down changes to parameters in the training process to improve stability. The DQN algorithm has two methods to achieve this: (1) experience replay and (2) infrequent weight updates. We will first look at experience replay.

3.2.3.2 Experience Replay

In reinforcement learning training samples are created in a sequence of interactions with the environment, and subsequent training states are strongly correlated to preceding states. There is a tendency to train the network on too many samples of a certain kind or in a certain area, and other parts of the state space remain under-explored. Furthermore, through function approximation and bootstrapping, some behavior may be forgotten. When an agent reaches a new level in a game that is different from previous levels, the agent may forget how to play the other level.

We can reduce correlation—and the local minima they cause—by adding a small amount of supervised learning. To break correlations and to create a more diverse set of training examples, DQN uses *experience replay*. Experience replay introduces a replay buffer [46], a cache of previously explored states, from which it samples training states at random.[5] Experience replay stores the last N examples in the replay memory and samples uniformly when performing updates. A typical number for N is 10^6 [83]. By using a buffer, a dynamic dataset from which recent training examples are sampled, we train states from a more diverse set, instead of only from the most recent one. The goal of experience replay is to increase the independence of subsequent training examples. The next state to be trained on is no longer a direct successor of the current state, but one somewhere in a long history of previous states. In this way the replay buffer spreads out the learning over more previously seen states, breaking temporal correlations between samples. DQN's replay buffer (1) improves coverage and (2) reduces correlation.

DQN treats all examples equal, old, and recent alike. A form of importance sampling might differentiate between important transitions, as we will see in the next section.

Note that, curiously, training by experience replay is a form of off-policy learning, since the target parameters are different from those used to generate the sample. Off-policy learning is one of the three elements of the deadly triad, and we find that stable learning can actually be improved by a special form of one of its problems.

Experience replay works well in Atari [54]. However, further analysis of replay buffers has pointed to possible problems. Zhang et al. [83] study the deadly triad with experience replay and find that larger networks resulted in more instabilities, but also that longer multi-step returns yielded fewer unrealistically high reward values. In Sect. 3.2.4 we will see many further enhancements to DQN-like algorithms.

---

[5] Originally experience replay is, as so much in artificial intelligence, a biologically inspired mechanism [47, 51, 59].

3.2.3.3 Infrequent Updates of Target Weights

The second improvement in DQN is *infrequent weight updates,* introduced in the 2015 paper on DQN [54]. The aim of this improvement is to reduce divergence that is caused by frequent updates of weights of the target Q-value. Again, the aim is to improve the stability of the network optimization by improving the stability of the Q-target in the loss function.

Every n updates, the network Q is cloned to obtain target network $\hat{Q}$, which is used for generating the targets for the following n updates to Q. In the original DQN implementation, a single set of network weights θ are used, and the network is trained on a moving loss target. Now, with infrequent updates the weights of the target network change much slower than those of the behavior policy, improving the stability of the Q-targets.

The second network improves the stability of Q-learning, where normally an update to $Q_\theta(s_t, a_t)$ also changes the target at each time step, quite possibly leading to oscillations and divergence of the policy. Generating the targets using an older set of parameters adds a delay between the time an update to Q_θ is made and the time the update changes the targets, making oscillations less likely.

3.2.3.4 Hands On: DQN and Breakout Gym Example

To get some hands-on experience with DQN, we will now have a look at how DQN can be used to play the Atari game Breakout.

The field of deep reinforcement learning is an open field where most codes of algorithms are freely shared on GitHub and where test environments are available. The most widely used environment is Gym, in which benchmarks such as ALE and MuJoCo can be found, see also Appendix C. The open availability of the software allows for easy replication, and, importantly, for further improvement of the methods. Let us have a closer look at the code of DQN, to experience how it works.

The DQN papers come with source code. The original DQN code from [54] is available at Atari DQN.[6] This code is the original code, in the programming language Lua, which may be interesting to study, if you are familiar with this language. A modern reference implementation of DQN, with further improvements, is in the (stable) baselines.[7] The RL Baselines Zoo even provides a collection of pretrained agents, at Zoo [22, 62].[8] The Network Zoo is especially useful if your desired application happens to be in the Zoo, to prevent long training times.

---

[6] https://github.com/kuz/DeepMind-Atari-Deep-Q-Learner.

[7] https://stable-baselines.readthedocs.io/en/master/index.html.

[8] https://github.com/araffin/rl-baselines-zoo.

3.2.3.5   Install Stable Baselines

The environment is only half of the reinforcement learning experiment, and we also need an agent algorithm to learn the policy. OpenAI also provides implementations of *agent* algorithms, called the Baselines, at the Gym GitHub repository Baselines.[9] Most algorithms that are covered in this book are present. You can download them, study the code, and experiment to gain an insight into their behavior.

In addition to OpenAI's Baselines, there is *Stable* Baselines, a fork of the OpenAI algorithms; it has more documentation and other features. It can be found at Stable Baselines,[10] and the documentation is at docs.[11]

The stable release from the Stable Baselines is installed by typing

```
pip install stable-baselines
```

or

```
pip install stable-baselines[mpi]
```

if support for OpenMPI is desired (a parallel message passing implementation for cluster computers). A very quick check to see if everything works is to run the PPO trainer from Listing 3.4. PPO is a policy-based algorithm that will be discussed in the next chapter in Sect. 4.2.5. The Cartpole should appear again but should now learn to stabilize for a brief moment.

3.2.3.6   The DQN Code

After having studied tabular Q-learning on Taxi in Sect. 2.2.4.14, let us now see how the network-based DQN works in practice. Listing 3.5 illustrates how easy it is to use the Stable Baselines implementation of DQN on the Atari Breakout environment. (See Sect. 2.2.4.2 for installation instructions of Gym.)

After you have run the DQN code and seen that it works, it is worthwhile to study how the code is implemented. Before you dive into the Python implementation of Stable Baselines, let us look at the pseudocode to refresh how the elements of DQN work together. See Listing 3.6. In this pseudocode we follow the 2015 version of DQN [54]. (The 2013 version of DQN did not use the target network [53].)

---

[9] https://github.com/openai/baselines.

[10] https://github.com/hill-a/stable-baselines.

[11] https://stable-baselines.readthedocs.io/en/master/.

```
1  import gym
2
3  from stable_baselines.common.policies import MlpPolicy
4  from stable_baselines.common.vec_env import DummyVecEnv
5  from stable_baselines import PPO2
6
7  env = gym.make('CartPole-v1')
8
9  model = PPO2(MlpPolicy, env, verbose=1)
10 model.learn(total_timesteps=10000)
11
12 obs = env.reset()
13 for i in range(1000):
14     action, _states = model.predict(obs)
15     obs, rewards, dones, info = env.step(action)
16     env.render()
```

Listing 3.4 Running stable baseline PPO on the Gym Cartpole environment

```
1  from stable_baselines.common.atari_wrappers import make_atari
2  from stable_baselines.deepq.policies import MlpPolicy, CnnPolicy
3  from stable_baselines import DQN
4
5  env = make_atari('BreakoutNoFrameskip-v4')
6
7  model = DQN(CnnPolicy, env, verbose=1)
8  model.learn(total_timesteps=25000)
9
10 obs = env.reset()
11 while True:
12     action, _states = model.predict(obs)
13     obs, rewards, dones, info = env.step(action)
14     env.render()
```

Listing 3.5 Deep Q-network Atari breakout example with Stable Baselines

DQN is based on Q-learning, with as extra a replay buffer and a target network to improve stability and convergence. First, at the start of the code, the replay buffer is initialized to empty, and the weights of the Q-network and the separate Q-target network are initialized. The state s is set to the start state.

Next is the optimization loop that runs until convergence. At the start of each iteration an action is selected at the state s, following an ϵ-greedy approach. The action is executed in the environment, and the new state and the reward are stored in a tuple in the replay buffer. Then, we train the Q-network. A minibatch is sampled randomly from the replay buffer, and one gradient descent step is performed. For

```
1  def dqn:
2      initialize replay_buffer empty
3      initialize Q network with random weights
4      initialize Qt target network with random weights
5      set s = s0
6      while not convergence:
7          # DQN in Atari uses preprocessing; not shown
8          epsilon-greedy select action a in argmax(Q(s,a)) # action
               selection depends on Q (moving target)
9          sx,reward = execute action in environment
10         append (s,a,r,sx) to buffer
11         sample minibatch from buffer # break temporal correlation
12         take target batch R (when terminal) or Qt
13         do gradient descent step on Q # loss function uses target
               Qt network
```

Listing 3.6 Pseudocode for DQN, after [54]

this step the loss function is calculated with the separate Q-target network $\hat{Q}_\theta$, which is updated less frequently than the primary Q-network Q_θ. In this way the loss function

$$\mathcal{L}_t(\theta_t) = \mathbb{E}_{s,a\sim\rho(\cdot)}\left[\left(\mathbb{E}_{s'\sim\mathcal{E}}(r + \gamma \max_{a'} \hat{Q}_{\theta_{t-1}}(s',a')|s,a) - Q_{\theta_t}(s,a)\right)^2\right]$$

is more stable, causing better convergence; $\rho(s, a)$ is the behavior distribution over s and a, and $\mathcal{E}$ is the Atari emulator [53]. Sampling the minibatch reduces the correlation that is inherent in reinforcement learning between subsequent states.

Conclusion

In summary, DQN was able to successfully learn end-to-end behavior policies for many different games (although similar and from the same benchmark set). Minimal prior knowledge was used to guide the system, and the agent only got to see the pixels and the game score. The same network architecture and procedure were used on each game; however, a network trained for one game could not be used to play another game.

The DQN achievement was an important milestone in the history of deep reinforcement learning. The main problems that were overcome by Mnih et al. [53] were training divergence and learning instability.

The nature of most Atari 2600 games is that they require eye–hand reflexes. The games have some strategic elements; credit assignment is mostly over a short term and can be learned with a surprisingly simple neural network. Most Atari games are more about immediate reflexes than about longer term reasoning. In this sense, the problem of playing Atari well is not unlike an image categorization problem: both problems are to find the right response that matches an input consisting of a set of pixels. Mapping pixels to categories is not that different from mapping pixels to joystick actions (see also the observations in [40]).

The Atari results have stimulated much subsequent research. Many blogs have been written on reproducing the result, which is not a straightforward task, requiring the finetuning of many hyperparameters [4].

3.2.4 Improving Exploration

The DQN results have spawned much activity among reinforcement learning researchers to improve training stability and convergence further, and many refinements have been devised, some of which we will review in this section.

Many of the topics that are covered by the enhancements are older ideas that work well in deep reinforcement learning. DQN applies random sampling of its replay buffer, and one of the first enhancements was prioritized sampling [68]. It was found that DQN, being an off-policy algorithm, typically overestimates action values (due to the max operation, Sect. 2.2.4.10). Double DQN addresses overestimation [79], and dueling DDQN introduces the advantage function to standardize action values [81]. Other approaches look at variance in addition to expected value, the effect of random noise on exploration was tested [21], and distributional DQN showed that networks that use probability distributions work better than networks that only use single point expected values [5].

In 2017 Hessel et al. [31] performed a large experiment that combined seven important enhancements. They found that the enhancements worked well together. The paper has become known as the Rainbow paper, since the major graph showing the cumulative performance over 57 Atari games of the seven enhancements is multi-colored (Fig. 3.5). Table 3.1 summarizes the enhancements, and this section provides an overview of the main ideas. The enhancements were tested on the same benchmarks (ALE, Gym), and most algorithm implementations can be found on the OpenAI Gym GitHub site in the baselines.[12]

3.2.4.1 Overestimation

Van Hasselt et al. introduce double deep Q-learning (DDQN) [79]. DDQN is based on the observation that Q-learning may overestimate action values. On the Atari 2600 games, DQN suffers from substantial overestimations. Remember that DQN uses Q-learning. Because of the max operation in Q-learning, this results in an

---

[12] https://github.com/openai/baselines.

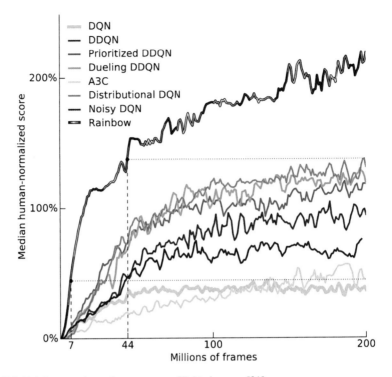

Fig. 3.5 Rainbow graph: performance over 57 Atari games [31]

Table 3.1 Deep value-based approaches

Name	Principle	Applicability	Effectiveness
DQN [53]	Replay buffer	Atari	Stable Q-learning
Double DQN [79]	De-overestimate values	DQN	Convergence
Prioritized experience [68]	Decorrelation	Replay buffer	Convergence
Distributional [5]	Probability distr.	Stable gradients	Generalization
Random noise [21]	Parametric noise	Stable gradients	More exploration

overestimation of the Q-value. To resolve this issue, DDQN uses the Q-Network to choose the action but uses the separate target Q-Network to evaluate the action. Let us compare the training target for DQN

$$y = r_{t+1} + \gamma Q_{\theta_t}(s_{t+1}, \arg\max_a Q_{\theta_t}(s_{t+1}, a))$$

with the training target for DDQN (the difference is a single ϕ)

$$y = r_{t+1} + \gamma Q_{\phi_t}(s_{t+1}, \arg\max_a Q_{\theta_t}(s_{t+1}, a)).$$

The DQN target uses the same set of weights θ_t twice, for selection and evaluation; the DDQN target uses a separate set of weights ϕ_t for evaluation, preventing overestimation due to the max operator. Updates are assigned randomly to either set of weights.

Earlier Van Hasselt et al. [28] introduced the double Q-learning algorithm in a tabular setting. The later paper shows that this idea also works with a large deep network. They report that the DDQN algorithm not only reduces the overestimations but also leads to better performance on several games. DDQN was tested on 49 Atari games and achieved about twice the average score of DQN with the same hyperparameters, and four times the average DQN score with tuned hyperparameters [79].

3.2.4.2   Prioritized Experience Replay

DQN samples uniformly over the entire history in the replay buffer, where Q-learning uses only the most recent (and important) state. It stands to reason to see if a solution in between these two extremes performs well.

Prioritized experience replay, or PEX, is such an attempt. It was introduced by Schaul et al. [68]. In the Rainbow paper PEX is combined with DDQN, and, as we can see, the blue line (with PEX) indeed outperforms the purple line.

In DQN experience, replay lets agents reuse examples from the past, although experience transitions are uniformly sampled, and actions are simply replayed at the same frequency that they were originally experienced, regardless of their significance. The PEX approach provides a framework for prioritizing experience. Important actions are replayed more frequently, and therefore, learning efficiency is improved. As measure of importance, standard proportional prioritized replay is used, with the absolute TD error to prioritize actions. Prioritized replay is used widely in value-based deep reinforcement learning. The measure can be computed in the distributional setting using the mean action values. In the Rainbow paper, all distributional variants prioritize actions by the Kullback–Leibler loss [31].

3.2.4.3   Advantage Function

The original DQN uses a single neural network as function approximator; DDQN (double deep Q-network) uses a separate target Q-network to evaluate an action. Dueling DDQN [81], also known as DDDQN, improves on this architecture by using two separate estimators: a value function and an advantage function

$$A(s, a) = Q(s, a) - V(s).$$

Advantage functions are related to the actor critic approach (see Chap. 4). An advantage function computes the difference between the value of an action and the value of the state. The function standardizes values on a baseline for the actions of a state [26]. Advantage functions provide better policy evaluation when many actions have similar values.

3.2.4.4 Distributional Methods

The original DQN learns a single value, which is the estimated mean of the state value. This approach does not take uncertainty into account. To remedy this, distributional Q-learning [5] learns a categorical probability distribution of discounted returns instead, increasing exploration. Bellemare et al. design a new distributional algorithm that applies Bellman's equation to the learning of distributions, a method called distributional DQN. Moerland et al. [56, 57] propose uncertain value networks. Interestingly, a link between the distributional approach and biology has been reported. Dabney et al. [14] showed correspondence between distributional reinforcement learning algorithms and the dopamine levels in mice, suggesting that the brain represents possible future rewards as a probability distribution.

3.2.4.5 Noisy DQN

Another distributional method is noisy DQN [21]. Noisy DQN uses stochastic network layers that add parametric noise to the weights. The noise induces randomness in the agent's policy, which increases exploration. The parameters that govern the noise are learned by gradient descent together with the remaining network weights. In their experiments the standard exploration heuristics for A3C (Sect. 4.2.4), DQN, and dueling agents (entropy reward and ϵ-greedy) were replaced with NoisyNet. The increased exploration yields substantially higher scores for Atari (dark red line).

3.3 Atari 2600 Environments

In their original 2013 workshop paper Mnih et al. [53] achieved human-level play for some of the games. Training was performed on 50 million frames in total on seven Atari games. The neural network performed better than an expert human player on Breakout, Enduro, and Pong. On Seaqest, Q*Bert, and Space Invaders, performance was far below that of a human. In these games a strategy must be found that extends over longer time periods. In their follow-up journal article two years later they were able to achieve human-level play for 49 of the 57 games that are in ALE [54] and performed better than human-level play in 29 of the 49 games.

Some of the games still proved difficult, notably games that require longer-range planning, where long stretches of the game do not give rewards, such as in Montezuma's Revenge, where the agent has to walk long distances and pick up a key to reach new rooms to enter new levels. In reinforcement learning terms, delayed credit assignment over long periods is hard. Toward the end of the book we will see Montezuma's Revenge again, when we discuss hierarchical reinforcement learning methods, in Chap 8. These methods are specifically developed to take large steps in the state space. The Go-Explore algorithm was able to solve Montezuma's Revenge [17, 18].

3.3.1 Network Architecture

End-to-end learning of challenging problems is computationally intensive. In addition to the two algorithmic innovations, the success of DQN is also due to the creation of a specialized efficient training architecture [54].

Playing the Atari games is a computationally intensive task for a deep neural network: the network trains a behavior policy directly from pixel frame input. Therefore, the training architecture contains reduction steps. To start with, the network consists of only three hidden layers (one fully connected, two convolutional), which is simpler than what is used in most supervised learning tasks.

The pixel images are high-resolution data. Since working with the full resolution of 210×160 pixels of 128 color values at 60 frames per second would be computationally too intensive, the images are reduced in resolution. The 210×160 with a 128 color palette is reduced to gray scale and 110×84 pixels, which is further cropped to 84×84. The first hidden layer convolves 16 8×8 filters with stride 4 and ReLU neurons. The second hidden layer convolves 32 4×4 filters with stride 2 and ReLU neurons. The third hidden layer is fully connected and consists of 256 ReLU neurons. The output layer is also fully connected with one output per action (18 joystick actions). The outputs correspond to the Q-values of the individual action. Figure 3.6 shows the architecture of DQN. The network receives the change

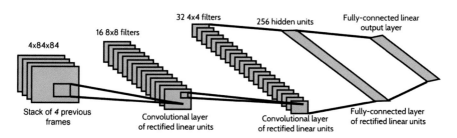

Fig. 3.6 DQN architecture [34]

in game score as a number from the emulator, and derivative updates are mapped to $\{-1, 0, +1\}$ to indicate decrease, no change, or improvement of the score (the Huber loss [4]).

To reduce computational demands further, frame skipping is employed. Only one in every 3–4 frames was used, depending on the game. To take game history into account, the net takes as input the last four resulting frames. This allows movement to be seen by the net. As optimizer, RMSprop is used [65]. A variant of ϵ-greedy is used, which starts with an ϵ of 1.0 (fully exploring) going down to 0.1 (90% exploiting).

3.3.2 Benchmarking Atari

To end the Atari story, we discuss two final algorithms. Of the many value-based model-free deep reinforcement learning algorithms that have been developed, one more algorithm that we discuss is R2D2 [39], because of its performance. R2D2 is not part of the Rainbow experiments but is a significant further improvement of the algorithms. R2D2 stands for Recurrent Replay Distributed DQN. It is built upon prioritized distributed replay and 5-step double Q-learning. Furthermore, it uses a dueling network architecture and an LSTM layer after the convolutional stack. Details about the architecture can be found in [27, 81]. The LSTM uses the recurrent state to exploit long-term temporal dependencies, which improve performance. The authors also report that the LSTM allows for better representation learning. R2D2 achieved good results on all 57 Atari games [39].

A more recent benchmark achievement has been published as Agent57. Agent57 is the first program that achieves a score higher than the human baseline on all 57 Atari 2600 games from ALE. It uses a controller that adapts the long- and short-term behavior of the agent, training for a range of policies, from very exploitative to very explorative, depending on the game [2].

Conclusion
Progress has come a long way since the replay buffer of DQN. Performance has been improved greatly in value-based model-free deep reinforcement learning, and now super-human performance in all 57 Atari games of ALE has been achieved. Many enhancements that improve coverage, correlation, and convergence have been developed. The presence of a clear benchmark was instrumental for progress so that researchers could clearly see which ideas worked and why. The earlier mazes and navigation games, OpenAI's Gym [9], and especially the ALE [6], have enabled this progress.

In the next chapter we will look at the other main branch of model-free reinforcement learning: policy-based algorithms. We will see how they work and that they are well suited for a different kind of application, with continuous action spaces.

Summary and Further Reading

This has been the first chapter in which we have seen deep reinforcement learning algorithms learn complex, high-dimensional, tasks. We end with a summary and pointers to the literature.

Summary

The methods that have been discussed in the previous chapter were exact, tabular methods. Most interesting problems have large state spaces that do not fit into memory. Feature learning identifies states by their common features. Function values are not calculated exactly, but are approximated, with deep learning.

Much of the recent success of reinforcement learning is due to deep learning methods. For reinforcement learning a problem arises when states are approximated. Since in reinforcement learning the next state is determined by the previous state, algorithms may get stuck in local minima or run in circles when values are shared with different states.

Another problem is training convergence. Supervised learning has a static dataset, and training targets are also static. In reinforcement learning the loss function targets depend on the parameters that are being optimized. This causes further instability. DQN caused a breakthrough by showing that with a replay buffer and a separate, more stable, target network, enough stability could be found for DQN to converge and learn how to play Atari arcade games.

Many further improvements to increase stability have been found. The Rainbow paper implements some of these improvements and finds that they are complementary and together achieve very strong play.

Further Reading

Deep learning revolutionized reinforcement learning. A comprehensive overview of the field is provided by Dong et al. [16]. For more on deep learning, see Goodfellow et al. [23], a book with much detail on deep learning; a major journal article is [45]. A brief survey is [1]. Also see Appendix B.

In 2013 the Arcade Learning Environment was presented [6, 49]. Experimenting with reinforcement learning was made even more accessible with OpenAI's Gym [9], with clear and easy to use Python bindings.

Deep learning versions of value-based tabular algorithms suffer from convergence and stability problems [77], yet the idea that stable deep reinforcement learning might be practical took hold with [29, 66]. Zhang et al. [83] study the deadly triad with experience replay. Deep gradient TD methods were proven to converge for evaluating a fixed policy [8]. Riedmiller et al. relaxed the fixed control policy in neural fitted Q-learning algorithm (NFQ) [64]. NFQ builds on work on stable function approximation [19, 24] and experience replay [46], and more recently on least-squares policy iteration [43]. In 2013 the first DQN paper appeared, showing results on a small number of Atari games [53] with the replay buffer to reduce temporal correlations. In 2015 the follow-up Nature paper reported results in more games [54], with a separate target network to improve training convergence. A well-known overview paper is the Rainbow paper [31, 38].

The use of benchmarks is of great importance for reproducible reinforcement learning experiments [30, 35, 36, 41]. For TensorFlow and Keras, see [12, 22].

Exercises

We will end this chapter with some questions to review the concepts that we have covered. Next are programming exercises to get some more exposure on how to use the deep reinforcement learning algorithms in practice.

Questions

Below are some questions to check your understanding of this chapter. Each question is a closed question where a simple, single sentence answer is expected.

1. What is Gym?
2. What are the Stable Baselines?
3. The loss function of DQN uses the Q-function as target. What is a consequence?
4. Why is the exploration/exploitation trade-off central in reinforcement learning?
5. Name one simple exploration/exploitation method.
6. What is bootstrapping?
7. Describe the architecture of the neural network in DQN.
8. Why is deep reinforcement learning more susceptible to unstable learning than deep supervised learning?
9. What is the deadly triad?
10. How does function approximation reduce stability of Q-learning?
11. What is the role of the replay buffer?

12. How can correlation between states lead to local minima?
13. Why should the coverage of the state space be sufficient?
14. What happens when deep reinforcement learning algorithms do not converge?
15. How large is the state space of chess estimated to be? 10^{47}, 10^{170} or 10^{1685}?
16. How large is the state space of Go estimated to be? 10^{47}, 10^{170} or 10^{1685}?
17. How large is the state space of StarCraft estimated to be? 10^{47}, 10^{170} or 10^{1685}?
18. What does the rainbow in the Rainbow paper stand for, and what is the main message?
19. Mention three Rainbow improvements that are added to DQN.

Exercises

Let us now start with some exercises. If you have not done so already, install Gym, PyTorch[13] or TensorFlow and Keras (see Sects. 2.2.4.2 and B.3.3.1 or go to the TensorFlow page).[14] Be sure to check the right versions of Python, Gym, TensorFlow, and the Stable Baselines to make sure that they work well together. The exercises below are designed to be done with Keras:

1. *DQN* Implement DQN from the Stable Baselines on Breakout from Gym. Turn off dueling and priorities. Find out what the values are for α, the training rate, for ϵ, the exploration rate, what kind of neural network architecture is used, what the replay buffer size is, and how frequently the target network is updated.
2. *Hyperparameters* Change all those hyperparameters, up, and down, and note the effect on training speed, and the training outcome: how good is the result? How sensitive is performance to hyperparameter optimization?
3. *Cloud* Use different computers, experiment with GPU versions to speed up training, consider Colab, AWS, or another cloud provider with fast GPU (or TPU) machines.
4. *Gym* Go to Gym and try different problems. For what kind of problems does DQN work, what are characteristics of problems for which it works less well?
5. *Stable Baselines* Go to the Stable Baselines and implement different agent algorithms. Try dueling algorithms, prioritized experience replay, but also other algorithm, such as actor critic or policy-based. (These algorithms will be explained in the next chapter.) Note their performance.
6. *TensorBoard* With TensorBoard you can follow the training process as it progresses. TensorBoard works on log files. Try TensorBoard on a Keras exercise and follow different training indicators. Also try TensorBoard on the Stable Baselines and see which indicators you can follow.

---

[13] https://pytorch.org.

[14] https://www.tensorflow.org.

7. *Checkpointing* Long training runs in Keras need checkpointing, to save valuable computations in case of a hardware or software failure. Create a large training job, and set up checkpointing. Test everything by interrupting the training, and try to re-load the pretrained checkpoint to restart the training where it left off.

References

1. Kai Arulkumaran, Marc Peter Deisenroth, Miles Brundage, and Anil Anthony Bharath. Deep reinforcement learning: A brief survey. *IEEE Signal Processing Magazine*, 34(6):26–38, 2017. 93
2. Adrià Puigdomènech Badia, Bilal Piot, Steven Kapturowski, Pablo Sprechmann, Alex Vitvit-skyi, Daniel Guo, and Charles Blundell. Agent57: Outperforming the Atari human benchmark. *arXiv preprint arXiv:2003.13350*, 2020. 92
3. Leemon Baird. Residual algorithms: Reinforcement learning with function approximation. In *Machine Learning Proceedings 1995*, pages 30–37. Elsevier, 1995. 72, 80
4. OpenAI Baselines. DQN https://openai.com/blog/openai-baselines-dqn/, 2017. 87, 92
5. Marc G Bellemare, Will Dabney, and Rémi Munos. A distributional perspective on reinforcement learning. In *International Conference on Machine Learning*, pages 449–458, 2017. 87, 88, 90
6. Marc G Bellemare, Yavar Naddaf, Joel Veness, and Michael Bowling. The Arcade Learning Environment: An evaluation platform for general agents. *Journal of Artificial Intelligence Research*, 47:253–279, 2013. 71, 73, 92, 94
7. Christopher Berner, Greg Brockman, Brooke Chan, Vicki Cheung, Przemyslaw Debiak, Christy Dennison, David Farhi, Quirin Fischer, Shariq Hashme, Christopher Hesse, Rafal Józefowicz, Scott Gray, Catherine Olsson, Jakub Pachocki, Michael Petrov, Henrique Pondé de Oliveira Pinto, Jonathan Raiman, Tim Salimans, Jeremy Schlatter, Jonas Schneider, Szymon Sidor, Ilya Sutskever, Jie Tang, Filip Wolski, and Susan Zhang. Dota 2 with large scale deep reinforcement learning. *arXiv preprint arXiv:1912.06680*, 2019. 75
8. Shalabh Bhatnagar, Doina Precup, David Silver, Richard S Sutton, Hamid R Maei, and Csaba Szepesvári. Convergent temporal-difference learning with arbitrary smooth function approximation. In *Advances in Neural Information Processing Systems*, pages 1204–1212, 2009. 81, 94
9. Greg Brockman, Vicki Cheung, Ludwig Pettersson, Jonas Schneider, John Schulman, Jie Tang, and Wojciech Zaremba. OpenAI Gym. *arXiv preprint arXiv:1606.01540*, 2016. 92, 94
10. Murray Campbell, A Joseph Hoane Jr, and Feng-Hsiung Hsu. Deep Blue. *Artificial Intelligence*, 134(1–2):57–83, 2002. 70
11. Kumar Chellapilla and David B Fogel. Evolving neural networks to play checkers without relying on expert knowledge. *IEEE Transactions on Neural Networks*, 10(6):1382–1391, 1999. 81
12. François Chollet. Keras. https://keras.io, 2015. 94
13. Christopher Clark and Amos Storkey. Teaching deep convolutional neural networks to play Go. arxiv preprint. *arXiv preprint arXiv:1412.3409*, 1, 2014. 81
14. Will Dabney, Zeb Kurth-Nelson, Naoshige Uchida, Clara Kwon Starkweather, Demis Hassabis, Rémi Munos, and Matthew Botvinick. A distributional code for value in dopamine-based reinforcement learning. *Nature*, pages 1–5, 2020. 90
15. Jia Deng, Wei Dong, Richard Socher, Li-Jia Li, Kai Li, and Li Fei-Fei. ImageNet: A large-scale hierarchical image database. In *2009 IEEE Conference on Computer Vision and Pattern Recognition*, pages 248–255. IEEE, 2009. 71
16. Hao Dong, Zihan Ding, and Shanghang Zhang. *Deep Reinforcement Learning*. Springer, 2020. 93

17. Adrien Ecoffet, Joost Huizinga, Joel Lehman, Kenneth O Stanley, and Jeff Clune. Go-Explore: a new approach for hard-exploration problems. *arXiv preprint arXiv:1901.10995*, 2019. 91

18. Adrien Ecoffet, Joost Huizinga, Joel Lehman, Kenneth O Stanley, and Jeff Clune. First return, then explore. *Nature*, 590(7847):580–586, 2021. 91

19. Damien Ernst, Pierre Geurts, and Louis Wehenkel. Tree-based batch mode reinforcement learning. *Journal of Machine Learning Research*, 6:503–556, April 2005. 94

20. Li Fei-Fei, Jia Deng, and Kai Li. ImageNet: Constructing a large-scale image database. *Journal of Vision*, 9(8):1037–1037, 2009. 71

21. Meire Fortunato, Mohammad Gheshlaghi Azar, Bilal Piot, Jacob Menick, Ian Osband, Alex Graves, Vlad Mnih, Remi Munos, Demis Hassabis, Olivier Pietquin, Charles Blundell, and Shane Legg. Noisy networks for exploration. In *International Conference on Learning Representations*, 2018. 87, 88, 90

22. Aurélien Géron. *Hands-on machine learning with Scikit-Learn and TensorFlow: concepts, tools, and techniques to build intelligent systems*. O'Reilly Media, Inc., 2019. 83, 94

23. Ian Goodfellow, Yoshua Bengio, and Aaron Courville. *Deep Learning*. MIT Press, Cambridge, 2016. 76, 93

24. Geoffrey J Gordon. Stable function approximation in dynamic programming. In *Machine Learning Proceedings 1995*, pages 261–268. Elsevier, 1995. 94

25. Geoffrey J Gordon. *Approximate solutions to Markov decision processes*. Carnegie Mellon University, 1999. 72, 80

26. Ivo Grondman, Lucian Busoniu, Gabriel AD Lopes, and Robert Babuska. A survey of actor-critic reinforcement learning: Standard and natural policy gradients. *IEEE Transactions on Systems, Man, and Cybernetics, Part C (Applications and Reviews)*, 42(6):1291–1307, 2012. 90

27. Audrunas Gruslys, Will Dabney, Mohammad Gheshlaghi Azar, Bilal Piot, Marc Bellemare, and Remi Munos. The reactor: A fast and sample-efficient actor-critic agent for reinforcement learning. In *International Conference on Learning Representations*, 2018. 92

28. Hado V Hasselt. Double Q-learning. In *Advances in Neural Information Processing Systems*, pages 2613–2621, 2010. 89

29. Nicolas Heess, David Silver, and Yee Whye Teh. Actor-critic reinforcement learning with energy-based policies. In *European Workshop on Reinforcement Learning*, pages 45–58, 2013. 81, 81, 94

30. Peter Henderson, Riashat Islam, Philip Bachman, Joelle Pineau, Doina Precup, and David Meger. Deep reinforcement learning that matters. In *Thirty-Second AAAI Conference on Artificial Intelligence*, 2018. 94

31. Matteo Hessel, Joseph Modayil, Hado Van Hasselt, Tom Schaul, Georg Ostrovski, Will Dabney, Dan Horgan, Bilal Piot, Mohammad Azar, and David Silver. Rainbow: Combining improvements in deep reinforcement learning. In *AAAI*, pages 3215–3222, 2018. 87, 88, 89, 94

32. Feng-Hsiung Hsu. *Behind Deep Blue: Building the computer that defeated the world chess champion*. Princeton University Press, 2004. 70

33. Feng-Hsiung Hsu, Thomas Anantharaman, Murray Campbell, and Andreas Nowatzyk. A grandmaster chess machine. *Scientific American*, 263(4):44–51, 1990. 75

34. Jonathan Hui. RL—DQN Deep Q-network https://medium.com/@jonathan_hui/rl-dqn-deep-q-network-e207751f7ae4. Medium post. 91

35. Matthew Hutson. Artificial Intelligence faces reproducibility crisis. *Science*, 359:725–726, 2018. 94

36. Riashat Islam, Peter Henderson, Maziar Gomrokchi, and Doina Precup. Reproducibility of benchmarked deep reinforcement learning tasks for continuous control. *arXiv preprint arXiv:1708.04133*, 2017. 94

37. Max Jaderberg, Wojciech M. Czarnecki, Iain Dunning, Luke Marris, Guy Lever, Antonio Garcia Castañeda, Charles Beattie, Neil C. Rabinowitz, Ari S. Morcos, Avraham Ruderman, Nicolas Sonnerat, Tim Green, Louise Deason, Joel Z. Leibo, David Silver, Demis Hassabis, Koray Kavukcuoglu, and Thore Graepel. Human-level performance in 3D multiplayer games

with population-based reinforcement learning. *Science*, 364(6443):859–865, 2019. 75, 75

38. Niels Justesen, Philip Bontrager, Julian Togelius, and Sebastian Risi. Deep learning for video game playing. *IEEE Transactions on Games*, 12(1):1–20, 2019. 94

39. Steven Kapturowski, Georg Ostrovski, John Quan, Remi Munos, and Will Dabney. Recurrent experience replay in distributed reinforcement learning. In *International Conference on Learning Representations*, 2018. 92, 92

40. Andrej Karpathy. Deep reinforcement learning: Pong from pixels. http://karpathy.github.io/2016/05/31/rl/. Andrej Karpathy Blog, 2016. 86

41. Khimya Khetarpal, Zafarali Ahmed, Andre Cianflone, Riashat Islam, and Joelle Pineau. Re-evaluate: Reproducibility in evaluating reinforcement learning algorithms. In *Reproducibility in Machine Learning Workshop, ICML*, 2018. 94

42. Alex Krizhevsky, Ilya Sutskever, and Geoffrey E Hinton. ImageNet classification with deep convolutional neural networks. In *Advances in Neural Information Processing Systems*, pages 1097–1105, 2012. 71

43. Michail G Lagoudakis and Ronald Parr. Least-squares policy iteration. *Journal of Machine Learning Research*, 4:1107–1149, Dec 2003. 94

44. Sascha Lange and Martin Riedmiller. Deep auto-encoder neural networks in reinforcement learning. In *The 2010 International Joint Conference on Neural Networks (IJCNN)*, pages 1–8. IEEE, 2010. 81

45. Yann LeCun, Yoshua Bengio, and Geoffrey Hinton. Deep learning. *Nature*, 521(7553):436, 2015. 93

46. Long-Ji Lin. Self-improving reactive agents based on reinforcement learning, planning and teaching. *Machine Learning*, 8(3–4):293–321, 1992. 82, 94

47. Long-Ji Lin. Reinforcement learning for robots using neural networks. Technical report, Carnegie-Mellon Univ Pittsburgh PA School of Computer Science, 1993. 81, 82

48. Hao Liu and Pieter Abbeel. Hybrid discriminative-generative training via contrastive learning. *arXiv preprint arXiv:2007.09070*, 2020. 79

49. Marlos C Machado, Marc G Bellemare, Erik Talvitie, Joel Veness, Matthew Hausknecht, and Michael Bowling. Revisiting the Arcade Learning Environment: Evaluation protocols and open problems for general agents. *Journal of Artificial Intelligence Research*, 61:523–562, 2018. 94

50. Hamid Reza Maei, Csaba Szepesvári, Shalabh Bhatnagar, and Richard S Sutton. Toward off-policy learning control with function approximation. In *International Conference on Machine Learning*, 2010. 81

51. James L McClelland, Bruce L McNaughton, and Randall C O'Reilly. Why there are complementary learning systems in the hippocampus and neocortex: insights from the successes and failures of connectionist models of learning and memory. *Psychological Review*, 102(3):419, 1995. 82

52. Francisco S Melo and M Isabel Ribeiro. Convergence of Q-learning with linear function approximation. In *2007 European Control Conference (ECC)*, pages 2671–2678. IEEE, 2007. 78

53. Volodymyr Mnih, Koray Kavukcuoglu, David Silver, Alex Graves, Ioannis Antonoglou, Daan Wierstra, and Martin Riedmiller. Playing Atari with deep reinforcement learning. *arXiv preprint arXiv:1312.5602*, 2013. 71, 78, 78, 78, 78, 81, 81, 84, 86, 86, 88, 90, 94

54. Volodymyr Mnih, Koray Kavukcuoglu, David Silver, Andrei A. Rusu, Joel Veness, Marc G. Bellemare, Alex Graves, Martin A. Riedmiller, Andreas Fidjeland, Georg Ostrovski, Stig Petersen, Charles Beattie, Amir Sadik, Ioannis Antonoglou, Helen King, Dharshan Kumaran, Daan Wierstra, Shane Legg, and Demis Hassabis. Human-level control through deep reinforcement learning. *Nature*, 518(7540):529–533, 2015. 71, 80, 81, 81, 82, 83, 83, 84, 86, 91, 91, 94

55. Thomas Moerland. Continuous Markov decision process and policy search. Lecture notes for the course reinforcement learning, Leiden University, 2021. 78

56. Thomas M Moerland, Joost Broekens, and Catholijn M Jonker. Efficient exploration with double uncertain value networks. *arXiv preprint arXiv:1711.10789*, 2017. 90

57. Thomas M Moerland, Joost Broekens, and Catholijn M Jonker. The potential of the return distribution for exploration in RL. *arXiv preprint arXiv:1806.04242*, 2018. 90

58. Martin Müller. Computer Go. *Artificial Intelligence*, 134(1–2):145–179, 2002. 75

59. Joseph O'Neill, Barty Pleydell-Bouverie, David Dupret, and Jozsef Csicsvari. Play it again: reactivation of waking experience and memory. *Trends in Neurosciences*, 33(5):220–229, 2010. 82

60. Santiago Ontanón, Gabriel Synnaeve, Alberto Uriarte, Florian Richoux, David Churchill, and Mike Preuss. A survey of real-time strategy game AI research and competition in StarCraft. *IEEE Transactions on Computational Intelligence and AI in Games*, 5(4):293–311, 2013. 75, 75

61. Aske Plaat. *Learning to Play: Reinforcement Learning and Games*. Springer Verlag, Heidelberg, https://learningtoplay.net, 2020. 70

62. Matthias Plappert. Keras-RL. https://github.com/keras-rl/keras-rl, 2016. 83

63. Jordan B Pollack and Alan D Blair. Why did TD-Gammon work? In *Advances in Neural Information Processing Systems*, pages 10–16, 1997. 81

64. Martin Riedmiller. Neural fitted Q iteration—first experiences with a data efficient neural reinforcement learning method. In *European Conference on Machine Learning*, pages 317–328. Springer, 2005. 81, 94

65. Sebastian Ruder. An overview of gradient descent optimization algorithms. *arXiv preprint arXiv:1609.04747*, 2016. 92

66. Brian Sallans and Geoffrey E Hinton. Reinforcement learning with factored states and actions. *Journal of Machine Learning Research*, 5:1063–1088, Aug 2004. 81, 81, 94

67. Jonathan Schaeffer. *One Jump Ahead: Computer Perfection at Checkers*. Springer Science & Business Media, 2008. 70

68. Tom Schaul, John Quan, Ioannis Antonoglou, and David Silver. Prioritized experience replay. In *International Conference on Learning Representations*, 2016. 87, 88, 89

69. Nicol N Schraudolph, Peter Dayan, and Terrence J Sejnowski. Temporal difference learning of position evaluation in the game of Go. In *Advances in Neural Information Processing Systems*, pages 817–824, 1994. 81

70. David Silver, Julian Schrittwieser, Karen Simonyan, Ioannis Antonoglou, Aja Huang, Arthur Guez, Thomas Hubert, Lucas Baker, Matthew Lai, Adrian Bolton, Yutian Chen, Timothy Lillicrap, Fan Hui, Laurent Sifre, George van den Driessche, Thore Graepel, and Demis Hassabis. Mastering the game of Go without human knowledge. *Nature*, 550(7676):354, 2017. 81

71. Ilya Sutskever and Vinod Nair. Mimicking Go experts with convolutional neural networks. In *International Conf. on Artificial Neural Networks*, pages 101–110. Springer, 2008. 81

72. Richard S Sutton and Andrew G Barto. *Reinforcement learning, an Introduction, Second Edition*. MIT Press, 2018. 72, 77, 80

73. Gerald Tesauro. Neurogammon wins Computer Olympiad. *Neural Computation*, 1(3):321–323, 1989. 81

74. Gerald Tesauro. TD-Gammon: A self-teaching backgammon program. In *Applications of Neural Networks*, pages 267–285. Springer, 1995. 81, 81

75. Gerald Tesauro. Temporal difference learning and TD-Gammon. *Communications of the ACM*, 38(3):58–68, 1995. 81

76. John Tromp. Number of legal Go states. http://tromp.github.io/go/legal.html, 2016. 75

77. John N Tsitsiklis and Benjamin Van Roy. Analysis of temporal-difference learning with function approximation. In *Advances in Neural Information Processing Systems*, pages 1075–1081, 1997. 72, 80, 94

78. Hado Van Hasselt, Yotam Doron, Florian Strub, Matteo Hessel, Nicolas Sonnerat, and Joseph Modayil. Deep reinforcement learning and the deadly triad. *arXiv:1812.02648*, 2018. 81

79. Hado Van Hasselt, Arthur Guez, and David Silver. Deep reinforcement learning with Double Q-Learning. In *AAAI*, volume 2, page 5. Phoenix, AZ, 2016. 87, 87, 88, 89

80. Oriol Vinyals, Igor Babuschkin, Wojciech M. Czarnecki, Michaël Mathieu, Andrew Dudzik, Junyoung Chung, David H. Choi, Richard Powell, Timo Ewalds, Petko Georgiev, Junhyuk

Oh, Dan Horgan, Manuel Kroiss, Ivo Danihelka, Aja Huang, Laurent Sifre, Trevor Cai, John P. Agapiou, Max Jaderberg, Alexander Sasha Vezhnevets, Rémi Leblond, Tobias Pohlen, Valentin Dalibard, David Budden, Yury Sulsky, James Molloy, Tom Le Paine, Çaglar Gülçehre, Ziyu Wang, Tobias Pfaff, Yuhuai Wu, Roman Ring, Dani Yogatama, Dario Wünsch, Katrina McKinney, Oliver Smith, Tom Schaul, Timothy P. Lillicrap, Koray Kavukcuoglu, Demis Hassabis, Chris Apps, and David Silver. Grandmaster level in Starcraft II using multi-agent reinforcement learning. *Nature*, 575(7782):350–354, 2019. 75

81. Ziyu Wang, Tom Schaul, Matteo Hessel, Hado Hasselt, Marc Lanctot, and Nando Freitas. Dueling network architectures for deep reinforcement learning. In *International Conference on Machine Learning*, pages 1995–2003, 2016. 87, 89, 92

82. Christopher JCH Watkins. *Learning from Delayed Rewards*. PhD thesis, King's College, Cambridge, 1989. 77, 77

83. Shangtong Zhang and Richard S Sutton. A deeper look at experience replay. *arXiv preprint arXiv:1712.01275*, 2017. 82, 82, 94

Chapter 4
Policy-Based Reinforcement Learning

Some of the most successful applications of deep reinforcement learning have a continuous action space, such as applications in robotics, self-driving cars, and real-time strategy games.

The previous chapters introduced value-based reinforcement learning. Value-based methods find the policy in a two-step process. First they find the best action value of a state, for which then the accompanying actions are found (by means of arg max). This works in environments with discrete actions, where the highest valued action is clearly separate from the next best action. Examples of continuous action spaces are robot arms that can move over arbitrary angles or poker bets that can be any monetary value. In these action spaces, value-based methods become unstable and arg max is not appropriate.

Another approach works better: policy-based methods. Policy-based methods do not use a separate value function but find the policy directly. They start with a policy function, which they then improve, episode by episode, with policy gradient methods. Policy-based methods are applicable to more domains than value-based methods. They work well with deep neural networks and gradient learning; they are some of the most popular methods of deep reinforcement learning, and this chapter introduces you to them.

We start by looking at applications with continuous action spaces. Next, we look at policy-based agent algorithms. We will introduce basic policy search algorithms and the policy gradient theorem. We will also discuss algorithms that combine value-based and policy-based approaches: the so-called actor critic algorithms. At the end of the chapter, we discuss larger environments for policy-based methods in more depth, where we will discuss progress in visuo-motor robotics and locomotion environments.

The chapter concludes with exercises, a summary, and pointers to further reading.

© The Author(s), under exclusive license to Springer Nature Singapore Pte Ltd. 2022   101
A. Plaat, *Deep Reinforcement Learning*,
https://doi.org/10.1007/978-981-19-0638-1_4

Core Concepts

- Policy gradient
- Bias–variance trade-off; actor critic

Core Problem

- Find a low-variance continuous action policy directly

Core Algorithms

- REINFORCE (Algorithm 4.2)
- Asynchronous advantage actor critic (Algorithm 4.4)
- Proximal policy optimization (Sect. 4.2.5)

Jumping Robots

One of the most intricate problems in robotics is learning to walk or, more generally, how to perform locomotion. Much work has been put into making robots walk, run, and jump. A video of a simulated robot that taught itself to jump over an obstacle course can be found on YouTube[1] [18].

Learning to walk is a challenge that takes human infants months to master. (Cats and dogs are quicker.) Teaching robots to walk is a challenging problem that is studied extensively in artificial intelligence and engineering. Movies abound on the Internet of robots that try to open doors, and fall over, or just try to stand upright, and still fall over.[2]

Locomotion of legged robots is a difficult sequential decision problem. For each leg, many different joints are involved. They must be actuated in the right order, turned with the right force, over the right duration, to the right angle. Most of these angles, forces, and durations are continuous. The algorithm has to decide how many degrees, Newtons, and seconds, constitute the optimal policy. All these actions are continuous quantities. Robot locomotion is a difficult problem, which is studied frequently in policy-based deep reinforcement learning.

---

[1] https://www.youtube.com/watch?v=hx_bgoTF7bs.

[2] See, for example, https://www.youtube.com/watch?v=g0TaYhjpOfo.

4.1 Continuous Problems

In this chapter, our actions are continuous and stochastic. We will discuss both of these aspects and some of the challenges they pose. We will start with continuous action policies.

4.1.1 Continuous Policies

In the previous chapter, we discussed environments with large state spaces. We will now move our attention to action spaces. The action spaces of the problems that we have seen so far—Grid worlds, mazes, and high-dimensional Atari games— were actually action spaces that were small and discrete—we could walk north, east, west, south, or we could choose from 9 joystick movements. In board games such as chess, the action space is larger but still discrete. When you move your pawn to e4, you do not move it to e4½.

In this chapter the problems are different. Steering a self-driving car requires turning the steering wheel a certain angle, duration, and angular velocity, to prevent jerky movements. Throttle movements should also be smooth and continuous. Actuation of robot joints is continuous, as we mentioned in the introduction of this chapter. An arm joint can move 1 degree, 2 degrees, or 90 or 180 degrees or anything in between.

An action in a continuous space is not one of a set of discrete choices, such as $\{N, E, W, S\}$, but rather a value over a continuous range, such as $[0, 2\pi]$ or $\mathbb{R}^+$; the number of possible values is infinite. How can we find the optimum value in an infinite space in a finite amount of time? Trying out all possible combinations of setting joint 1 to x degrees and applying force y in motor 2 will take infinitely long time. A solution could be to discretize the actions, although that introduces potential quantization errors.

When actions are not discrete, the arg max operation cannot be used to identify "the" best action, and value-based methods are no longer sufficient. Policy-based methods find suitable continuous or stochastic policies directly, without the intermediate step of a value function and the need for the arg max operation to construct the final policy.

4.1.2 Stochastic Policies

We will now turn to the modeling of stochastic policies.

When a robot moves its hand to open a door, it must judge the distance correctly. The robot may make a small error, and it may fail (as many movie clips show).[3] Stochastic environments cause stability problems for value-based methods [29]. Small perturbations in Q-values may lead to large changes in the policy of value-based methods. Convergence can typically only be achieved at slow learning rates, to smooth out the randomness. A stochastic policy (a target distribution) does not suffer from this problem. Stochastic policies have another advantage. By their nature they perform exploration, without the need to separately code ϵ-greediness or other exploration methods, since a stochastic policy returns a distribution over actions.

Policy-based methods find suitable stochastic policies directly. A potential disadvantage of purely episodic policy-based methods is that they are high variance; they may find local optima instead of global optima and converge slower than value-based methods. Newer (actor critic) methods, such as A3C, TRPO, and PPO, were designed to overcome these problems. We will discuss these algorithms later in this chapter.

Before we will explain these policy-based agent algorithms, we will have a closer look at some of the applications for which they are needed.

4.1.3 Environments: Gym and MuJoCo

Robotic experiments play an important role in reinforcement learning. However, because of the cost associated with real-world robotics experiments, in reinforcement learning, often simulated robotics systems are used. This is especially important in model-free methods, which tend to have a high sample complexity (real robots wear down when trials run in the millions). These software simulators model behavior of the robot and the effects on the environment, using physics models. This prevents the expense of real experiments with real robots, although some precision is lost to modeling error. Two well-known physics models are MuJoCo [48] and PyBullet [5]. They can be used easily via the Gym environment.

4.1.3.1 Robotics

Most robotic applications are more complicated than the classics such as mazes, Mountain car, and Cart pole. Robotic control decisions involve more joints, directions of travel, and degrees of freedom than a single cart that moves in one dimension. Typical problems involve learning of visuo-motor skills (eye–hand coordination, grasping) or learning of different locomotion gaits of multi-legged "animals." Some examples of grasping and walking are illustrated in Fig. 4.1.

---

[3] Even worse, when a robot thinks it stands still, it may actually be in the process of falling over (and, of course, robots cannot think, they only wished they could).

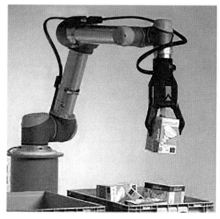

Fig. 4.1 Robot grasping and gait [30]

The environments for these actions are unpredictable to a certain degree: they require reactions to disturbances such as bumps in the road or the moving of objects in a scene.

4.1.3.2 Physics Models

Simulating robot motion involves modeling forces, acceleration, velocity, and movement. It also includes modeling mass and elasticity for bouncing balls, tactile/grasping mechanics, and the effect of different materials. A physics mechanics model needs to simulate the result of actions in the real world. Among the goals of such a simulation is to model grasping, locomotion, gaits, and walking and running (see also Sect. 4.3.1).

The simulations should be accurate. Furthermore, since model-free learning algorithms often involve millions of actions, it is important that the physics simulations are fast. Many different physics environments for model-based robotics have been created, among them Bullet, Havok, ODE, and PhysX, see [9] for a comparison. Of the models, MuJoCo [48] and PyBullet [5] are the most popular in reinforcement learning; especially MuJoCo is used in many experiments.

Although MuJoCo calculations are deterministic, the initial state of environments is typically randomized, resulting in an overall non-deterministic environment. Despite many code optimizations in MuJoCo, simulating physics is still an expensive proposition. Most MuJoCo experiments in the literature therefore are based on stick-like entities, which simulate limited motions, in order to limit the computational demands.

Figures 4.2 and 4.3 illustrate a few examples of some of the common Gym/MuJoCo problems that are often used in reinforcement learning: Ant, Half-Cheetah, and Humanoid.

Fig. 4.2 Gym MuJoCo Ant and Half-Cheetah [4]

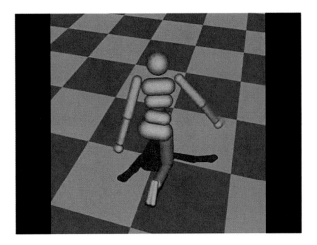

Fig. 4.3 Gym MuJoCo humanoid

4.1.3.3   Games

In real-time video games and certain card games, the decisions are also continuous.
For example, in some variants of poker, the size of monetary bets can be any amount,
which makes the action space quite large (although strictly speaking still discrete).
In games such as StarCraft and Capture the Flag, aspects of the physical world are
modeled, and movement of agents can vary in duration and speed. The environment
for these games is also stochastic: some information is hidden for the agent. This
increases the size of the state space greatly. We will discuss these games in Chap. 7
when we discuss multi-agent methods.

Table 4.1 Policy-Based Algorithms: REINFORCE, Asynchronous Advantage actor critic, Deep Deterministic Policy Gradient, Trust Region Policy Optimization, Proximal Policy Optimization, and Soft Actor Critic

Name	Approach	Ref
REINFORCE	Policy gradient optimization	[49]
A3C	Distributed actor critic	[35]
DDPG	Derivative of continuous action function	[29]
TRPO	Dynamically sized step size	[38]
PPO	Improved TRPO, first order	[40]
SAC	Variance-Based actor critic for robustness	[15]

4.2 Policy-Based Agents

Now that we have discussed the problems and environments that are used with policy-based methods, it is time to see how policy-based algorithms work. Policy-based methods are a popular approach in model-free deep reinforcement learning. Many algorithms have been developed that perform well. Table 4.1 lists some of the better known algorithms that will be covered in this chapter.

We will first provide an intuitive explanation of the idea behind the basic policy-based approach. Then we will discuss some of the theory behind it, as well as advantages and disadvantages of the basic policy-based approach. Most of these disadvantages are alleviated by the actor critic method, which is discussed next.

Let us start with the basic idea behind policy-based methods.

4.2.1 Policy-Based Algorithm: REINFORCE

Policy-based approaches learn a parameterized policy, which selects actions without consulting a value function.[4] In policy-based methods, the policy function is represented directly, allowing policies to select a continuous action, something that is difficult to do in value-based methods.

> **The Supermarket:** To build some intuition on the nature of policy-based methods, let us think back again at the supermarket navigation task, which we used in Chap. 2. In this navigation problem, we can try to assess our current distance to the supermarket with the Q-value function, as we have

(continued)

---

[4] Policy-based methods may use a value function to *learn* the policy parameters θ but do not use it for *action selection*.

done before. The Q-value assesses the distance of each direction to take; it tells us how far each action is from the goal. We can then use this distance function to find our path.

In contrast, the policy-based alternative would be to ask a local the way, who tells us, for example, to go straight and then left and then right at the Opera House and straight until we reach the supermarket on our left. The local just gave us a full path to follow, without having to infer which action was the closest and then use that information to determine the way to go. We can subsequently try to improve this full trajectory.

Let us see how we can optimize such a direct policy directly, without the intermediate step of the Q-function. We will develop a first, generic, policy-based algorithm to see how the pieces fit together. The explanation will be intuitive in nature.

The basic framework for policy-based algorithms is straightforward. We start with a parameterized policy function π_θ. We first (1) initialize the parameters θ of the policy function, (2) sample a new trajectory τ, (3) if τ is a good trajectory, increase the parameters θ toward τ, otherwise decrease them, and (4) keep going until convergence. Algorithm 4.1 provides a framework in pseudocode. Please note the similarity with the codes in the previous chapter (Listings 3.1–3.3), and especially the deep learning algorithms, where we also optimized function parameters in a loop.

The policy is represented by a set of parameters θ (these can be the weights in a neural network). Together, the parameters θ map the states S to action probabilities A. When we are given a set of parameters, how should we adjust them to improve the policy? The basic idea is to randomly sample a new policy, and if it is better, adjust the parameters a bit in the direction of this new policy (and away if it is worse). Let us see in more detail how this idea works.

To know which policy is best, we need some kind of measure of its quality. We denote the quality of the policy that is defined by the parameters as $J(\theta)$. It is natural to use the value function of the start state as our measure of quality

$$J(\theta) = V^\pi(s_0).$$

We wish to maximize $J(\cdot)$. When the parameters are differentiable, then all we need to do is to find a way to improve the gradient

$$\nabla_\theta J(\theta) = \nabla_\theta V^\pi(s_0)$$

of this expression to maximize our objective function $J(\cdot)$.

Policy-based methods apply gradient-based optimization, using the derivative of the objective to find the optimum. Since we are maximizing, we apply gradient

---

Algorithm 4.1 Gradient ascent optimization

Input: a differentiable objective $J(\theta)$, learning rate $\alpha \in \mathbb{R}^+$, threshold $\epsilon \in \mathbb{R}^+$
Initialization: randomly initialize θ in $\mathbb{R}^d$
repeat
    Sample trajectory τ and compute gradient ∇_θ
    $\theta \leftarrow \theta + \alpha \cdot \nabla_\theta J(\theta)$
until $\nabla_\theta J(\theta)$ converges below ϵ
return parameters θ

---

ascent. In each time step t of the algorithm, we perform the following update:

$$\theta_{t+1} = \theta_t + \alpha \cdot \nabla_\theta J(\theta)$$

for learning rate $\alpha \in \mathbb{R}^+$ and performance objective J, see the gradient ascent algorithm in Algorithm 4.1.

Remember that $\pi_\theta(a|s)$ is the probability of taking action a in state s. This function π is represented by a neural network θ, mapping states S at the input side of the network to action probabilities on the output side of the network. The parameters θ determine the mapping of our function π. Our goal is to update the parameters so that π_θ becomes the optimal policy. The better the action a is, the more we want to increase the parameters θ.

If we now would know, by some magical way, the optimal action $a^\star$, then we could use the gradient to push each parameter θ_t, $t \in$ trajectory, of the policy, in the direction of the optimal action, as follows:

$$\theta_{t+1} = \theta_t + \alpha \nabla \pi_{\theta_t}(a^\star|s).$$

Unfortunately, we do not know which action is best. We can, however, take a sample trajectory and use estimates of the value of the actions of the sample. This estimate can use the regular $\hat{Q}$ function from the previous chapter, or the discounted return function, or an advantage function (to be introduced shortly). Then, by multiplying the push of the parameters (the probability) with our estimate, we get

$$\theta_{t+1} = \theta_t + \alpha \hat{Q}(s,a) \nabla \pi_{\theta_t}(a|s).$$

A problem with this formula is that not only are we going to push harder on actions with a high value but also more often, because the policy $\pi_{\theta_t}(a|s)$ is the probability of action a in state s. Good actions are thus doubly improved, which may cause instability. We can correct by dividing by the general probability:

$$\theta_{t+1} = \theta_t + \alpha \hat{Q}(s,a) \frac{\nabla \pi_{\theta_t}(a|s)}{\pi_\theta(a|s)}.$$

Algorithm 4.2 Monte Carlo policy gradient (REINFORCE) [49]

Input: A differentiable policy $\pi_\theta(a|s)$, learning rate $\alpha \in \mathbb{R}^+$, threshold $\epsilon \in \mathbb{R}^+$
Initialization: Initialize parameters θ in $\mathbb{R}^d$
repeat
    Generate full trace $\tau = \{s_0, a_0, r_0, s_1, .., s_T\}$ following $\pi_\theta(a|s)$
    for $t \in 0, \ldots, T-1$ **do**   ▷ Do for each step of the episode
        $R \leftarrow \sum_{k=t}^{T-1} \gamma^{k-t} \cdot r_k$ ▷ Sum Return from trace
        $\theta \leftarrow \theta + \alpha R \nabla_\theta \log \pi_\theta(a_t|s_t)$ ▷ Adjust parameters
    end for
until $\nabla_\theta J(\theta)$ converges below ϵ
return Parameters θ

In fact, we have now almost arrived at the classic policy-based algorithm, REIN-FORCE, introduced by Williams in 1992 [49]. In this algorithm our formula is expressed in a way that is reminiscent of a logarithmic cross-entropy loss function. We can arrive at such a log-formulation by using the basic fact from calculus that

$$\nabla \log f(x) = \frac{\nabla f(x)}{f(x)}.$$

Substituting this formula into our equation, we arrive at

$$\theta_{t+1} = \theta_t + \alpha \hat{Q}(s, a) \nabla_\theta \log \pi_\theta(a|s).$$

This formula is indeed the core of REINFORCE, the prototypical policy-based algorithm, which is shown in full in Algorithm 4.2, with discounted cumulative reward.

To summarize, the REINFORCE formula pushes the parameters of the policy in the direction of the better action (multiplied proportionally by the size of the estimated action value) to know which action is best.

We have arrived at a method to improve a policy that can be used directly to indicate the action to take. The method works, regardless whether the action is discrete, continuous, or stochastic. It works directly, without having to go through intermediate value or argmax functions to find it. Algorithm 4.2 shows the full algorithm, which is called *Monte Carlo policy gradient*. The algorithm is called *Monte Carlo* because it samples a trajectory.

Online and Batch

The versions of gradient ascent (Algorithm 4.1) and REINFORCE (Algorithm 4.2) that we show update the parameters inside the innermost loop. All updates are performed as the time steps of the trajectory are traversed. This method is called the *online* approach. When multiple processes work in parallel to update data, the online approach makes sure that information is used as soon as it is known.

The policy gradient algorithm can also be formulated in *batch*-fashion: all gradients are summed over the states and actions, and the parameters are updated at the end of the trajectory. Since parameter updates can be expensive, the batch approach can be more efficient. An intermediate form that is frequently applied in practice is to work with *mini-batches*, trading off computational efficiency for information efficiency.

Let us now take a step back and look at the algorithm and assess how well it works.

4.2.2 Bias–Variance Trade-Off in Policy-Based Methods

Now that we have seen the principles behind a policy-based algorithm, let us see how policy-based algorithms work in practice and compare advantages and disadvantages of the policy-based approach.

Let us start with the advantages. First of all, parameterization is at the core of policy-based methods, making them a good match for deep learning. For value-based methods, deep learning had to be retrofitted, giving rise to complications as we saw in Sect. 3.2.3. Second, policy-based methods can easily find stochastic policies; value-based methods find deterministic policies. Due to their stochastic nature, policy-based methods naturally explore, without the need for methods such as ϵ-greedy, or more involved methods, which may require tuning to work well. Third, policy-based methods are effective in large or continuous action spaces. Small changes in θ lead to small changes in π and to small changes in state distributions (they are smooth). Policy-based algorithms do not suffer (as much) from convergence and stability issues that are seen in arg max-based algorithms in large or continuous action spaces.

On the other hand, there are disadvantages to the episodic Monte Carlo version of the REINFORCE algorithm. Remember that REINFORCE generates a full random episode in each iteration, before it assesses the quality. (Value-based methods use a reward to select the next action in each time step of the episode.) Because of this, the policy-based method is low bias, since full random trajectories are generated. However, they are also high variance, since the full trajectory is generated randomly (whereas value-based method uses the value for guidance at each selection step). What are the consequences? First, policy evaluation of full trajectories has low sample efficiency and high variance. As a consequence, policy improvement happens infrequently, leading to slow convergence compared to value-based methods. Second, this approach often finds a local optimum, since convergence to the global optimum takes too long.

Much research has been performed to address the high variance of the episode-based vanilla policy gradient [2, 13, 25, 26]. The enhancements that have been found have greatly improved performance, so much so that policy-based approaches—such as A3C, PPO, SAC, and DDPG—have become favorite model-free reinforcement learning algorithms for many applications. The enhancements to reduce high variance that we discuss are:

- *Actor critic* introduces within-episode value-based critics based on temporal difference value bootstrapping.

- *Baseline subtraction* introduces an advantage function to lower variance.
- *Trust regions* reduce large policy parameter changes.
- *Exploration* is crucial to get out of local minima and for more robust result; high entropy action distributions are often used.

Let us have a look at these enhancements.

4.2.3   Actor Critic Bootstrapping

The actor critic approach combines value-based elements with the policy-based method. The actor stands for the action, or policy-based, approach, and the critic stands for the value-based approach [43].

Action selection in episodic REINFORCE is random and hence low bias. However, variance is high, since the full episode is sampled (the size and direction of the update can strongly vary between different samples). The actor critic approach is designed to combine the advantage of the value-based approach (low variance) with the advantage of the policy-based approach (low bias). Actor critic methods are popular because they work well. It is an active field where many different algorithms have been developed.

The variance of policy methods can originate from two sources: (1) high variance in the cumulative reward estimate and (2) high variance in the gradient estimate. For both problems, a solution has been developed: bootstrapping for better reward estimates and baseline subtraction to lower the variance of gradient estimates. Both of these methods use the learned value function, which we denote by $V_\phi(s)$. The value function can use a separate neural network, with separate parameters ϕ, or it can use a value head on top of the actor parameters θ. In this case the actor and the critic share the lower layers of the network, and the network has two separate top heads: a policy and a value head. We will use ϕ for the parameters of the value function, to discriminate them from the policy parameters θ.

Temporal Difference Bootstrapping
To reduce the variance of the policy gradient, we can increase the number of traces M that we sample. However, the possible number of different traces is exponential in the length of the trace for a given stochastic policy, and we cannot afford to sample them all for one update. In practice, the number of sampled traces M is small, sometimes even $M = 1$, updating the policy parameters from a single trace. The return of the trace depends on many random action choices; the update has high variance. A solution is to use a principle that we know from temporal difference learning, to bootstrap the value function step by step. Bootstrapping uses the value function to compute intermediate n-step values per episode, trading off variance for bias. The n-step values are in-between full-episode Monte Carlo and single-step temporal difference targets.

Algorithm 4.3 Actor critic with bootstrapping

Input: A policy $\pi_\theta(a|s)$, a value function $V_\phi(s)$
An estimation depth n, learning rate α, number of episodes M
Initialization: Randomly initialize θ and ϕ
repeat
    for $i \in 1, \ldots, M$ **do**
        Sample trace $\tau = \{s_0, a_0, r_0, s_1, \ldots, s_T\}$ following $\pi_\theta(a|s)$
        for $t \in 0, \ldots, T-1$ **do**
            $\hat{Q}_n(s_t, a_t) = \sum_{k=0}^{n-1} \gamma^k r_{t+k} + \gamma^n V_\phi(s_{t+n})$ ▷ n-step target
        end for
    end for
    $\phi \leftarrow \phi - \alpha \cdot \nabla_\phi \sum_t \left(\hat{Q}_n(s_t, a_t) - V_\phi(s_t) \right)^2$ ▷ Descent value loss
    $\theta \leftarrow \theta + \alpha \cdot \sum_t [\hat{Q}_n(s_t, a_t) \cdot \nabla_\theta \log \pi_\theta(a_t|s_t)]$ ▷ Ascent policy gradient
until $\nabla_\theta J(\theta)$ converges below ϵ
return Parameters θ

We can use bootstrapping to compute an n-step target

$$\hat{Q}_n(s_t, a_t) = \sum_{k=0}^{n-1} r_{t+k} + V_\phi(s_{t+n}),$$

and we can then update the value function, for example, on a squared loss

$$\mathcal{L}(\phi|s_t, a_t) = \left(\hat{Q}_n(s_t, a_t) - V_\phi(s_t) \right)^2$$

and update the policy with the standard policy gradient but with that (improved) value $\hat{Q}_n$

$$\nabla_\theta \mathcal{L}(\theta|s_t, a_t) = \hat{Q}_n(s_t, a_t) \cdot \nabla_\theta \log \pi_\theta(a_t|s_t).$$

We are now using the value function prominently in the algorithm, which is parameterized by a separate set of parameters, denoted by ϕ; the policy parameters are still denoted by θ. The use of both policy and value is what gives the actor critic approach its name.

An example algorithm is shown in Algorithm 4.3. When we compare this algorithm with Algorithm 4.2, we see how the policy gradient ascent update now uses the n-step $\hat{Q}_n$ value estimate instead of the trace return R. We also see that this time the parameter updates are in batch mode, with separate summations.

4.2.4 Baseline Subtraction with Advantage Function

Another method to reduce the variance of the policy gradient is by baseline subtraction. Subtracting a baseline from a set of numbers reduces the variance but

---

Algorithm 4.4 Actor critic with bootstrapping and baseline subtraction

---

Input: A policy $\pi_\theta(a|s)$, a value function $V_\phi(s)$
An estimation depth n, learning rate α, number of episode M
Initialization: Randomly initialize θ and ϕ
while not converged **do**
    for $i = 1, \ldots, M$ **do**
        Sample trace $\tau = \{s_0, a_0, r_0, s_1, .., s_T\}$ following $\pi_\theta(a|s)$
        for $t = 0, \ldots, T - 1$ **do**
            $\hat{Q}_n(s_t, a_t) = \sum_{k=0}^{n-1} \gamma^k r_{t+k} + \gamma^n V_\phi(s_{t+n})$ $\triangleright$ n-step target
            $\hat{A}_n(s_t, a_t) = \hat{Q}_n(s_t, a_t) - V_\phi(s_t)$ $\triangleright$ Advantage
        end for
    end for
    $\phi \leftarrow \phi - \alpha \cdot \nabla_\phi \sum_t \left(\hat{A}_n(s_t, a_t)\right)^2$ $\triangleright$ Descent Advantage loss
    $\theta \leftarrow \theta + \alpha \cdot \sum_t [\hat{A}_n(s_t, a_t) \cdot \nabla_\theta \log \pi_\theta(a_t|s_t)]$ $\triangleright$ Ascent policy gradient
end while
return Parameters θ

---

leaves the expectation unaffected. Assume, in a given state with three available actions, which we sample action returns of 65, 70, and 75, respectively. Policy gradient will then try to push the probability of each action up, since the return for each action is positive. The above method may lead to a problem, since we are pushing all actions up (only somewhat harder on one of them). It might be better if we only push up on actions that are higher than the average (action 75 is higher than the average of 70 in this example) and push down on actions that are below average (65 in this example). We can do so through *baseline subtraction*.

The most common choice for the baseline is the value function. When we subtract the value V from a state–action value estimate Q, the function is called the *advantage function*:

$$A(s_t, a_t) = Q(s_t, a_t) - V(s_t).$$

The A function subtracts the value of the state s from the state–action value. It now estimates how much better a particular action is compared to the expectation of a particular state.

We can combine baseline subtraction with any bootstrapping method to estimate the cumulative reward $\hat{Q}(s_t, a_t)$. We compute

$$\hat{A}_n(s_t, a_t) = \hat{Q}_n(s_t, a_t) - V_\phi(s_t)$$

and update the policy with

$$\nabla_\theta \mathcal{L}(\theta|s_t, a_t) = \hat{A}_n(s_t, a_t) \cdot \nabla_\theta \log \pi_\theta(a_t|s_t).$$

We have now seen the ingredients to construct a full actor critic algorithm. An example algorithm is shown in Algorithm 4.4.

Generic Policy Gradient Formulation

With these two ideas, we can formulate an entire spectrum of policy gradient methods, depending on the type of cumulative reward estimate that they use. In general, the policy gradient estimator takes the following form, where we now introduce a new target Ψ_t that we sample from the trajectories τ:

$$\nabla_\theta J(\theta) = \mathbb{E}_{\tau_0 \sim p_\theta(\tau_0)} \left[\sum_{t=0}^{n} \Psi_t \nabla_\theta \log \pi_\theta(a_t|s_t) \right]$$

There is a variety of potential choices for Ψ_t, based on the use of bootstrapping and baseline subtraction:

$$\Psi_t = \hat{Q}_{MC}(s_t, a_t) = \sum_{i=t}^{\infty} \gamma^i \cdot r_i \qquad\qquad \text{Monte Carlo target}$$

$$\Psi_t = \hat{Q}_n(s_t, a_t) = \sum_{i=t}^{n-1} \gamma^i \cdot r_i + \gamma^n V_\theta(s_n) \qquad\qquad \text{bootstrap (n-step target)}$$

$$\Psi_t = \hat{A}_{MC}(s_t, a_t) = \sum_{i=t}^{\infty} \gamma^i \cdot r_i - V_\theta(s_t) \qquad\qquad \text{baseline subtraction}$$

$$\Psi_t = \hat{A}_n(s_t, a_t) = \sum_{i=t}^{n-1} \gamma^i \cdot r_i + \gamma^n V_\theta(s_n) - V_\theta(s_t) \qquad \text{baseline + bootstrap}$$

$$\Psi_t = Q_\phi(s_t, a_t) \qquad\qquad \text{Q-value approximation}$$

actor critic algorithms are among the most popular model-free reinforcement learning algorithms in practice, due to their good performance. After having discussed relevant theoretical background, it is time to look at how actor critic can be implemented in a practical high-performance algorithm. We will start with A3C.

Asynchronous Advantage Actor Critic

Many high-performance implementations are based on the actor critic approach. For large problems, the algorithm is typically parallelized and implemented on a large cluster computer. A well-known parallel algorithm is asynchronous advantage actor critic (A3C). A3C is a framework that uses asynchronous (parallel and distributed) gradient descent for optimization of deep neural network controllers [35].

There is also a non-parallel version of A3C, the synchronous variant A2C [50]. Together they popularized this approach to actor critic methods. Figure 4.4 shows the distributed architecture of A3C [22]; Algorithm 4.5 shows the pseudocode, from Mnih et al. [35]. The A3C network will estimate both a value function $V_\phi(s)$ and an advantage function $A_\phi(s, a)$, as well as a policy function $\pi_\theta(a|s)$. In the experiments

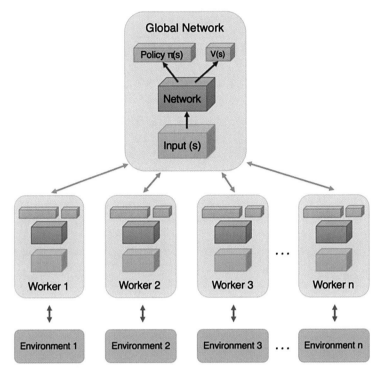

Fig. 4.4 A3C network [22]

on Atari [35], the neural networks were separate fully connected policy and value heads at the top (orange in Fig. 4.4), followed by joint convolutional networks (blue). This network architecture is replicated over the distributed workers. Each of these workers is run on a separate processor thread and is synced with global parameters from time to time.

A3C improves on classic REINFORCE in the following ways: it uses an advantage actor critic design, it uses deep learning, and it makes efficient use of parallelism in the training stage. The gradient accumulation step at the end of the code can be considered as a parallelized reformulation of minibatch-based stochastic gradient update: the values of ϕ or θ are adjusted in the direction of each training thread independently. A major contribution of A3C comes from its parallelized and asynchronous architecture: multiple actor-learners are dispatched to separate instantiations of the environment; they all interact with the environment and collect experience and asynchronously push their gradient updates to a central target network (just as DQN).

It was found that the parallel actor-learners have a stabilizing effect on training. A3C surpassed the previous state-of-the-art on the Atari domain and succeeded on a wide variety of continuous motor control problems as well as on a new task of navigating random 3D mazes using high-resolution visual input [35].

Algorithm 4.5 Asynchronous advantage actor critic pseudocode for each actor-learner thread [35]

Input: Assume global shared parameter vectors θ and ϕ and global shared counter $T = 0$
Assume thread-specific parameter vectors θ' and ϕ'
Initialize thread step counter $t \leftarrow 1$
repeat
    Reset gradients: $d\theta \leftarrow 0$ and $d\phi \leftarrow 0$.
    Synchronize thread-specific parameters $\theta' = \theta$ and $\phi' = \phi$
    $t_{start} = t$
    Get state s_t
    repeat
        Perform a_t according to policy $\pi(a_t | s_t; \theta')$
        Receive reward r_t and new state s_{t+1}
        $t \leftarrow t + 1$
        $T \leftarrow T + 1$
    until terminal s_t **or** $t - t_{start} == t_{max}$
    $R = \begin{cases} 0 & \text{for terminal } s_t \\ V(s_t, \phi') & \text{for non-terminal } s_t \text{ // Bootstrap from last state} \end{cases}$
    for $i \in \{t - 1, \dots, t_{start}\}$ **do**
        $R \leftarrow r_i + \gamma R$
        Accumulate gradients wrt θ': $d\theta \leftarrow d\theta + \nabla_{\theta'} \log \pi(a_i | s_i; \theta')(R - V(s_i; \phi'))$
        Accumulate gradients wrt ϕ': $d\phi \leftarrow d\phi + \partial \left(R - V(s_i; \phi') \right)^2 / \partial \phi'$
    end for
    Perform asynchronous update of θ using $d\theta$ and of ϕ using $d\phi$.
until $T > T_{max}$

4.2.5 Trust Region Optimization

Another important approach to further reduce the variance of policy methods is the trust region approach. Trust region policy optimization (TRPO) aims to further reduce the high variability in the policy parameters, by using a special loss function with an additional constraint on the optimization problem [38].

A naive approach to speed up an algorithm is to try to increase the step size of hyperparameters, such as the learning rate, and the policy parameters. This approach will fail to uncover solutions that are hidden in finer grained trajectories, and the optimization will converge to local optima. For this reason, the step size should not be too large. A less naive approach is to use an adaptive step size that depends on the *output* of the optimization progress.

Trust regions are used in general optimization problems to constrain the update size [42]. The algorithms work by computing the quality of the approximation; if it is still good, then the trust region is expanded. Alternatively, the region can be shrunk if the divergence of the new and current policy is getting large.

Schulman et al. [38] introduced trust region policy optimization (TRPO) based on this ideas, trying to take the largest possible parameter improvement step on a policy, without accidentally causing performance to collapse.

To this end, as it samples policies, TRPO compares the old and the new policy:

$$\mathcal{L}(\theta) = \mathbb{E}_t \left[\frac{\pi_\theta(a_t|s_t)}{\pi_{\theta_{\text{old}}}(a_t|s_t)} \cdot A_t \right].$$

In order to increase the learning step size, TRPO tries to maximize this loss function $\mathcal{L}$, subject to the constraint that the old and the new policies are not too far away. In TRPO, the Kullback–Leibler divergence[5] is used for this purpose:

$$\mathbb{E}_t[\text{KL}(\pi_{\theta_{\text{old}}}(\cdot|s_t), \pi_\theta(\cdot|s_t))] \le \delta.$$

TRPO scales to complex high-dimensional problems. Original experiments demonstrated its robust performance on simulated robotic Swimming, Hopping, Walking gaits, and Atari games. TRPO is commonly used in experiments and as a baseline for developing new algorithms. A disadvantage of TRPO is that it is a complicated algorithm that uses second-order derivatives; we will not cover the pseudocode here. Implementations can be found at Spinning Up[6] and Stable Baselines.[7]

Proximal policy optimization (PPO) [40] was developed as an improvement of TRPO. PPO has some of the benefits of TRPO, but it is simpler to implement, is more general, has better empirical sample complexity, and has better run-time complexity. It is motivated by the same question as TRPO, to take the largest possible improvement step on a policy parameter without causing performance collapse.

There are two variants of PPO: PPO-Penalty and PPO-Clip. PPO-Penalty approximately solves a KL-constrained update (like TRPO) but merely penalizes the KL-divergence in the objective function instead of making it a hard constraint. PPO-Clip does not use a KL-divergence term in the objective and has no constraints either. Instead it relies on clipping in the objective function to remove incentives for the new policy to get far from the old policy; it clips the difference between the old and the new policies within a fixed range $[1 - \epsilon, 1 + \epsilon] \cdot A_t$.

While simpler than TRPO, PPO is still a complicated algorithm to implement, and we omit the code here. The authors of PPO provide an implementation as a baseline.[8] Both TRPO and PPO are on-policy algorithms. Hsu et al. [20] reflect on design choices of PPO.

---

[5] The Kullback–Leibler divergence is a measure of distance between probability distributions [3, 28].

[6] https://spinningup.openai.com.

[7] https://stable-baselines.readthedocs.io.

[8] https://openai.com/blog/openai-baselines-ppo/#ppo.

4.2.6 Entropy and Exploration

A problem in many deep reinforcement learning experiments where only a fraction of the state space is sampled is brittleness: the algorithms get stuck in local optima, and different choices for hyperparameters can cause large differences in performance. Even a different choice for seed for the random number generator can cause large differences in performance for many algorithms.

For large problems, exploration is important, in value-based and policy-based approaches alike. We must provide the incentive to sometimes try an action which currently seems suboptimal [36]. Too little exploration results in brittle, local, and optima.

When we learn a *deterministic* policy $\pi_\theta(s) \rightarrow a$, we can manually add exploration noise to the behavior policy. In a continuous action space, we can use Gaussian noise, while in a discrete action space we can use Dirichlet noise [27]. For example, in a 1D continuous action space, we could use

$$\pi_{\theta,\text{behavior}}(a|s) = \pi_\theta(s) + \mathcal{N}(0, \sigma),$$

where $\mathcal{N}(\mu, \sigma)$ is the Gaussian (normal) distribution with hyperparameters mean $\mu = 0$ and standard deviation σ; σ is our exploration hyperparameter.

Soft Actor Critic

When we learn a *stochastic* policy $\pi(a|s)$, then exploration is already partially ensured due to the stochastic nature of our policy. For example, when we predict a Gaussian distribution, then simply sampling from this distribution will already induce variation in the chosen actions.

$$\pi_{\theta,\text{behavior}}(a|s) = \pi_\theta(a|s).$$

However, when there is not sufficient exploration, a potential problem is the collapse of the policy distribution. The distribution then becomes too narrow, and we lose the exploration pressure that is necessary for good performance.

Although we could simply add additional noise, another common approach is to use *entropy regularization* (see Sect. A.2 for details). We then add an additional penalty to the loss function, which enforces the entropy H of the distribution to stay larger. Soft actor critic (SAC) is a well-known algorithm that focuses on exploration [15, 16].[9] SAC extends the policy gradient equation to

$$\theta_{t+1} = \theta_t + R \cdot \nabla_\theta \log \pi_\theta(a_t|s_t) + \eta \nabla_\theta H[\pi_\theta(\cdot|s_t)],$$

where $\eta \in \mathbb{R}^+$ is a constant that determines the amount of entropy regularization. SAC ensures that we will move $\pi_\theta(a|s)$ to the optimal policy, while also ensuring that the policy stays as wide as possible (trading off the two against each other).

---

[9] https://github.com/haarnoja/sac.

Entropy is computed as $H = -\sum_i p_i \log p_i$, where p_i is the probability of being in state i; in SAC, entropy is the negative log of the stochastic policy function $-\log \pi_\theta(a|s)$.

High-entropy policies favor exploration. First, the policy is incentivized to explore more widely, while giving up on clearly unpromising avenues. Second, with improved exploration comes improved learning speed.

Most policy-based algorithms (including A3C, TRPO, and PPO) only optimize for the expected value. By including entropy explicitly in the optimization goal, SAC is able to increase the stability of outcome policies, achieving stable results for different random seeds and reducing the sensitivity to hyperparameter settings. Including entropy into the optimization goal has been studied widely, see, for example, [14, 23, 37, 47, 52].

A further element that SAC uses to improve stability and sample efficiency is a replay buffer. Many policy-based algorithms are on-policy learners (including A3C, TRPO, and PPO). In on-policy algorithms, each policy improvement uses feedback on actions according to the most recent version of the behavior policy. On-policy methods converge well but tend to require many samples to do so. In contrast, many value-based algorithms are off-policy: each policy improvement can use feedback collected at any earlier point during training, regardless of how the behavior policy was acting to explore the environment at the time when the feedback was obtained. The replay buffer is such a mechanism, breaking out of local maxima. Large replay buffers cause off-policy behavior, not only improving sample efficiency by learning from behavior of the past but also potentially causing convergence problems. Like DQN, SAC has overcome these problems and achieves stable off-policy performance.

4.2.7 Deterministic Policy Gradient

Actor critic approaches improve the policy-based approach with various value-based ideas and with good results. Another method to join policy and value approaches is to use a learned value function as a differentiable target to optimize the policy against—we let the policy follow the value function [36]. An example is the *deterministic policy gradient* [41]. Imagine we collect data D and train a value network $Q_\phi(s, a)$. We can then attempt to optimize the parameters θ of a deterministic policy by optimizing the prediction of the value network:

$$J(\theta) = \mathbb{E}_{s \sim D}\left[\sum_{t=0}^{n} Q_\phi(s, \pi_\theta(s))\right],$$

which by the chain rule gives the following gradient expression:

$$\nabla_\theta J(\theta) = \sum_{t=0}^{n} \nabla_a Q_\phi(s, a) \cdot \nabla_\theta \pi_\theta(s).$$

In essence, we first train a state–action value network based on sampled data and then *let the policy follow the value network*, by simply chaining the gradients. Thereby, we push the policy network in the direction of those actions a that increase the value network prediction, toward actions that perform better.

Lillicrap et al. [29] present deep deterministic policy gradient (DDPG). It is based on DQN, with the purpose of applying it to continuous action functions. In DQN, if the optimal action-value function $Q^\star(s, a)$ is known, then the optimal action $a^\star(s)$ can be found via $a^\star(s) = \arg\max_a Q^\star(s, a)$. DDPG uses the derivative of a continuous function $Q(s, a)$ with respect to the action argument to efficiently approximate $\max_a Q(s, a)$. DDPG is also based on the algorithms, deterministic policy gradients (DPGs) [41] and neurally fitted Q-learning with continuous actions (NFQCAs) [17], two actor critic algorithms.

Algorithm 4.6 DDPG algorithm [29]

Randomly initialize critic network $Q_\phi(s, a)$ and actor $\pi_\theta(s)$ with weights ϕ and θ.
Initialize target network Q' and π' with weights $\phi' \leftarrow \phi, \theta' \leftarrow \theta$
Initialize replay buffer R
for episode = 1, M **do**
    Initialize a random process $\mathcal{N}$ for action exploration
    Receive initial observation state s_1
    for t = 1, T **do**
        Select action $a_t = \pi_\theta(s_t) + \mathcal{N}_t$ according to the current policy and exploration noise
        Execute action a_t and observe reward r_t and observe new state s_{t+1}
        Store transition (s_t, a_t, r_t, s_{t+1}) in R
        Sample a random minibatch of N transitions (s_i, a_i, r_i, s_{i+1}) from R
        Set $y_i = r_i + \gamma Q_{\phi'}(s_{i+1}, \pi_{\theta'}(s_{i+1}))$
        Update critic by minimizing the loss: $\mathcal{L} = \frac{1}{N} \sum_i (y_i - Q_\phi(s_i, a_i))^2$
        Update the actor policy using the sampled policy gradient:

$$\nabla_\theta J \approx \frac{1}{N} \sum_i \nabla_a Q_\phi(s, a)|_{s=s_i, a=\mu(s_i)} \nabla_\theta \pi_\theta(s)|_{s_i}$$

        Update the target networks:

$$\phi' \leftarrow \tau\phi + (1 - \tau)\phi'$$

$$\theta' \leftarrow \tau\theta + (1 - \tau)\theta'$$

    end for
end for

The pseudocode of DDPG is shown in Algorithm 4.6. DDPG has been shown to work well on simulated physics tasks, including classic problems such as Cartpole, Gripper, Walker, and Car driving, being able to learn policies directly from raw pixel inputs. DDPG is off-policy and uses a replay buffer and a separate target network to achieve stable deep reinforcement learning (just as DQN).

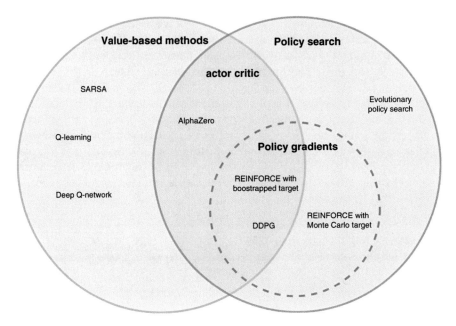

Fig. 4.5 Value-based, policy-based, and actor critic methods [36].

DDPG is a popular actor critic algorithm. Annotated pseudocode and efficient implementations can be found at Spinning Up[10] and Stable Baselines[11] in addition to the original paper [29].

Conclusion

We have seen quite some algorithms that combine the policy and value approach, and we have discussed possible combinations of these building blocks to construct working algorithms. Figure 4.5 provides a conceptual map of how the different approaches are related, including two approaches that will be discussed in later chapters (AlphaZero and evolutionary approaches).

Researchers have constructed many algorithms and performed experiments to see when they perform best. Quite a number of actor critic algorithms have been developed. Working high-performance Python implementations can be found on GitHub in the Stable Baselines.[12]

---

[10] https://spinningup.openai.com.

[11] https://stable-baselines.readthedocs.io.

[12] https://stable-baselines.readthedocs.io/en/master/guide/quickstart.html.

4.2.8 Hands On: PPO and DDPG MuJoCo Examples

OpenAI's Spinning Up provides a tutorial on policy gradient algorithms, complete with TensorFlow and PyTorch versions of REINFORCE to learn Gym's Cartpole.[13] with the TensorFlow code[14] or PyTorch.[15]

Now that we have discussed these algorithms, let us see how they work in practice, to get a feeling for the algorithms and their hyperparameters. MuJoCo is the most frequently used physics simulator in policy-based learning experiments. Gym, the (Stable) Baselines, and Spinning Up allow us to run any mix of learning algorithms and experimental environments. You are encouraged to try these experiments yourself.

Please be warned, however, that attempting to install all necessary pieces of software may invite a minor version-hell. Different versions of your operating system, of Python, of GCC, of Gym, of the Baselines, of TensorFlow or PyTorch, and of MuJoCo all need to line up before you can see beautiful images of moving arms, legs, and jumping humanoids. Unfortunately, not all of these versions are backward-compatible, specifically the switch from Python 2 to 3 and from TensorFlow 1 to 2 introduced incompatible language changes.

Getting everything to work may be an effort and may require switching machines, operating systems, and languages, but you should really try. This is the disadvantage of being part of one of the fastest moving fields in machine learning research. If things do not work with your current operating system and Python version, in general a combination of Linux Ubuntu (or macOS), Python 3.7, TensorFlow 1 or PyTorch, Gym, and the Baselines may be a good idea to start with. Search on the GitHub repositories or Stack Overflow when you get error messages. Sometimes it will be necessary to use an older version, or make some changes to include paths or library paths.

If everything works, then both Spinning Up and the Baselines provide convenient scripts that facilitate mixing and matching algorithms and environments from the command line.

For example, to run Spinup's PPO on MuJoCo's Walker environment, with a 32×32 hidden layer, the following command line does the job:

```
python -m spinup.run ppo
--hid "[32,32]" --env Walker2d-v2 --exp_name mujocotest
```

---

[13] Tutorial: https://spinningup.openai.com/en/latest/spinningup/rl_intro3.html#deriving-the-simplest-policy-gradient.

[14] TensorFlow: https://github.com/openai/spinningup/blob/master/spinup/examples/tf1/pg_math/1_simple_pg.py.

[15] PyTorch: https://github.com/openai/spinningup/blob/master/spinup/examples/pytorch/pg_math/1_simple_pg.py.

To train DDPG from the Baselines on the Half-Cheetah, the command is

```
python -m base-lines.run
--alg=ddpg --env=HalfCheetah-v2 --num_timesteps=1e6
```

All hyperparameters can be controlled via the command line, providing for a flexible way to run experiments. A final example command line is

```
python scripts/all_plots.py
-a ddpg -e HalfCheetah Ant Hopper Walker2D -f logs/
-o logs/ddpg_results
```

The Stable Baselines site explains what this command line does.

4.3   Locomotion and Visuo-Motor Environments

We have seen many different policy-based reinforcement learning algorithms that can be used in agents with continuous action spaces. Let us have a closer look at the environments that they have been used in and how well they perform.

Policy-based methods, and especially the actor critic policy/value hybrid, work well for many problems, both with discrete and with continuous action spaces. Policy-based methods are often tested on complex high-dimensional robotics applications [24]. Let us have a look at the kind of environments that have been used to develop PPO, A3C, and the other algorithms.

Two application categories are robot locomotion and visuo-motor interaction. These two problems have drawn many researchers, and many new algorithms have been devised, some of which were able to learn impressive performance. For each of the two problems, we will discuss a few results in more detail.

4.3.1   Locomotion

One of the problems of locomotion of legged entities is the problem of learning gaits. Humans, with two legs, can walk, run, and jump, among others. Dogs and horses, with four legs, have other gaits, where their legs may move in even more interesting patterns, such as the trot, canter, pace, and gallop. The challenges that we pose robots are often easier. Typical reinforcement learning tasks are for a one-legged robot to learn to jump, for biped robots to walk and jump, and for a quadruped to get to learn to use its multitude of legs in any coordinated fashion that results in forward moving. Learning such policies can be quite computationally expensive, and a curious simulated virtual animal has emerged that is cheaper to simulate: the

Fig. 4.6 Humanoid standing
up [39]

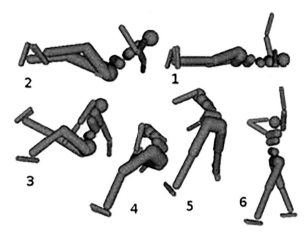

two-legged *half-cheetah*, whose task is to run forward. We have already seen some
of these robotic creatures in Figs. 4.2 and 4.3.

The first approach that we will discuss is by Schulman et al. [39]. They report
experiments where human-like bipeds and quadrupeds must learn to stand up and
learn running gaits. These are challenging 3D locomotion tasks that were formerly
attempted with hand-crafted policies. Figure 4.6 shows a sequence of states.

The challenge in these situations is actually somewhat spectacular: the agent is
only provided with a positive reward for moving forward; based on nothing more,
it has to learn to control all its limbs by itself, through trial and error; and no hint is
given on how to control a leg or what its purpose is. These results are best watched
in the movies that have been made[16] about the learning process.

The authors use an advantage actor critic algorithm with trust regions. The
algorithm is fully model-free, and learning with simulated physics was reported to
take one to 2 weeks of real time. Learning to walk is quite a complicated challenge,
as the movies illustrate. They also show the robot learning to scale an obstacle run
all by itself.

In another study, Heess et al. [18] report on end-to-end learning of complex
robot locomotion from pixel input to (simulated) motor actuation. Figure 4.7 shows
how a walker scales an obstacle course and Fig. 4.8 shows a time lapse of how
a quadruped traverses a course. Agents learned to run, jump, crouch, and turn as
the environment required, without explicit reward shaping or other hand-crafted
features. For this experiment, a distributed version of PPO was used. Interestingly,
the researchers stress that the use of a rich—varied, difficult—environment helps to
promote learning of complex behavior, which is also robust across a range of tasks.

---

[16] Such as the movie from the start of this chapter: https://www.youtube.com/watch?v=
hx_bgoTF7bs.

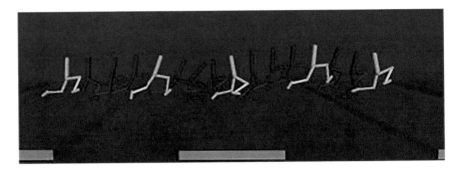

Fig. 4.7 Walker obstacle course [18]

Fig. 4.8 Quadruped obstacle course [18]

4.3.2   Visuo-Motor Interaction

Most experiments in "end-to-end" learning of robotic locomotion are set up so that the input is received directly from features that are derived from the states as calculated by the simulation software. A step further toward real-world interaction is to learn directly from camera pixels. We then model eye–hand coordination in visuo-motor interaction tasks, and the state of the environment has to be inferred from camera or other visual means and then be translated in joint (muscle) actuations.

Visuo-motor interaction is a difficult task, requiring many techniques to work together. Different environments have been introduced to test algorithms. Tassa et al. report on benchmarking efforts in robot locomotion with MuJoCo [45], introducing the DeepMind control suite, a suite of environments consisting of different MuJoCo control tasks (see Fig. 4.9). The authors also present baseline implementations of learning agents that use A3C, DDPG, and D4PG (distributional distributed deep deterministic policy gradients—an algorithm that extends DDPG).

In addition to learning from state derived features, results are presented where the agent learns from 84×84 pixel information, in a simulated form of visuo-motor

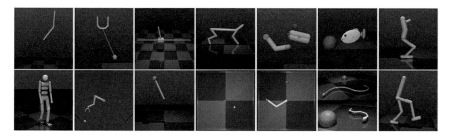

Fig. 4.9 DeepMind Control Suite. Top: Acrobot, Ball-in-cup, Cartpole, Cheetah, Finger, Fish, Hopper. Bottom: Humanoid, Manipulator, Pendulum, Point-mass, Reacher, Swimmer (6 and 15 links), and Walker [45]

interaction. The DeepMind control suite is especially designed for further research in the field [31–34, 46]. Other environment suites are Meta-World [51], Surreal [10], and RLbench [21].

Visuo-motor interaction is a challenging problem that remains an active area of research.

4.3.3 Benchmarking

Benchmarking efforts are of great importance in the field [7]. Henderson et al. [19] published an influential study of the sensitivity of outcomes to different hyperparameter settings and the influence of non-determinism, by trying to reproduce many published works in the field. They find large variations in outcomes, and in general that reproducibility of results is problematic. They conclude that *without significance metrics and tighter standardization of experimental reporting, it is difficult to determine whether improvements over the prior state-of-the-art are meaningful* [19]. Further studies confirmed these findings [1],[17] and today more works are being published with code, hyperparameters, and environments.

Taking inspiration from the success of the Arcade Learning Environment in game playing, benchmark suites of continuous control tasks with high state and action dimensionality have been introduced [7, 12].[18] The tasks include 3D humanoid locomotion, tasks with partial observations, and tasks with hierarchical structure. The locomotion tasks are Swimmer, Hopper, Walker, Half-Cheetah, Ant, simple Humanoid, and full Humanoid, with the goal being to move forward as fast as possible. These are difficult tasks because of the high degree of freedom of movement. Partial observation is achieved by adding noise or leaving out certain

---

[17] https://github.com/google-research/rliable.

[18] The suite in the paper is called RLlab. A newer version of the suite is named Garage. See also Appendix C.

parts of the regular observations. The hierarchical tasks consist of low-level tasks such as learning to move and a high-level task such as finding the way out of a maze.

Summary and Further Reading

This chapter is concerned with the second kind of model-free algorithms: policy-based methods. We summarize what we have learned and provide pointers to further reading.

Summary

Policy-based model-free methods are some of the most popular methods of deep reinforcement learning. For large continuous action spaces, indirect value-based methods are not well suited because of the use of the arg max function to recover the best action to go with the value. Where value-based methods work step by step, vanilla policy-based methods roll out a full future trajectory or episode. Policy-based methods work with a parameterized current policy, which is well suited for a neural network as policy function approximator.

After the full trajectory has been rolled out, the reward and the value of the trajectory is calculated and the policy parameters are updated, using gradient ascent. Since the value is only known at the end of an episode, classic policy-based methods have a higher variance than value-based methods and may converge to a local optimum. The best known classic policy method is called REINFORCE.

Actor critic methods add a value network to the policy network, to achieve the benefits of both approaches. To reduce variance, n-step temporal difference bootstrapping can be added, and a baseline value can be subtracted, so that we get the so-called *advantage* function (which subtracts the value of the parent state from the action values of the future states, bringing their expected value closer to zero). Well-known actor critic methods are A3C, DDPG, TRPO, PPO, and SAC.[19] A3C features an asynchronous (parallel, distributed) implementation, DDPG is an actor critic version of DQN for continuous action spaces, TRPO and PPO use trust regions to achieve adaptive step sizes in nonlinear spaces, and SAC optimizes for the expected value and entropy of the policy. Benchmark studies have shown that the performance of these actor critic algorithm is as good as or better than value-based methods [7, 19].

---

[19] Asynchronous advantage actor critic, deep deterministic policy gradients, trust region policy optimization, proximal policy optimization, soft actor critic.

Robot learning is among the most popular applications for policy-based method. Model-free methods have low sample efficiency, and to prevent the cost of wear after millions of samples, most experiments use a physics simulation as environment, such as MuJoCo. Two main application areas are locomotion (learning to walk, learning to run) and visuo-motor interaction (learning directly from camera images of one's own actions).

Further Reading

Policy-based methods have been an active research area for some time. Their natural suitability for deep function approximation for robotics applications and other applications with continuous action spaces has spurred a large interest in the research community. The classic policy-based algorithm is Williams' REINFORCE [49], which is based on the policy gradient theorem, see [44]. Our explanation is based on [6, 8, 11]. Joining policy and value-based methods as we do in actor critic are discussed in Barto et al. [2]. Mnih et al. [35] introduce a modern efficient parallel implementation named A3C. After the success of DQN, a version for the continuous action space of policy-based methods was introduced as DDPG by Lillicrap et al. [29]. Schulman et al. have worked on trust regions, yielding efficient popular algorithms TRPO [38] and PPO [40].

Important benchmark studies of policy-based methods are Duan et al. [7] and Henderson et al. [19]. These papers have stimulated reproducibility in reinforcement learning research.

Software environments that are used in testing policy-based methods are MuJoCo [48] and PyBullet [5]. Gym [4] and the DeepMind control suite [45] incorporate MuJoCo and provide an easy to use Python interface. An active research community has emerged around the DeepMind control suite.

Exercises

We have come to the end of this chapter, and it is time to test our understanding with questions, exercises, and a summary.

Questions

Below are some quick questions to check your understanding of this chapter. For each question, a simple single sentence answer is sufficient.

1. Why are value-based methods difficult to use in continuous action spaces?

2. What is MuJoCo? Can you name a few example tasks?
3. What is an advantage of policy-based methods?
4. What is a disadvantage of full-trajectory policy-based methods?
5. What is the difference between actor critic and vanilla policy-based methods?
6. How many parameter sets are used by actor critic? How can they be represented in a neural network?
7. Describe the relation between Monte Carlo REINFORCE, n-step methods, and temporal difference bootstrapping.
8. What is the advantage function?
9. Describe a MuJoCo task that methods such as PPO can learn to perform well.
10. Give two actor critic approaches to further improve upon bootstrapping and advantage functions, which are used in high-performing algorithms such as PPO and SAC.
11. Why is learning robot actions from image input hard?

Exercises

Let us now look at programming exercises. If you have not already done so, install MuJoCo or PyBullet, and install the DeepMind control suite.[20] We will use agent algorithms from the Stable Baselines. Furthermore, browse the examples directory of the DeepMind control suite on GitHub, and study the Colab notebook.

1. *REINFORCE.* Go to the Medium blog[21] and reimplement REINFORCE. You can choose PyTorch or TensorFlow/Keras, in which case you will have to improvise. Run the algorithm on an environment with a discrete action space, and compare with DQN. Which works better? Run in an environment with a continuous action space. Note that Gym offers a discrete and a continuous version of Mountain Car.
2. *Algorithms.* Run REINFORCE on a Walker environment from the Baselines. Run DDPG, A3C, and PPO. Run them for different time steps. Make plots. Compare training speed and outcome quality. Vary hyperparameters to develop an intuition for their effect.
3. *Suite.* Explore the DeepMind control suite. Look around and see what environments have been provided and how you can use them. Consider extending an environment. What learning challenges would you like to introduce? First do a survey of the literature that has been published about the DeepMind control suite.

---

[20] https://github.com/deepmind/dm_control.

[21] https://medium.com/@ts1829/policy-gradient-reinforcement-learning-in-pytorch-df1383ea0baf.

References

1. Rishabh Agarwal, Max Schwarzer, Pablo Samuel Castro, Aaron Courville, and Marc G Bellemare. Deep reinforcement learning at the edge of the statistical precipice. *arXiv preprint arXiv:2108.13264*, 2021. 127
2. Andrew G Barto, Richard S Sutton, and Charles W Anderson. Neuronlike adaptive elements that can solve difficult learning control problems. *IEEE Transactions on Systems, Man, and Cybernetics*, (5):834–846, 1983. 111, 129
3. Christopher M Bishop. *Pattern Recognition and Machine Learning*. Information science and statistics. Springer Verlag, Heidelberg, 2006. 118
4. Greg Brockman, Vicki Cheung, Ludwig Pettersson, Jonas Schneider, John Schulman, Jie Tang, and Wojciech Zaremba. OpenAI Gym. *arXiv preprint arXiv:1606.01540*, 2016. 106, 129
5. Erwin Coumans and Yunfei Bai. Pybullet, a python module for physics simulation for games, robotics and machine learning. http://pybullet.org, 2016–2019. 104, 105, 129
6. Mohit Deshpande. Deep RL policy methods https://mohitd.github.io/2019/01/20/deep-rl-policy-methods.html. 129
7. Yan Duan, Xi Chen, Rein Houthooft, John Schulman, and Pieter Abbeel. Benchmarking deep reinforcement learning for continuous control. In *International Conference on Machine Learning*, pages 1329–1338, 2016. 127, 127, 128, 129
8. Adrien Lucas Ecofet. An intuitive explanation of policy gradient https://towardsdatascience.com/an-intuitive-explanation-of-policy-gradient-part-1-reinforce-aa4392cbfd3c. 129
9. Tom Erez, Yuval Tassa, and Emanuel Todorov. Simulation tools for model-based robotics: Comparison of Bullet, Havok, MuJoCo, Ode and Physx. In *2015 IEEE International Conference on Robotics and Automation (ICRA)*, pages 4397–4404. IEEE, 2015. 105
10. Linxi Fan, Yuke Zhu, Jiren Zhu, Zihua Liu, Orien Zeng, Anchit Gupta, Joan Creus-Costa, Silvio Savarese, and Li Fei-Fei. Surreal: Open-source reinforcement learning framework and robot manipulation benchmark. In *Conference on Robot Learning*, pages 767–782, 2018. 127
11. Vincent François-Lavet, Peter Henderson, Riashat Islam, Marc G Bellemare, and Joelle Pineau. An introduction to deep reinforcement learning. *Foundations and Trends in Machine Learning*, 11(3-4):219–354, 2018. 129
12. The garage contributors. Garage: A toolkit for reproducible reinforcement learning research. https://github.com/rlworkgroup/garage, 2019. 127
13. Ivo Grondman, Lucian Busoniu, Gabriel AD Lopes, and Robert Babuska. A survey of actor-critic reinforcement learning: Standard and natural policy gradients. *IEEE Transactions on Systems, Man, and Cybernetics, Part C (Applications and Reviews)*, 42(6):1291–1307, 2012. 111
14. Tuomas Haarnoja, Haoran Tang, Pieter Abbeel, and Sergey Levine. Reinforcement learning with deep energy-based policies. In *International Conference on Machine Learning*, pages 1352–1361. PMLR, 2017. 120
15. Tuomas Haarnoja, Aurick Zhou, Pieter Abbeel, and Sergey Levine. Soft actor-critic: Off-policy maximum entropy deep reinforcement learning with a stochastic actor. In *International Conference on Machine Learning*, pages 1861–1870. PMLR, 2018. 107, 119
16. Tuomas Haarnoja, Aurick Zhou, Kristian Hartikainen, George Tucker, Sehoon Ha, Jie Tan, Vikash Kumar, Henry Zhu, Abhishek Gupta, Pieter Abbeel, and Sergey Levine. Soft actor-critic algorithms and applications. *arXiv preprint arXiv:1812.05905*, 2018. 119
17. Roland Hafner and Martin Riedmiller. Reinforcement learning in feedback control. *Machine Learning*, 84(1-2):137–169, 2011. 121
18. Nicolas Heess, Dhruva TB, Srinivasan Sriram, Jay Lemmon, Josh Merel, Greg Wayne, Yuval Tassa, Tom Erez, Ziyu Wang, SM Eslami, Martin Riedmiller, and David Silver. Emergence of locomotion behaviours in rich environments. *arXiv preprint arXiv:1707.02286*, 2017. 102, 125, 126, 126
19. Peter Henderson, Riashat Islam, Philip Bachman, Joelle Pineau, Doina Precup, and David Meger. Deep reinforcement learning that matters. In *Thirty-Second AAAI Conference on Artificial Intelligence*, 2018. 127, 127, 128, 129

20. Chloe Ching-Yun Hsu, Celestine Mendler-Dünner, and Moritz Hardt. Revisiting design choices in proximal policy optimization. *arXiv preprint arXiv:2009.10897*, 2020. 118
21. Stephen James, Zicong Ma, David Rovick Arrojo, and Andrew J Davison. RLbench: the robot learning benchmark & learning environment. *IEEE Robotics and Automation Letters*, 5(2):3019–3026, 2020. 127
22. Arthur Juliani. Simple reinforcement learning with TensorFlow part 8: Asynchronous actor-critic agents (A3C) https://medium.com/emergent-future/simple-reinforcement-learning-with-tensorflow-part-8-asynchronous-actor-critic-agents-a3c-c88f72a5e9f2, 2016. 115, 116
23. Hilbert J Kappen. Path integrals and symmetry breaking for optimal control theory. *Journal of statistical mechanics: theory and experiment*, 2005(11):P11011, 2005. 120
24. Jens Kober, J Andrew Bagnell, and Jan Peters. Reinforcement learning in robotics: A survey. *The International Journal of Robotics Research*, 32(11):1238–1274, 2013. 124
25. Vijay R Konda and John N Tsitsiklis. Actor–critic algorithms. In *Advances in Neural Information Processing Systems*, pages 1008–1014, 2000. 111
26. Vijaymohan R Konda and Vivek S Borkar. Actor–Critic-type learning algorithms for Markov Decision Processes. *SIAM Journal on Control and Optimization*, 38(1):94–123, 1999. 111
27. Samuel Kotz, Narayanaswamy Balakrishnan, and Norman L Johnson. *Continuous Multivariate Distributions, Volume 1: Models and Applications*. John Wiley & Sons, 2004. 119
28. Solomon Kullback and Richard A Leibler. On information and sufficiency. *The Annals of Mathematical Statistics*, 22(1):79–86, 1951. 118
29. Timothy P Lillicrap, Jonathan J Hunt, Alexander Pritzel, Nicolas Heess, Tom Erez, Yuval Tassa, David Silver, and Daan Wierstra. Continuous control with deep reinforcement learning. In *International Conference on Learning Representations*, 2016. 104, 107, 121, 121, 122, 129
30. Vince Martinelli. How robots autonomously see, grasp, and pick. https://www.therobotreport.com/grasp-sight-picking-evolve-robots/, 2019. 105
31. Josh Merel, Arun Ahuja, Vu Pham, Saran Tunyasuvunakool, Siqi Liu, Dhruva Tirumala, Nicolas Heess, and Greg Wayne. Hierarchical visuomotor control of humanoids. In *International Conference on Learning Representations*, 2019. 127
32. Josh Merel, Diego Aldarondo, Jesse Marshall, Yuval Tassa, Greg Wayne, and Bence Ölveczky. Deep neuroethology of a virtual rodent. In *International Conference on Learning Representations*, 2020.
33. Josh Merel, Leonard Hasenclever, Alexandre Galashov, Arun Ahuja, Vu Pham, Greg Wayne, Yee Whye Teh, and Nicolas Heess. Neural probabilistic motor primitives for humanoid control. In *International Conference on Learning Representations*, 2019.
34. Josh Merel, Yuval Tassa, Dhruva TB, Sriram Srinivasan, Jay Lemmon, Ziyu Wang, Greg Wayne, and Nicolas Heess. Learning human behaviors from motion capture by adversarial imitation. *arXiv preprint arXiv:1707.02201*, 2017. 127
35. Volodymyr Mnih, Adria Puigdomenech Badia, Mehdi Mirza, Alex Graves, Timothy Lillicrap, Tim Harley, David Silver, and Koray Kavukcuoglu. Asynchronous methods for deep reinforcement learning. In *International Conference on Machine Learning*, pages 1928–1937, 2016. 107, 115, 115, 116, 116, 117, 129
36. Thomas Moerland. Continuous Markov decision process and policy search. Lecture notes for the course reinforcement learning, Leiden University, 2021. 119, 120, 122
37. Ofir Nachum, Mohammad Norouzi, Kelvin Xu, and Dale Schuurmans. Bridging the gap between value and policy based reinforcement learning. In *Advances in Neural Information Processing Systems*, pages 2775–2785, 2017. 120
38. John Schulman, Sergey Levine, Pieter Abbeel, Michael Jordan, and Philipp Moritz. Trust region policy optimization. In *International Conference on Machine Learning*, pages 1889–1897, 2015. 107, 117, 117, 129
39. John Schulman, Philipp Moritz, Sergey Levine, Michael Jordan, and Pieter Abbeel. High-dimensional continuous control using generalized advantage estimation. In *International Conference on Learning Representations*, 2016. 125, 125
40. John Schulman, Filip Wolski, Prafulla Dhariwal, Alec Radford, and Oleg Klimov. Proximal policy optimization algorithms. *arXiv preprint arXiv:1707.06347*, 2017. 107, 118, 129

41. David Silver, Guy Lever, Nicolas Heess, Thomas Degris, Daan Wierstra, and Martin Ried-miller. Deterministic policy gradient algorithms. In *International Conference on Machine Learning*, pages 387–395, 2014. 120, 121

42. Wenyu Sun and Ya-Xiang Yuan. *Optimization Theory and Methods: Nonlinear Programming*, volume 1. Springer Science & Business Media, 2006. 117

43. Richard S Sutton and Andrew G Barto. *Reinforcement learning, An Introduction, Second Edition*. MIT Press, 2018. 112

44. Richard S Sutton, David A McAllester, Satinder P Singh, and Yishay Mansour. Policy gradient methods for reinforcement learning with function approximation. In *Advances in Neural Information Processing Systems*, pages 1057–1063, 2000. 129

45. Yuval Tassa, Yotam Doron, Alistair Muldal, Tom Erez, Yazhe Li, Diego de Las Casas, David Budden, Abbas Abdolmaleki, Josh Merel, Andrew Lefrancq, Timothy Lillicrap, and Martin Riedmiller. DeepMind control suite. *arXiv preprint arXiv:1801.00690*, 2018. 126, 127, 129

46. Yuval Tassa, Saran Tunyasuvunakool, Alistair Muldal, Yotam Doron, Siqi Liu, Steven Bohez, Josh Merel, Tom Erez, Timothy Lillicrap, and Nicolas Heess. dm_control: Software and tasks for continuous control. *arXiv preprint arXiv:2006.12983*, 2020. 127

47. Emanuel Todorov. Linearly-solvable Markov decision problems. In *Advances in Neural Information Processing Systems*, pages 1369–1376, 2007. 120

48. Emanuel Todorov, Tom Erez, and Yuval Tassa. MuJoCo: A physics engine for model-based control. In *IEEE/RSJ International Conference on Intelligent Robots and Systems (IROS)*, pages 5026–5033, 2012. 104, 105, 129

49. Ronald J Williams. Simple statistical gradient-following algorithms for connectionist reinforcement learning. *Machine Learning*, 8(3-4):229–256, 1992. 107, 110, 110, 129

50. Yuhuai Wu, Elman Mansimov, Roger B Grosse, Shun Liao, and Jimmy Ba. Scalable trust-region method for deep reinforcement learning using Kronecker-factored approximation. In *Advances in Neural Information Processing Systems*, pages 5279–5288, 2017. 115

51. Tianhe Yu, Deirdre Quillen, Zhanpeng He, Ryan Julian, Karol Hausman, Chelsea Finn, and Sergey Levine. Meta-world: A benchmark and evaluation for multi-task and meta reinforcement learning. In *Conference on Robot Learning*, pages 1094–1100. PMLR, 2020. 127

52. Brian D Ziebart, Andrew L Maas, J Andrew Bagnell, and Anind K Dey. Maximum entropy inverse reinforcement learning. In *AAAI*, volume 8, pages 1433–1438. Chicago, IL, USA, 2008. 120

Chapter 5
Model-Based Reinforcement Learning

The previous chapters discussed model-*free* methods, and we saw their success in video games and simulated robotics. In model-free methods, the agent updates a policy directly from the feedback that the environment provides on its actions. The environment performs the state transitions and calculates the reward. A disadvantage of deep model-free methods is that they can be slow to train; for stable convergence or low variance, often millions of environment samples are needed before the policy function converges to a high-quality optimum.

In contrast, with model-*based* methods, the agent first builds its own internal transition model from the environment feedback. The agent can then use this local transition model to find out about the effect of actions on states and rewards. The agent can use a planning algorithm to play *what-if* games and generate policy updates, all without causing any state changes in the environment. This approach promises higher quality at lower sample complexity. Generating policy updates from the internal model is called planning or *imagination*.

Model-based methods update the policy indirectly: the agent first learns a local transition model from the environment, which the agent then uses to update the policy. Indirectly learning the policy function has two consequences. On the positive side, as soon as the agent has its own model of the state transitions of the world, it can learn the best policy for free, without further incurring the cost of acting in the environment. Model-based methods thus may have a lower sample complexity. The downside is that the learned transition model may be inaccurate, and the resulting policy may be of low quality. No matter how many samples can be taken for free from the model, if the agent's local transition model does not reflect the environment's real transition model, then the locally learned policy function will not work in the environment. Thus, dealing with uncertainty and model bias are important elements in model-based reinforcement learning.

The idea to first learn an internal representation of the environment's transition function has been conceived many years ago, and transition models have been

© The Author(s), under exclusive license to Springer Nature Singapore Pte Ltd. 2022   135
A. Plaat, *Deep Reinforcement Learning*,
https://doi.org/10.1007/978-981-19-0638-1_5

implemented in many different ways. Models can be tabular, or they can be based on various kinds of deep learning, as we will see.

This chapter will start with an example showing how model-based methods work. Next, we describe in more detail different kinds of model-based approaches: approaches that focus on *learning* an accurate model and approaches for *planning* with an imperfect model. Finally, we describe application environments for which model-based methods have been used in practice, to see how well the approaches perform.

The chapter is concluded with exercises, a summary, and pointers to further reading.

Core Concepts

- Imagination
- Uncertainty models
- World models and latent models
- Model-predictive control
- Deep end-to-end planning and learning

Core Problem

- Learn and use accurate transition models for high-dimensional problems

Core Algorithms

- Dyna-Q (Algorithm 5.3)
- Ensembles and model-predictive control (Algorithms 5.4 and 5.6)
- Value prediction networks (Algorithm 5.5)
- Value iteration networks (Sect. 5.2.2.2)

Building a Navigation Map

To illustrate the basic concepts of model-based reinforcement learning, we return to the supermarket example.

Let us compare how model-free and model-based methods find their way to the supermarket in a new city.[1] In this example we will use value-based Q-learning; our policy $\pi(s, a)$ will be derived directly from the $Q(s, a)$ values with arg max, and writing Q is in this sense equivalent to writing π.

Model-free Q-learning: the agent picks the start state s_0 and uses (for example) an ϵ-greedy behavior policy on the action-value function $Q(s, a)$ to select the next action. The environment then executes the action, computes the next state s' and reward r, and returns these to the agent. The agent updates its action-value function $Q(s, a)$ with the familiar update rule

$$Q(s, a) \leftarrow Q(s, a) + \alpha[r + \gamma \max_a Q(s', a) - Q(s, a)].$$

The agent repeats this procedure until the values in the Q-function no longer change greatly.

Thus we pick our start location in the city, perform one walk along a block in an ϵ-greedy direction, and record the reward and the new state at which we arrive. We use the information to update the policy, and from our new location, we walk again in an ϵ-greedy direction using the policy. If we find the supermarket, we start over again, trying to find a shorter path, until our policy values no longer change (this may take many environment interactions). Then the best policy is the path with the shortest distances.

Model-based planning and learning: the agent uses the $Q(s, a)$ function as behavior policy as before to sample the new state and reward from the environment and to update the policy (Q-function). In addition, however, the agent will record the new state and reward in a local transition $T_a(s, s')$ and reward function $R_a(s, s')$. Because the agent now has these local entries, we can also sample from our local functions to update the policy. We can choose sample from the (expensive) environment transition function or from the (cheap) local transition function. There is a caveat with sampling locally, however. The local functions may contain fewer entries—or only high-variance entries—especially in the early stages, when few environment samples have been performed. The usefulness of the local functions increases as more environment samples are performed.

Thus, we now have a local map on which to record the new states and rewards. We will use this map to peek, as often as we like and at no cost, at a location on that map, to update the Q-function. As more environment samples come in, the map will have more and more locations for which a distance to the supermarket is recorded. When glances at the map do not improve the policy anymore, we have to walk in the environment again and, as before, update the map and the policy.

---

[1] We use distance to the supermarket as negative reward, in order to formulate this as a distance minimization problem, while still being able to reason in our familiar reward maximization setting.

Fig. 5.1 Direct and indirect
reinforcement learning [85]

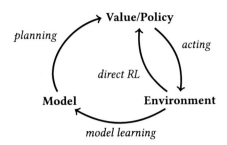

In conclusion, model-free finds all policy updates outside the agent, from the
environment feedback, and model-based also[2] uses policy updates from within the
agent, using information from its local map (see Fig. 5.1). In both methods all
updates to the policy are ultimately derived from the environment feedback; model-
based offers a different way to use the information to update the policy, a way that
may be more information-efficient, by keeping information from each sample within
the agent transition model and reusing that information.

5.1 Dynamics Models of High-Dimensional Problems

The application environments for model-based reinforcement learning are the same
as for model-free; our goal, however, is to solve larger and more complex problems
in the same amount of time, by virtue of the lower sample complexity and, as it
were, a deeper understanding of the environment.

Transition Model and Knowledge Transfer
The principle of model-based learning is as follows. Where model-free methods
sample the environment to learn the state-to-*action* policy function $\pi(s, a)$ based
on action rewards, model-based methods sample the environment to learn the state-
to-*state* transition function $T_a(s, s')$ based on action rewards. Once the accuracy of
this local transition function is good enough, the agent can sample from this local
function to improve the policy $\pi(s, a)$ as often as it likes, without incurring the cost
of actual environment samples. In the model-based approach, the agent builds its
own local state-to-state transition (and reward) model of the environment, so that,
in theory at least, it does not need the environment anymore.

This brings us to another reason for the interest in model-based methods. For
sequential decision problems, knowing the transition function is a natural way of

---

[2] One option is to only update the policy from the agent's internal transition model and not by the
environment samples anymore. However, another option is to keep using the environment samples
to also update the policy in the model-free way. Sutton's Dyna [83] approach is a well-known
example of this last hybrid approach. Compare also Figs. 5.2 and 5.4.

capturing *the essence* of how the environment works—π gives the next action, and T gives the next state.

This is useful, for example, when we switch to a related environment. When the transition function of the environment is known by the agent, then the agent can be adapted quickly, without having to learn a whole new policy by sampling the environment. When a good local transition function of the domain is known by the agent, then new, but related, problems might be solved efficiently. Hence, model-based reinforcement learning may contribute to efficient transfer learning (see Chap. 9).

Sample Efficiency
The sample efficiency of an agent algorithm tells us how many environment samples it needs for the policy to reach a certain accuracy.

To achieve high sample efficiency, model-based methods learn a dynamics model. Learning high-accuracy high-capacity models of high-dimensional problems requires a high number of training examples, to prevent overfitting (see Sect. B.2.7). Thus, reducing overfitting in learning the transition model would negate (some of) the advantage of the low sample complexity that model-based learning of the policy function achieves. Constructing accurate deep transition models can be difficult in practice, and for many complex sequential decision problems the best results are often achieved with model-free methods, although deep model-based methods are becoming stronger (see, for example, Wang et al. [88]).

5.2 Learning and Planning Agents

The promise of model-based reinforcement learning is to find a high-accuracy behavior policy at a low cost, by building a local model of the world. This will only work if the learned transition model provides accurate predictions and if the extra cost of planning with the model is reasonable.

Let us see which solutions have been developed for deep model-based reinforcement learning. In Sect. 5.3, we will have a closer look at the performance in different environments. First, in this section, we will look at four different algorithmic approaches and at a classic approach: Dyna's tabular imagination.

Tabular Imagination
A classic approach is Dyna [83], which popularized the idea of model-based reinforcement learning. In Dyna, environment samples are used in a hybrid model-free/model-based manner, to train the transition model, use planning to improve the policy, while also training the policy function directly.

Why is Dyna a hybrid approach? Strict model-based methods update the policy only by planning using the agent's transition model, see Algorithm 5.1. In Dyna, however, in addition to using the agent's transition model, *also* the environment's transition model is used to update the policy function (see Fig. 5.2 and Algorithm 5.2). Thus we get a hybrid approach combining model-based and model-free

---

Algorithm 5.1 Strict learned dynamics model

---

repeat
    Sample environment E to generate data $D = (s, a, r', s')$
    Use D to learn $M = T_a(s, s')$, $R_a(s, s')$ ▷ learning
    for $n = 1, \ldots, N$ **do**
        Use M to update policy $\pi(s, a)$ ▷ planning
    end for
until π converges

Fig. 5.2 Hybrid model-based imagination

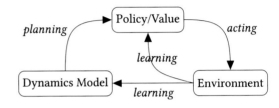

Algorithm 5.2 Hybrid model-based imagination

---

repeat
    Sample env E to generate data $D = (s, a, r', s')$
    Use D to update policy $\pi(s, a)$ ▷ learning
    Use D to learn $M = T_a(s, s')$, $R_a(s, s')$ ▷ learning
    for $n = 1, \ldots, N$ **do**
        Use M to update policy $\pi(s, a)$ ▷ planning
    end for
until π converges

---

learning. This hybrid model-based planning is called *imagination* because looking ahead with the agent's own dynamics model resembles imagining environment samples outside the real environment inside the "mind" of the agent. In this approach the imagined samples augment the real (environment) samples at no sample cost.[3]

Imagination is a mix of model-based and model-free reinforcement learning. Imagination performs regular direct reinforcement learning, where the environment is sampled with actions according to the behavior policy, and the feedback is used to update the same behavior policy. Imagination also uses the environment sample to update the dynamics model $\{T_a, R_a\}$. This extra model is also sampled and provides extra updates to the behavior policy, in between the model-free updates.

The diagram in Fig. 5.2 shows how sample feedback is used both for updating the policy directly and for updating the model, which then updates the policy, by planning "imagined" feedback. In Algorithm 5.2, the general imagination approach is shown as pseudocode.

---

[3] The term *imagination* is used somewhat loosely in the field. In a strict sense, imagination refers only to updating the policy from the internal model by planning. In a wider sense, imagination refers to hybrid schemes where the policy is updated from both the internal model and the environment. Sometimes the term *dreaming* is used for agents imagining environments.

Algorithm 5.3 Dyna-Q [83]

Initialize $Q(s, a) \to \mathbb{R}$ randomly
Initialize $M(s, a) \to \mathbb{R} \times S$ randomly   ▷ Model
repeat
    Select $s \in S$ randomly
    $a \leftarrow \pi(s)$ ▷ $\pi(s)$ can be ϵ-greedy(s) based on Q
    $(s', r) \leftarrow E(s, a)$ ▷ Learn new state and reward from environment
    $Q(s, a) \leftarrow Q(s, a) + \alpha \cdot [r + \gamma \cdot \max_{a'} Q(s', a') - Q(s, a)]$
    $M(s, a) \leftarrow (s', r)$
    for $n = 1, \ldots, N$ **do**
        Select $\hat{s}$ and $\hat{a}$ randomly
        $(s', r) \leftarrow M(\hat{s}, \hat{a})$ ▷ Plan imagined state and reward from model
        $Q(\hat{s}, \hat{a}) \leftarrow Q(\hat{s}, \hat{a}) + \alpha \cdot [r + \gamma \cdot \max_{a'} Q(s', a') - Q(\hat{s}, \hat{a})]$
    end for
until Q converges

Sutton's Dyna-Q [83, 85], which is shown in more detail in Algorithm 5.3, is a concrete implementation of the imagination approach. Dyna-Q uses the Q-function as behavior policy $\pi(s)$ to perform ϵ-greedy sampling of the environment. It then updates this policy with the reward and an explicit model M. When the model M has been updated, it is used N times by planning with random actions to update the Q-function. The pseudocode shows the learning steps (from environment E) and N planning steps (from model M). In both cases, the Q-function state–action values are updated. The best action is then derived from the Q-values as usual.

Thus, we see that the number of updates to the policy can be increased without more environment samples. By choosing the value for N, we can tune how many of the policy updates will be environment samples and how many will be model samples. In the larger problems that we will see later in this chapter, the ratio of environment-to-model samples is often set at, for example, 1 : 1000, greatly reducing sample complexity. The questions then become, of course, *how good is the model* and *how far is the resulting policy from a model-free baseline?*

Hands On: Imagining Taxi Example
It is time to illustrate how Dyna-Q works with an example. For that, we turn to one of our favorites, the Taxi world.

Let us see what the effect of imagining with a model can be. Please refer to Fig. 5.3. We use our simple maze example, the Taxi maze, with zero imagination ($N = 0$) and with large imagination ($N = 50$). Let us assume that the reward at all states returned by the environment is 0, except for the goal, where the reward is $+1$. In states, the usual actions are present (north, east, west, and south), except at borders or walls.

When $N = 0$, Dyna-Q performs exactly Q-learning, randomly sampling action rewards, building up the Q-function, and using the Q-values following the ϵ-greedy policy for action selection. The purpose of the Q-function is to act as a vessel of information to find the goal. How does our vessel get filled with

Fig. 5.3 Taxi world [44]

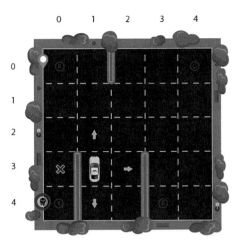

information? Sampling starts off randomly, and the Q-values fill slowly, since the reward landscape is flat or sparse: only the goal state returns $+1$, and all other states return 0. In order to fill the Q-values with actionable information on where to find the goal, first the algorithm must be lucky enough to choose a state next to the goal, including the appropriate action to reach the goal. Only then the first useful reward information is found and the first non-zero step toward finding the goal can be entered into the Q-function. We conclude that, with $N = 0$, the Q-function is filled up slowly, due to sparse rewards.

What happens when we turn on planning? When we set N to a high value, such as 50, we perform 50 planning steps for each learning step. As we can see in the algorithm, the model is built alongside the Q-function, from environment returns. As long as the Q-function is still fully zero, then planning with the model will also be useless. But as soon as one goal entry is entered into Q and M, then planning will start to shine: it will perform 50 planning samples on the M-model, probably finding the goal information and possibly building up an entire trajectory filling states in the Q-function with actions toward the goal.

In a way, the model-based planning amplifies any useful reward information that the agent has learned from the environment and plows it back quickly into the policy function. The policy is learned much quicker, with fewer environment samples.

Reversible Planning and Irreversible Learning

Model-free methods sample the environment and learn the policy function $\pi(s, a)$ directly, in one step. Model-based methods sample the environment to learn the policy indirectly, using a dynamics model $\{T_a, R_a\}$ (as we see in Figs. 5.1 and 5.4 and in Algorithm 5.1).

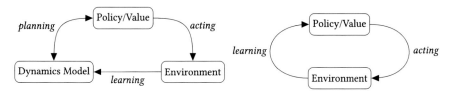

Fig. 5.4 Model-based (left) and model-free (right). Learning changes the environment state irreversibly (single arrow), and planning changes the agent state reversibly (undo, double arrow)

Table 5.1 Difference between Planning and Learning

	Planning	Learning
Transition model in	Agent	Environment
Agent can undo	Yes	No
State is	Reversible by agent	Irreversible by agent
Dynamics	Backtrack	Forward only
Data structure	Tree	Path
New state	In agent	Sample from environment
Reward	By agent	Sample from environment
Synonyms	Imagination, simulation	Sampling, rollout

It is useful to step back for a moment to consider the place of learning and planning algorithms in the reinforcement learning paradigm. Please refer to Table 5.1 for a summary of differences between planning and learning.

Planning with an internal transition model is *reversible*. When the agent uses its own transition model to perform local actions on a local state, then the actions can be *undone*, since the agent applied them to a copy in its own memory [61].[4] Because of this local state memory, the agent can return to the old state, reversing the local state change caused by the local action that it has just performed. The agent can then try an alternative action (which it can also reverse). The agent can use tree-traversal methods to traverse the state space, backtracking to try other states.

In contrast to planning, *learning* is done when the agent does not have access to its own transition function $T_a(s, s')$. The agent can get reward information by sampling real actions in the environment. These actions are not played out inside the agent but executed in the actual environment; they are irreversible and cannot be undone by the agent. Learning uses actions that irreversibly change the state of the environment. Learning does not permit backtracking; learning algorithms learn a policy by repeatedly sampling the environment.

Note the similarity between learning and planning: learning samples rewards from the external environment and planning from the internal model; both use the samples to update the policy function $\pi(s, a)$.

---

[4] In our dreams, we can undo our actions, play *what-if*, and imagine alternative realities.

Four Types of Model-Based Methods

In model-based reinforcement learning, the challenge is to learn deep high-dimensional transition models from limited data. Our methods should be able to account for model uncertainty and plan over these models to achieve policy and value functions that perform as well or better than model-free methods. Let us look in more detail at specific model-based reinforcement learning methods to see how this can be achieved.

Over the years, many different approaches for high-accuracy high-dimensional model-based reinforcement learning have been devised. Following [69], we group the methods into four main approaches. We start with two approaches for learning the model and then two approaches for planning, using the model. For each, we will take a few representative papers from the literature that we describe in more depth. After we have done so, we will look at their performance in different environments. But let us start with the methods for learning a deep model first.

5.2.1 Learning the Model

In model-based approaches, the transition model is learned from sampling the environment. If this model is not accurate, then planning will not improve the value or policy function, and the method will perform worse than model-free methods. When the learning/planning ratio is set to $1/1000$, as it is in some experiments, inaccuracy in the models will reveal itself quickly in a low accuracy policy function.

Much research has focused on achieving high accuracy dynamics models for high-dimensional problems. Two methods to achieve better accuracy are *uncertainty modeling* and *latent models*. We will start with uncertainty modeling.

5.2.1.1 Modeling Uncertainty

The variance of the transition model can be reduced by increasing the number of environment samples, but there are also other approaches that we will discuss. A popular approach for smaller problems is to use Gaussian processes, where the dynamics model is learned by giving an estimate of the function and of the uncertainty around the function with a covariance matrix on the entire dataset [5]. A Gaussian model can be learned from few data points, and the transition model can be used to plan the policy function successfully. An example of this approach is the PILCO system, which stands for Probabilistic Inference for Learning Control [13, 14]. This system was effective on Cartpole and Mountain car but does not scale to larger problems.

We can also sample from a trajectory distribution optimized for cost and use that to train the policy, with a policy-based method [57]. Then we can optimize policies with the aid of locally linear models and a stochastic trajectory optimizer. This is the approach that is used in guided policy search (GPS), which has been shown to train

---

Algorithm 5.4 Planning with an ensemble of models [52]

Initialize policy π_θ and the models $\hat{m}_{\phi_1}, \hat{m}_{\phi_1}, \ldots, \hat{m}_{\phi_K}$ $\triangleright$ ensemble
Initialize an empty dataset D
repeat
    $D \leftarrow$ sample with π_θ from environment E
    Learn models $\hat{m}_{\phi_1}, \hat{m}_{\phi_1}, \ldots, \hat{m}_{\phi_K}$ using D $\triangleright$ ensemble
    repeat
        $D' \leftarrow$ sample with π_θ from $\{\hat{m}_{\phi_i}\}_{i=1}^{K}$
        Update π_θ with TRPO using D' $\triangleright$ planning
        Estimate performance of trajectories $\hat{\eta}_\tau(\theta, \phi_i)$ for $i = 1, \ldots, K$
    until performance converges
until π_θ performs well in environment E

---

complex policies with thousands of parameters, learning tasks in MuJoCo such as Swimming, Hopping, and Walking.

Another popular method to reduce variance in machine learning is the ensemble method. Ensemble methods combine multiple learning algorithms to achieve better predictive performance; for example, a random forest of decision trees often has better predictive performance than a single-decision tree [5, 68]. In deep model-based methods, the ensemble methods are used to estimate the variance and account for it during planning. A number of researchers have reported good results with ensemble methods on larger problems [10, 40]. For example, Chua et al. use an ensemble of probabilistic neural network models [9] in their approach named probabilistic ensembles with trajectory sampling (PETS). They report good results on high-dimensional simulated robotic tasks (such as Half-Cheetah and Reacher). Kurutach et al. [52] combine an ensemble of models with TRPO, in ME-TRPO.[5] In ME-TRPO, an ensemble of deep neural networks is used to maintain model uncertainty, while TRPO is used to control the model parameters. In the planner, each imagined step is sampled from the ensemble predictions (see Algorithm 5.4).

Uncertainty modeling tries to improve the accuracy of high-dimensional models by probabilistic methods. A different approach, specifically designed for high-dimensional deep models, is the latent model approach, which we will discuss next.

5.2.1.2 Latent Models

Latent models focus on dimensionality reduction of high-dimensional problems. The idea behind latent models is that in most high-dimensional environments some elements are less important, such as buildings in the background that never move and that have no relation with the reward. We can abstract these unimportant elements away from the model, reducing the effective dimensionality of the space.

---

[5] A video is available at https://sites.google.com/view/me-trpo. The code is at https://github.com/thanard/me-trpo. A blog post is at [38].

Latent models do so by learning to represent the elements of the input and the reward. Since planning and learning are now possible in a lower-dimensional latent space, the sampling complexity of learning from the latent models improves.

Even though latent model approaches are often complicated designs, many works have been published that show good results [29, 31–33, 41, 76, 81]. Latent models use multiple neural networks, as well as different learning and planning algorithms.

To understand this approach, we will briefly discuss one such latent-model approach: the Value prediction network (VPN) by Oh et al. [67].[6] VPN uses four differentiable functions, which are trained to predict the value [25], Fig. 5.5 shows how the core functions. The core idea in VPN is not to learn directly in the actual observation space but first to transform the state representations to a smaller latent representation model, also known as abstract model. The other functions, such as value, reward, and next state, then work on these smaller latent states, instead of on the more complex high-dimensional states. In this way, planning and learning occur in a space where states are encouraged only to contain the elements that influence value changes. Latent space is lower dimensional, and training and planning become more efficient.

The four functions in VPN are (1) an encoding function $f_{\theta_e}^{enc}$, (2) a reward function, (3) a value function, and (4) a transition function. All functions are parameterized with their own set of parameters. To distinguish these latent-based functions from the conventional observation-based functions R, V, T, they are denoted as $f_{\theta_r}^{reward}$, $f_{\theta_v}^{value}$, and $f_{\theta_t}^{trans}$.

- The encoding function $f_{\theta_e}^{enc} : s_{actual} \rightarrow s_{latent}$ maps the observation s_{actual} to the abstract state using neural network θ_e, such as a CNN for visual observations. This is the function that performs the dimensionality reduction.
- The latent reward function $f_{\theta_r}^{reward} : (s_{latent}, o) \rightarrow r, \gamma$ maps the latent state s and option o (a kind of action) to the reward and discount factor. If the option takes k primitive actions, the network should predict the discounted sum of the k immediate rewards as a scalar. (The role of options is explained in the paper [67].) The network also predicts option-discount factor γ for the number of steps taken by the option.
- The latent-value function $f_{\theta_v}^{value} : s_{latent} \rightarrow V_{\theta_v}(s_{latent})$ maps the abstract state to its value using a separate neural network θ_v. This value is the value of the latent state, not of the actual observation state $V(s_{actual})$.
- The latent-transition function $f_{\theta_t}^{trans} : (s_{latent}, o) \rightarrow s'_{latent}$ maps the latent state to the next latent state, depending also on the option.

Figure 5.5 shows how the core functions work together in the smaller latent space, with x the observed actual state and s the encoded latent state [67].

The figure shows a single rollout step, planning one step ahead. However, a model also allows looking further into the future, by performing multi-step rollouts. Of course, this requires a highly accurate model; otherwise the accumulated inaccura-

---

[6] See https://github.com/junhyukoh/value-prediction-network for the code.

Fig. 5.5 Architecture of latent model [67]

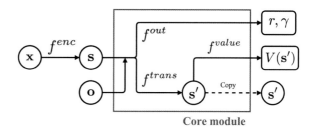

Algorithm 5.5 Multi-step planning [67]

---

function Q-PLAN(s, o, d)
    $r, \gamma, V(s'), s' \rightarrow f_\theta^{core}(s, o)$ ▷ Perform the four latent functions
    if $d = 1$ **then**
        return $r + \gamma V(s')$
    end if
    $A \leftarrow b$-best options based on $r + \gamma V_\theta(s')$ ▷ See paper for other expansion strategies
    for $o' \in A$ **do**
        $q_{o'} \leftarrow$ Q-Plan$(s', o', d - 1)$
    end for
    return $r + \gamma[\frac{1}{d} V(s') + \frac{d-1}{d} \max_{o' \in A} q_{o'}]$
end function

---

cies diminish the accuracy of the far-into-the-future lookahead. Algorithm 5.5 shows the pseudocode for a d-step planner for the value prediction network.

The networks are trained with n-step Q-learning and TD search [80]. Trajectories are generated with an ϵ-greedy policy using the planning algorithm from Algorithm 5.5. VPN achieved good results on Atari games such as Pacman and Seaquest, outperforming model-free DQN and outperforming observation-based planning in stochastic domains.

Another relevant approach is presented in a sequence of papers by Hafner et al. [31–33]. Their PlaNet and Dreamer approaches use latent models based on a Recurrent State Space Model (RSSM), which consists of a transition model, an observation model, a variational encoder, and a reward model, to improve consistency between one-step and multi-step predictions in latent space [7, 16, 45].

The latent-model approach reduces the dimensionality of the observation space. Dimensionality reduction is related to unsupervised learning (Sect. 1.2.2) and autoencoders (Sect. B.2.6.3). The latent-model approach is also related to world models, a term used by Ha and Schmidhuber [28, 29]. World models are inspired by the manner in which humans are thought to construct a mental model of the world in which we live. Ha et al. implement world models using generative recurrent neural networks that generate states for simulation using a variational autoencoder [48, 49] and a recurrent network. Their approach learns a compressed spatial and temporal representation of the environment. By using features extracted from the world model as inputs to the agent, a compact and simple policy can be trained to solve a task, and planning occurs in the compressed world. The term *world model* goes back to 1990, see Schmidhuber [72].

Latent models and world models achieve promising results and are, despite their complexity, an active area of research, see, for example [90]. In the next section, we will further discuss the performance of latent models, but we will first look at two methods for planning with deep transition models.

5.2.2   Planning with the Model

We have discussed in some depth methods to improve the accuracy of models. We will now switch from how to *create* deep models to how to *use* them. We will describe two planning approaches that are designed to be forgiving for models that contain inaccuracies. The planners try to reduce the impact of the inaccuracy of the model, for example, by planning ahead with a limited horizon and by re-learning and re-planning at each step of the trajectory. We will start with planning with a limited horizon.

5.2.2.1   Trajectory Rollouts and Model-Predictive Control

At each planning step, the local transition model $T_a(s) \rightarrow s'$ computes the new state, using the local reward to update the policy. Due to the inaccuracies of the internal model, planning algorithms that perform many steps will quickly accumulate model errors [26]. Full rollouts of long inaccurate trajectories are therefore problematic. We can reduce the impact of accumulated model errors by not planning too far ahead. For example, Gu et al. [26] perform experiments with locally linear models that roll out planning trajectories of length 5–10. This reportedly works well for MuJoCo tasks Gripper and Reacher.

In another experiment, Feinberg et al. [20] allow imagination to a fixed lookahead depth, after which value estimates are split into a near-future model-based component and a distant future model-free component (model-based value expansion, MVE). They experiment with horizons of 1, 2, and 10 and find that 10 generally performs best on typical MuJoCo tasks such as Swimmer, Walker, and Half-Cheetah. The sample complexity in their experiments is better than model-free methods such as DDPG [78]. Similarly good results are reported by others [40, 42], with a model horizon that is much shorter than the task horizon.

Model-Predictive Control
Taking the idea of shorter trajectories for planning than for learning further, we arrive at decision-time planning [55], also known as model-predictive control (MPC) [23, 53]. Model-predictive control is a well-known approach in process engineering, to control complex processes with frequent re-planning over a limited time horizon. Model-predictive control uses the fact that many real-world processes are approximately linear over a small operating range (even though they can be highly nonlinear over a longer range). In MPC, the model is optimized for a limited

time into the future, and then it is re-learned after each environment step. In this way small errors do not get a chance to accumulate and influence the outcome greatly. Related to MPC are other local re-planning methods. All try to reduce the impact of the use of an inaccurate model by not planning too far into the future and by updating the model frequently. Applications are found in the automotive industry and in aerospace, for example, for terrain-following and obstacle-avoidance algorithms [43].

MPC has been used in various deep model learning approaches. Both Finn et al. and Ebert et al. [18, 22] use a form of MPC in the planning for their visual foresight robotic manipulation system. The MPC part uses a model that generates the corresponding sequence of future frames based on an image to select the least-cost sequence of actions. This approach is able to perform multi-object manipulation, pushing, picking and placing, and cloth-folding tasks (which add the difficulty of material that changes shape as it is being manipulated).

Another approach is to use ensemble models for learning the transition model, with MPC for planning. PETS [9] uses probabilistic ensembles [54] for learning, based on cross-entropy methods (CEM) [6, 11]. In MPC fashion, only the first action from the CEM-optimized sequence is used, re-planning at every environment step. Many model-based approaches combine MPC and the ensemble method, as we will also see in the overview in Table 5.2 at the end of the next section. Algorithm 5.6 shows in pseudocode an example of model-predictive control (based on [63], only the model-based part is shown).[7]

MPC is a simple and effective planning method that is well-suited for use with inaccurate models, by restricting the planning horizon and by re-planning. It has also been used with success in combination with latent models [32, 41].

It is now time to look at the final method, which is a very different approach to planning.

5.2.2.2 End-to-End Learning and Planning-by-Network

Up until now, the *learning* of the dynamics model and its *use* is performed by separate algorithms. In the previous subsection, differentiable transition models were learned through backpropagation, and then the models were used by a conventional hand-crafted procedural planning algorithm, such as depth-limited search, with hand-coded selection and backup rules.

A trend in machine learning is to replace *all* hand-crafted algorithms by differentiable approaches, which are trained by example, end-to-end. These differentiable approaches often are more general and perform better than their hand-crafted versions.[8] We could ask the question if it would be possible to make the planning

---

[7] The code is at https://github.com/anagabandi/nn_dynamics.

[8] Note that here we use the term *end-to-end* to indicate the use of differentiable methods for the learning and use of a deep dynamics model—to replace hand-crafted planning algorithms to use

Table 5.2 Model-Based reinforcement learning approaches [69]

Name	Learning	Planning	Environment	Ref
PILCO	Uncertainty	Trajectory	Pendulum	[13]
iLQG	Uncertainty	MPC	Small	[87]
GPS	Uncertainty	Trajectory	Small	[56]
SVG	Uncertainty	Trajectory	Small	[34]
VIN	CNN	e2e	Mazes	[86]
VProp	CNN	e2e	Mazes	[64]
Planning	CNN/LSTM	e2e	Mazes	[27]
TreeQN	Latent	e2e	Mazes	[19]
I2A	Latent	e2e	Mazes	[70]
Predictron	Latent	e2e	Mazes	[81]
World Model	Latent	e2e	Car Racing	[29]
Local Model	Uncertainty	Trajectory	MuJoCo	[26]
Visual Foresight	Video Prediction	MPC	Manipulation	[22]
PETS	Ensemble	MPC	MuJoCo	[9]
MVE	Ensemble	Trajectory	MuJoCo	[20]
Meta-policy	Ensemble	Trajectory	MuJoCo	[10]
Policy Optim	Ensemble	Trajectory	MuJoCo	[40]
PlaNet	Latent	MPC	MuJoCo	[32]
Dreamer	Latent	Trajectory	MuJoCo	[31]
Plan2Explore	Latent	Trajectory	MuJoCo	[76]
L^3P	Latent	Trajectory	MuJoCo	[90]
Video prediction	Latent	Trajectory	Atari	[66]
VPN	Latent	Trajectory	Atari	[67]
SimPLe	Latent	Trajectory	Atari	[41]
Dreamer-v2	Latent	Trajectory	Atari	[33]
MuZero	Latent	e2e/MCTS	Atari/Go	[74]

phase differentiable as well? or to see if the planning rollouts can be implemented in a single computational model, the neural network?

At first sight, it may seem strange to think of a neural network as something that can perform planning and backtracking, since we often think of a neural network as a stateless mathematical function. Neural networks normally perform transformation and filter activities to achieve selection or classification. Planning consists of action selection and state unrolling. Note, however, that recurrent neural networks and LSTMs contain implicit state, making them a candidate to be used for planning (see

---

the learned model. Elsewhere, in supervised learning, the term *end-to-end* is used differently, to describe learning both features and their use from raw pixels for classification—to replace hand-crafted feature recognizers to preprocess the raw pixels and use in a hand-crafted machine learning algorithm.

Algorithm 5.6 Neural network dynamics for model-based deep reinforcement learning (based on [63])

Initialize the model $\hat{m}_\phi$
Initialize an empty dataset D
for $i = 1, \ldots, I$ **do**
    $D \leftarrow E_a$ ▷ sample action from environment
    Train $\hat{m}_\phi(s, a)$ on D minimizing the error by gradient descent
    for $t = 1, \ldots, T$ Horizon **do**   ▷ planning
        $A_t \leftarrow \hat{m}_\phi$ ▷ estimate optimal action sequence with finite MPC horizon
        Execute first action a_t from sequence A_t
        $D \leftarrow (s_t, a_t)$
    end for
end for

Sect. B.2.5). Perhaps it is not so strange to try to implement planning in a neural network. Let us have a look at attempts to perform planning with a neural network.

Tamar et al. [86] introduced value iteration networks (VINs), convolutional networks for planning in Grid worlds. A VIN is a differentiable multi-layer convolutional network that can execute the steps of a simple planning algorithm [65]. The core idea is that in a Grid world, value iteration can be implemented by a multi-layer convolutional network: each layer does a step of lookahead (refer back to Listing 2.1 for value iteration). The value iterations are rolled out in the network layers S with A channels, and the CNN architecture is shaped specifically for each problem task. Through backpropagation, the model learns the value iteration parameters including the transition function. The aim is to learn a general model, which can navigate in unseen environments.

Let us look in more detail at the value iteration algorithm. It is a simple algorithm that consists of a doubly nested loop over states and actions, calculating the sum of rewards $\sum_{s' \in S} T_a(s, s')(R_a(s, s') + \gamma V[s'])$ and a subsequent maximization operation $V[s] = \max_a(Q[s, a])$. This double loop is iterated to convergence. The insight is that each iteration can be implemented by passing the previous value function V_n and reward function R through a convolution layer and max-pooling layer. In this way, each channel in the convolution layer corresponds to the Q-function for a specific action—the innermost loop—and the convolution kernel weights correspond to the transitions. Thus by recurrently applying a convolution layer K times, K iterations of value iteration are performed.

The value iteration module is simply a neural network that has the capability of approximating a value iteration computation. Representing value iteration in this form makes learning the MDP parameters and functions natural—by backpropagating through the network, as in a standard CNN. In this way, the classic value iteration algorithm can be approximated by a neural network.

Why would we want to have a fully differentiable algorithm that can only give an approximation, if we have a perfectly good classic procedural implementation that can calculate the value function V exactly?

The reason is generalization. The exact algorithm only works for known transition probabilities. The neural network can learn $T(\cdot)$ when it is not given, from the environment, and it learns the reward and value functions at the same time. By learning all functions all at once in an end-to-end fashion, the dynamics and value functions might be better integrated than when a separately hand-crafted planning algorithm uses the results of a learned dynamics model. Indeed, reported results do indicate good generalization to unseen problem instances [86].

The idea of planning by gradient descent has existed for some time—actually, the idea of learning *all* functions by example has existed for some time—several authors explored learning approximations of dynamics in neural networks [39, 47, 73]. The VINs can be used for discrete and continuous path planning and have been tried in Grid world problems and natural language tasks.

Later work has extended the approach to other applications of more irregular shape, by adding abstraction networks [71, 81, 82]. The addition of latent models increases the power and versatility of end-to-end learning of planning and transitions even further. Let us look briefly in more detail at one such extension of VIN, to illustrate how latent models and planning go together. TreeQN by Farquhar et al. [19] is a fully differentiable model learner and planner, using observation abstraction so that the approach works on applications that are less regular than mazes.

TreeQN consists of five differentiable functions, four of which we have seen in the previous section in value prediction networks [67]; Fig. 5.5 on page 141

- The encoding function consists of a series of convolutional layers that embed the actual state in a lower-dimensional state $s_{latent} \leftarrow f_{\theta_e}^{enc}(s_{actual})$.
- The transition function uses a fully connected layer per action to calculate the next-state representation $s'_{latent} \leftarrow f_{\theta_t}^{trans}(s_{latent}, a_i)_{i=0}^{I}$.
- The reward function predicts the immediate reward for every action $a_i \in A$ in state s_{latent} using a ReLU layer $r \leftarrow f_{\theta_r}^{reward}(s'_{latent})$.
- The value function of a state is estimated with a vector of weights $V(s_{latent}) \leftarrow w^\top s_{latent} + b$.
- The backup function applies a softmax function[9] recursively to calculate the tree backup value $b(x) \leftarrow \sum_{i=0}^{I} x_i \, \text{softmax}(x)_i$.

These functions together can learn a model and can also execute n-step Q-learning, to use the model to update a policy. Further details can be found in [19] and the GitHub code.[10] TreeQN has been applied on games such as box-pushing and some Atari games and outperformed model-free DQN.

A limitation of VIN is that the tight connection between problem domain, iteration algorithm, and network architecture limited the applicability to other problems. Another system that addresses this limitation of Predictron. Like TreeQN,

---

[9] The softmax function normalizes an input vector of real numbers to a probability distribution $[0, 1]$; $p_\theta(y|x) = \text{softmax}(f_\theta(x)) = \frac{e^{f_\theta(x)}}{\sum_k e^{f_{\theta,k}(x)}}$.

[10] See https://github.com/oxwhirl/treeqn for the code of TreeQN.

the Predictron [81] introduces an abstract model to reduce this limitation. As in VPN, the latent model consists of four differentiable components: a representation model, a next-state model, a reward model, and a discount model. The goal of the abstract model in Predictron is to facilitate value prediction (not state prediction) or prediction of pseudo-reward functions that can encode special events, such as *staying alive* or *reaching the next room*. The planning part rolls forward its internal model k steps. Unlike VPN, Predictron uses joint parameters. The Predictron has been applied to procedurally generated mazes and a simulated pool domain. In both cases, it outperformed model-free algorithms.

End-to-end model-based learning and planning is an active area of research. Challenges include understanding the relation between planning and learning [1, 24], achieving performance that is competitive with classical planning algorithms and with model-free methods, and generalizing the class of applications. In Sect. 5.3, more methods will be shown.

Conclusion

In the previous sections, we have discussed two methods to reduce the inaccuracy of the model and two methods to reduce the impact of the use of an inaccurate model. We have seen a range of different approaches to model-based algorithms. Many of the algorithms were developed recently. Deep model-based reinforcement learning is an active area of research.

Ensembles and MPC have improved the performance of model-based reinforcement learning. The goal of latent or world models is to learn the essence of the domain, reducing the dimensionality, and for end-to-end, to also include the planning part in the learning. Their goal is generalization in a fundamental sense. Model-free learns a policy of which action to take in each state. Model-based methods learn the transition model, from state (via action) to state. Model-free teaches you how to best respond to actions in your world, and model-based helps you to understand your world. By learning the transition model (and possibly even how to best plan with it), it is hoped that new generalization methods can be learned.

The goal of model-based methods is to get to know the environment so intimately that the sample complexity can be reduced while staying close to the solution quality of model-free methods. A second goal is that the generalization power of the methods improves so much and that new classes of problems can be solved. The literature is rich and contains many experiments of these approaches on different environments. Let us now look at the environments to see if we have succeeded.

5.3 High-Dimensional Environments

We have now looked in some detail at approaches for deep model-based reinforcement learning. Let us now change our perspective from the agent to the environment and look at the kinds of environments that can be solved with these approaches.

5.3.1   Overview of Model-Based Experiments

The main goal of model-based reinforcement learning is to learn the transition model accurately—not just the policy function that finds the best action, but the function that finds the next state. By learning the full essence of the environment, a substantial reduction of sample complexity can be achieved. Also, the hope is that the model allows us to solve new classes of problems. In this section we will try to answer the question if these approaches have succeeded.

The answer to this question can be measured in training time and in run-time performance. For *performance*, most benchmark domains provide easily measurable quantities, such as the score in an Atari game. For model-based approaches, the scores achieved by state-of-the-art model-free algorithms such as DQN, DDPG, PPO, SAC, and A3C are a useful baseline. For *training time*, the reduction in sample complexity is an obvious choice. However, many model-based approaches use a fixed hyperparameter to determine the relation between external environment samples and internal model samples (such as 1 : 1000). Then the number of time steps needed for high performance to be reached becomes an important measure, and this is indeed published by most authors. For model-free methods, we often see time steps in the millions per training run and sometimes even billions. With so many time steps, it becomes quite important how much processing each time step takes. For model-based methods, individual time steps may take longer than for model-free, since more processing for learning and planning has to be performed. In the end, wall-clock time is important, and this is also often published.

There are two additional questions. First we are interested in knowing whether a model-based approach allows new types of problems to be solved, which could not be solved by model-free methods. Second is the question of brittleness. In many experiments, the numerical results are quite sensitive to different settings of hyperparameters (including the random seeds). This is the case in many model-free and model-based results [35]. However, when the transition model is accurate, the variance may diminish, and some model-based approaches might be more robust.

Table 5.2 lists 26 experiments with model-based methods [69]. In addition to the name, the table provides an indication of the type of model learning that the agent uses, of the type of planning, and of the application environment in which it was used. The categories in the table are described in the previous section, where e2e means end-to-end.

In the table, the approaches are grouped by environment. At the top are smaller applications such as mazes and navigation tasks. In the middle are larger MuJoCo tasks. At the bottom are high-dimensional Atari tasks. Let us look in more depth at the three groups of environments: small navigation, robotics, and Atari games.

5.3.2 Small Navigation Tasks

We see that a few approaches use smaller 2D Grid world navigation tasks, such as mazes, or block puzzles, such as Sokoban and Pacman. Grid world tasks are some of the oldest problems in reinforcement learning, and they are used frequently to test out new ideas. Tabular imagination approaches such as Dyna, and some latent model and end-to-end learning and planning, have been evaluated with these environments. They typically achieve good results, since the problems are of moderate complexity.

Grid world navigation problems are quintessential sequential decision problems. Navigation problems are typically low dimensional, and no visual recognition is involved; transition functions are easy to learn.

Navigation tasks are also used for latent model and end-to-end learning. Three latent model approaches in Table 5.2 use navigation problems. I2A deals with model imperfections by introducing a latent model, based on Chiappa et al. [8] and Buesing et al. [7]. I2A is applied to Sokoban and Mini-Pacman by [7, 70]. Performance compares favorably to model-free learning and to planning algorithms (MCTS).

Value iteration networks introduced the concept of end-to-end differentiable learning and planning [65, 86], after [39, 47, 73]. Through backpropagation, the model learns to perform value iteration. The aim to learn a general model that can navigate in unseen environments was achieved, although different extensions were needed for more complex environments.

5.3.3 Robotic Applications

Next, we look at papers that use MuJoCo to model continuous robotic problems. Robotic problems are high-dimensional problems with continuous action spaces. MuJoCo is used by most experiments in this category to simulate the physical behavior of robotic movement and the environment.

Uncertainty modeling with ensembles and MPC re-planning try to reduce or contain inaccuracies. The combination of ensemble methods with MPC is well suited for robotic problems, as we have seen in individual approaches such as PILCO and PETS.

Robotic applications are more complex than Grid worlds; model-free methods can take many time steps to find good policies. It is important to know if model-based methods succeed in reducing sample complexity in these problems. When we have a closer look at how well uncertainty modeling and MPC succeed at achieving our first goal, we find a mixed picture.

A benchmark study by Wang et al. [88] looked into the performance of ensemble methods and model-predictive control on MuJoCo tasks. It finds that these methods mostly find good policies and do so in significantly fewer time steps than model-free methods, typically in 200k time steps versus 1 million for model-free. So, it would appear that the lower sample complexity is achieved. However, they also note that per time step, the more complex model-based methods perform more processing than the simpler model-free methods. Although the sample complexity may be lower, the wall-clock time is not, and model-free methods such as PPO and SAC are still much faster for many problems. Furthermore, the score that the policy achieves varies greatly for different problems and is sensitive to different hyperparameter values.

Another finding is that in some experiments with a large number of time steps, the performance of model-based methods plateaus well below model-free performance, and the performance of the model-based methods themselves differs substantially. There is a need for further research in deep model-based methods, especially into robustness of results. More benchmarking studies are needed that compare different methods.

5.3.4 *Atari Game Applications*

Some experiments in Table 5.2 use the Arcade Learning Environment (ALE). ALE features high-dimensional inputs and provides one of the most challenging environments of the table. Especially, latent models choose Atari games to showcase their performance, and some do indeed achieve impressive results, in that they are able to solve new problems, such as playing all 57 Atari games well (Dreamer-v2) [33] and learning the rules of Atari and chess (MuZero) [74].

Hafner et al. published the papers *Dream to control: learning behaviors by latent imagination* and *Dreamer v2* [31, 33]. Their work extends the work on VPN and PlaNet with more advanced latent models and reinforcement learning methods [32, 67]. Dreamer uses an actor critic approach to learn behaviors that consider rewards beyond the horizon. Values are backpropagated through the value model, similar to DDPG [59] and soft actor critic [30].

An important advantage of model-based reinforcement learning is that it can generalize to unseen environments with similar dynamics [76]. The Dreamer experiments showed that latent models are indeed more robust to unseen environments than model-free methods. Dreamer is tested with applications from the DeepMind control suite (Sect. 4.3.2).

Value prediction networks are another latent approach. They outperform model-free DQN on mazes and Atari games such as Seaquest, QBert, Krull, and Crazy Climber. Taking the development of end-to-end learner/planners such as VPN and Predictron further is the work on MuZero [25, 37, 74]. In MuZero, a new architecture is used to learn the transition functions for a range of different games, from Atari to the board games chess, shogi, and Go. MuZero learns the transition model for all games from interaction with the environment.[11] The MuZero model includes different modules: a representation, dynamics, and prediction function. Like AlphaZero, MuZero uses a refined version of MCTS for planning (see Sect. 6.2.1.2 in the next chapter). This MCTS planner is used in a self-play training loop for policy improvement. MuZero's achievements are impressive: it is able to learn the rules of Atari games as well as board games, learning to play the games from scratch, in conjunction with learning the rules of the games. The MuZero achievements have created follow-up work to provide more insight into the relationship between actual and latent representations and to reduce the computational demands [1, 4, 12, 24, 25, 36, 75, 89].

Latent models reduce observational dimensionality to a smaller model to perform planning in latent space. End-to-end learning and planning is able to learn new problems—the second of our two goals: it is able to learn to generalize navigation tasks and to learn the rules of chess and Atari. These are new problems, which are out of reach for model-free methods (although the sample complexity of MuZero is quite large).

Conclusion

In deep model-based reinforcement learning, benchmarks drive progress. We have seen good results with ensembles and local re-planning in continuous problems and with latent models in discrete problems. In some applications, both the first goal, of better sample complexity, and the other goals, of learning new applications and reducing brittleness, are achieved.

The experiments used many different environments within the ALE and MuJoCo suites, from hard to harder. In the next two chapters, we will study multi-agent problems, where we encounter a new set of benchmarks, with a state space of many combinations, including hidden information and simultaneous actions. These provide even more complex challenges for deep reinforcement learning methods.

5.3.5 *Hands On: PlaNet Example*

Before we go to the next chapter, let us take a closer look at how one of these methods achieves efficient learning of a complex high-dimensional task. We will look at PlaNet, a well-documented project by Hafner et al. [32]. Code is available,[12] scripts are available, videos are available, and a blog is available[13] inviting us to take the experiments further. The name of the work is *Learning latent dynamics from pixels*, which describes what the algorithm does: use high dimensional visual

---

[11] This is a somewhat unusual usage of board games. Most researchers use board games because the transition function is given (see next chapter). MuZero instead does *not* know the rules of chess, but starts from scratch learning the rules from interaction with the environment.

[12] https://github.com/google-research/planet.

[13] https://planetrl.github.io.

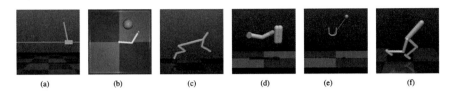

Fig. 5.6 Locomotion tasks of PlaNet [32]. (**a**) Cartpole. (**b**) Reacher. (**c**) Cheetah. (**d**) Finger. (**e**) Cup. (**f**) Walker

input, convert it to latent space, and plan in latent space to learn robot locomotion dynamics.

PlaNet solves continuous control tasks that include contact dynamics, partial observability, and sparse rewards. The applications used in the PlaNet experiments are (a) Cartpole, (b) Reacher, (c) Cheetah, (d) Finger, (e) Cup, and (f) Walker (see Fig. 5.6). The Cartpole task is a swing-up task, with a fixed viewpoint. The cart can be out of sight, requiring the agent to remember information from previous frames. The Finger spin task requires predicting the location of two separate objects and their interactions. The Cheetah task involves learning to run. It includes contacts of the feet with the ground that requires a model to predict multiple futures. The Cup task must catch a ball in a cup. It provides a sparse reward signal once the ball is caught, requiring accurate predictions far into the future. The Walker task involves a simulated robot that begins lying on the ground and must learn to stand up and then walk. PlaNet performs well on these tasks. On DeepMind control tasks, it achieves higher accuracy than an A3C or an D4PG agent. It reportedly does so using 5000% fewer interactions with the environment on average.

It is instructive to experiment with PlaNet. The code can be found on GitHub.[14] Scripts are available to run the experiments with simple one-line commands:

```
python3 -m planet.scripts.train --logdir /path/to/logdir
--params '{tasks:  [cheetah_run]}'
```

As usual, this does require having the right versions of the right libraries installed, which may be a challenge and may require some creativity on your part. The required versions are listed on the GitHub page. The blog also contains videos and pictures of what to expect, including comparisons to model-free baselines from the DeepMind control suite (A3C, D4PG).

The experiments show the viability of the idea to use rewards and values to compress actual states into lower-dimensional latent states and then plan with these latent states. Value-based compression reduces details in the high-dimensional

---

[14] https://github.com/google-research/planet.

actual states as noise that is not relevant to improve the value function [25]. To help understand how the actual state map to the latent states, see, for example [12, 46, 58].

Summary and Further Reading

This has been a diverse chapter. We will summarize the chapter and provide references for further reading.

Summary

Model-free methods sample the environment using the rewards to learn the policy function, providing actions for all states for an environment. Model-based methods use the rewards to learn the transition function and then use planning methods to sample the policy from this internal model. Metaphorically speaking, model-free learns how to *act* in the environment, and model-based learns how to *be* the environment. The learned transition model acts as a multiplier on the amount of information that is used from each environment sample. A consequence is that model-based methods have a lower sample complexity, although when the agent's transition model does not perfectly reflect the environment's transition function, the performance of the policy may be worse than a model-free policy (since that always uses the environment to sample from).

Another and perhaps more important aspect of the model-based approach is generalization. Model-based reinforcement learning builds a dynamics model of the domain. This model can be used multiple times, not only for new problem instances but also for related problem classes. By learning the transition and reward model, model-based reinforcement learning may be better at capturing the essence of a domain than model-free methods and thus be able to generalize to variations of the problem.

Imagination showed how to learn a model and use it to fill in extra samples based on the model (not the environment). For problems where tabular methods work, imagination can be many times more efficient than model-free methods.

When the agent has access to the transition model, it can apply reversible planning algorithms, in addition to one-way learning with samples. There is a large literature on backtracking and tree-traversal algorithms. Using a lookahead of more than one step can increase the quality of the reward even more. When the problem size increases or when we perform deep multi-step lookahead, the accuracy of the model becomes critical. For high-dimensional problems, high-capacity networks are used that require many samples to prevent overfitting. Thus a trade-off exists, to keep sample complexity low.

Methods such as PETS aim to take the uncertainty of the model into account in order to increase modeling accuracy. Model-predictive control methods replan at each environment step to prevent over-reliance on the accuracy of the model. Classical tabular approaches and Gaussian process approaches have been quite succesful in achieving low sample complexity for small problems [15, 50, 85].

Latent models observe that in many high-dimensional problems, the factors that influence changes in the value function are often lower dimensional. For example, the background scenery in an image may be irrelevant for the quality of play in a game and has no effect on the value. Latent models use an encoder to translate the high-dimensional actual state space into a lower-dimensional latent state space. Subsequent planning and value functions work on the (much smaller) latent space.

Finally, we considered end-to-end model-based algorithms. These fully differentiable algorithms not only learn the dynamics model but also learn the planning algorithm that uses the model. The work on value iteration networks [86] inspired recent work on end-to-end learning, where both the transition model and the planning algorithm are learned, end-to-end. Combined with latent models (or world models [29]), impressive results were achieved [81], and the model and planning accuracy was improved to the extent that tabula rasa self-learning of game-rules was achieved, in Muzero [74] for chess, shogi, Go, and Atari games.

Further Reading

Model-based reinforcement learning promises more sample efficient learning. The field has a long history. For exact tabular methods, Sutton's Dyna-Q is a classical approach that illustrates the basic concept of model-based learning [83, 84].

There is an extensive literature on the approaches that were discussed in this chapter. For uncertainty modeling, see [13, 14, 87] and for ensembles [9, 10, 22, 34, 40, 41, 52, 56, 57, 68]. For model-predictive control, see [3, 20, 23, 26, 51, 60, 63].

Latent models are an active field of research. Two of the earlier works are [66, 67], although the ideas go back to world models [29, 39, 47, 73]. Later, a sequence of PlaNet and Dreamer papers was influential [31–33, 76, 91].

The literature on end-to-end learning and planning is also extensive, starting with VIN [86], see [2, 19, 21, 27, 64, 74, 77, 79, 81].

As applications became more challenging, notably in robotics, other methods were developed, mostly based on uncertainty, see for surveys [15, 50]. Later, as high-dimensional problems became prevalent, latent and end-to-end methods were developed. The basis for the section on environments is an overview of recent model-based approaches [69]. Another survey is [62], and a comprehensive benchmark study is [88].

Exercises

Let us go to the Exercises.

Questions

Below are first some quick questions to check your understanding of this chapter. For each question, a simple single sentence answer is sufficient.

1. What is the advantage of model-based over model-free methods?
2. Why may the sample complexity of model-based methods suffer in high-dimensional problems?
3. Which functions are part of the dynamics model?
4. Mention four deep model-based approaches.
5. Do model-based methods achieve better sample complexity than model-free?
6. Do model-based methods achieve better performance than model-free?
7. In Dyna-Q, is the policy updated by two mechanisms: learning by sampling the environment and what other mechanism?
8. Why is the variance of ensemble methods lower than of the individual machine learning approaches that are used in the ensemble?
9. What does model-predictive control do and why is this approach suited for models with lower accuracy?
10. What is the advantage of planning with latent models over planning with actual models?
11. How are latent models trained?   ⟋
12. Mention four typical modules that constitute the latent model.
13. What is the advantage of end-to-end planning and learning?
14. Mention two end-to-end planning and learning methods.

Exercises

It is now time to introduce a few programming exercises. The main purpose of the exercises is to become more familiar with the methods that we have covered in this chapter. By playing around with the algorithms and trying out different hyperparameter settings, you will develop some intuition for the effect on performance and run time of the different methods.

The experiments may become computationally expensive. You may want to consider running them in the cloud, with Google Colab, Amazon AWS, or Microsoft Azure. They may have student discounts, and they will have the latest GPUs or TPUs for use with TensorFlow or PyTorch.

1. *Dyna.* Implement tabular Dyna-Q for the Gym Taxi environment. Vary the amount of planning N and see how performance is influenced.
2. *Keras.* Make a function approximation version of Dyna-Q and Taxi, with Keras. Vary the capacity of the network and the amount of planning. Compare against a pure model-free version, and note the difference in performance for different tasks and in computational demands.

3. *Planning.* In Dyna-Q, planning has so far been with single-step model samples. Implement a simple depth-limited multi-step lookahead planner, and see how performance is influenced for the different lookahead depths.

4. *MPC.* Read the paper by Nagabandi et al. [63] and download the code.[15] Acquire the right versions of the libraries, and run the code with the supplied scripts, just for the MB (model-based) versions. Note that plotting is also supported by the scripts. Run with different MPC horizons. Run with different ensemble sizes. What are the effects on performance and run time for the different applications?

5. *PlaNet.* Go to the PlaNet blog and read it (see previous section).[16] Go to the PlaNet GitHub site and download and install the code.[17] Install the DeepMind control suite,[18] and all necessary versions of the support libraries.

    Run Reacher and Walker in PlaNet, and compare against the model-free methods D4PG and A3C. Vary the size of the encoding network and note the effect on performance and run time. Now turn off the encoder and run with planning on actual states (you may have to change network sizes to achieve this). Vary the capacity of the latent model and of the value and reward functions. Also vary the amount of planning, and note its effect.

6. *End-to-end.* As you have seen, these experiments are computationally expensive. We will now turn to end-to-end planning and learning (VIN and MuZero). This exercise is also computationally expensive. Use small applications, such as small mazes and Cartpole. Find and download a MuZero implementation from GitHub and explore using the experience that you have gained from the previous exercises. Focus on gaining insight into the shape of the latent space. Try MuZero-General [17],[19] or a MuZero visualization [12] to get insight into latent space.[20] (This is a challenging exercise, suitable for a term project or thesis.)

References

1. Ankesh Anand, Jacob Walker, Yazhe Li, Eszter Vértes, Julian Schrittwieser, Sherjil Ozair, Théophane Weber, and Jessica B Hamrick. Procedural generalization by planning with self-supervised world models. *arXiv preprint arXiv:2111.01587*, 2021. 153, 157

2. Thomas Anthony, Zheng Tian, and David Barber. Thinking fast and slow with deep learning and tree search. In *Advances in Neural Information Processing Systems*, pages 5360–5370, 2017. 160

3. Kamyar Azizzadenesheli, Brandon Yang, Weitang Liu, Emma Brunskill, Zachary C Lipton, and Animashree Anandkumar. Surprising negative results for generative adversarial tree search. *arXiv preprint arXiv:1806.05780*, 2018. 160

---

[15] https://github.com/anagabandi/nn_dynamics.

[16] https://planetrl.github.io.

[17] https://github.com/google-research/planet.

[18] https://github.com/deepmind/dm_control.

[19] https://github.com/werner-duvaud/muzero-general.

[20] https://github.com/kaesve/muzero.

4. Mohammad Babaeizadeh, Mohammad Taghi Saffar, Danijar Hafner, Harini Kannan, Chelsea Finn, Sergey Levine, and Dumitru Erhan. Models, pixels, and rewards: Evaluating design trade-offs in visual model-based reinforcement learning. *arXiv preprint arXiv:2012.04603*, 2020. 157

5. Christopher M Bishop. *Pattern Recognition and Machine Learning*. Information science and statistics. Springer Verlag, Heidelberg, 2006. 144, 145

6. Zdravko I Botev, Dirk P Kroese, Reuven Y Rubinstein, and Pierre L'Ecuyer. The cross-entropy method for optimization. In *Handbook of Statistics*, volume 31, pages 35–59. Elsevier, 2013. 149

7. Lars Buesing, Theophane Weber, Sébastien Racaniere, SM Eslami, Danilo Rezende, David P Reichert, Fabio Viola, Frederic Besse, Karol Gregor, Demis Hassabis, and Daan Wierstra. Learning and querying fast generative models for reinforcement learning. *arXiv preprint arXiv:1802.03006*, 2018. 147, 155, 155

8. Silvia Chiappa, Sébastien Racaniere, Daan Wierstra, and Shakir Mohamed. Recurrent environment simulators. In *International Conference on Learning Representations*, 2017. 155

9. Kurtland Chua, Roberto Calandra, Rowan McAllister, and Sergey Levine. Deep reinforcement learning in a handful of trials using probabilistic dynamics models. In *Advances in Neural Information Processing Systems*, pages 4754–4765, 2018. 145, 149, 150, 160

10. Ignasi Clavera, Jonas Rothfuss, John Schulman, Yasuhiro Fujita, Tamim Asfour, and Pieter Abbeel. Model-based reinforcement learning via meta-policy optimization. In *2nd Annual Conference on Robot Learning, CoRL 2018, Zürich, Switzerland*, pages 617–629, 2018. 145, 150, 160

11. Pieter-Tjerk De Boer, Dirk P Kroese, Shie Mannor, and Reuven Y Rubinstein. A tutorial on the cross-entropy method. *Annals of Operations Research*, 134(1):19–67, 2005. 149

12. Joery A. de Vries, Ken S. Voskuil, Thomas M. Moerland, and Aske Plaat. Visualizing MuZero models. *arXiv preprint arXiv:2102.12924*, 2021. 157, 159, 162

13. Marc Deisenroth and Carl E Rasmussen. PILCO: a model-based and data-efficient approach to policy search. In *Proceedings of the 28th International Conference on Machine Learning (ICML-11)*, pages 465–472, 2011. 144, 150, 160

14. Marc Peter Deisenroth, Dieter Fox, and Carl Edward Rasmussen. Gaussian processes for data-efficient learning in robotics and control. *IEEE Transactions on Pattern Analysis and Machine Intelligence*, 37(2):408–423, 2013. 144, 160

15. Marc Peter Deisenroth, Gerhard Neumann, and Jan Peters. A survey on policy search for robotics. In *Foundations and Trends in Robotics 2*, pages 1–142. Now publishers, 2013. 160, 160

16. Andreas Doerr, Christian Daniel, Martin Schiegg, Duy Nguyen-Tuong, Stefan Schaal, Marc Toussaint, and Sebastian Trimpe. Probabilistic recurrent state-space models. *arXiv preprint arXiv:1801.10395*, 2018. 147

17. Werner Duvaud and Aurèle Hainaut. MuZero general: Open reimplementation of Muzero. https://github.com/werner-duvaud/muzero-general, 2019. 162

18. Frederik Ebert, Chelsea Finn, Sudeep Dasari, Annie Xie, Alex Lee, and Sergey Levine. Visual foresight: Model-based deep reinforcement learning for vision-based robotic control. *arXiv preprint arXiv:1812.00568*, 2018. 149

19. Gregory Farquhar, Tim Rocktäschel, Maximilian Igl, and SA Whiteson. TreeQN and ATreeC: Differentiable tree planning for deep reinforcement learning. In *International Conference on Learning Representations*, 2018. 150, 152, 152, 160

20. Vladimir Feinberg, Alvin Wan, Ion Stoica, Michael I Jordan, Joseph E Gonzalez, and Sergey Levine. Model-based value estimation for efficient model-free reinforcement learning. *arXiv preprint arXiv:1803.00101*, 2018. 148, 150, 160

21. Dieqiao Feng, Carla P Gomes, and Bart Selman. Solving hard AI planning instances using curriculum-driven deep reinforcement learning. *arXiv preprint arXiv:2006.02689*, 2020. 160

22. Chelsea Finn and Sergey Levine. Deep visual foresight for planning robot motion. In *2017 IEEE International Conference on Robotics and Automation (ICRA)*, pages 2786–2793. IEEE, 2017. 149, 150, 160

23. Carlos E Garcia, David M Prett, and Manfred Morari. Model predictive control: Theory and practice—a survey. *Automatica*, 25(3):335–348, 1989. 148, 160
24. Jean-Bastien Grill, Florent Altché, Yunhao Tang, Thomas Hubert, Michal Valko, Ioannis Antonoglou, and Rémi Munos. Monte-Carlo tree search as regularized policy optimization. In *International Conference on Machine Learning*, pages 3769–3778. PMLR, 2020. 153, 157
25. Christopher Grimm, André Barreto, Satinder Singh, and David Silver. The value equivalence principle for model-based reinforcement learning. In *Advances in Neural Information Processing Systems*, 2020. 146, 156, 157, 159
26. Shixiang Gu, Timothy Lillicrap, Ilya Sutskever, and Sergey Levine. Continuous deep Q-learning with model-based acceleration. In *International Conference on Machine Learning*, pages 2829–2838, 2016. 148, 148, 150, 160
27. Arthur Guez, Mehdi Mirza, Karol Gregor, Rishabh Kabra, Sébastien Racanière, Theophane Weber, David Raposo, Adam Santoro, Laurent Orseau, Tom Eccles, Greg Wayne, David Silver, and Timothy P. Lillicrap. An investigation of model-free planning. In *International Conference on Machine Learning*, pages 2464–2473, 2019. 150, 160
28. David Ha and Jürgen Schmidhuber. Recurrent world models facilitate policy evolution. In *Advances in Neural Information Processing Systems*, pages 2450–2462, 2018. 147
29. David Ha and Jürgen Schmidhuber. World models. *arXiv preprint arXiv:1803.10122*, 2018. 146, 147, 150, 160, 160
30. Tuomas Haarnoja, Aurick Zhou, Pieter Abbeel, and Sergey Levine. Soft actor-critic: Off-policy maximum entropy deep reinforcement learning with a stochastic actor. In *International Conference on Machine Learning*, pages 1861–1870. PMLR, 2018. 156
31. Danijar Hafner, Timothy Lillicrap, Jimmy Ba, and Mohammad Norouzi. Dream to control: Learning behaviors by latent imagination. In *International Conference on Learning Representations*, 2020. 146, 147, 150, 156, 160
32. Danijar Hafner, Timothy Lillicrap, Ian Fischer, Ruben Villegas, David Ha, Honglak Lee, and James Davidson. Learning latent dynamics for planning from pixels. In *International Conference on Machine Learning*, pages 2555–2565, 2019. 149, 150, 156, 157, 158
33. Danijar Hafner, Timothy Lillicrap, Mohammad Norouzi, and Jimmy Ba. Mastering Atari with discrete world models. In *International Conference on Learning Representations*, 2021. 146, 147, 150, 156, 156, 160
34. Nicolas Heess, Gregory Wayne, David Silver, Timothy Lillicrap, Tom Erez, and Yuval Tassa. Learning continuous control policies by stochastic value gradients. In *Advances in Neural Information Processing Systems*, pages 2944–2952, 2015. 150, 160
35. Peter Henderson, Riashat Islam, Philip Bachman, Joelle Pineau, Doina Precup, and David Meger. Deep reinforcement learning that matters. In *Thirty-Second AAAI Conference on Artificial Intelligence*, 2018. 154
36. Matteo Hessel, Ivo Danihelka, Fabio Viola, Arthur Guez, Simon Schmitt, Laurent Sifre, Theophane Weber, David Silver, and Hado van Hasselt. Muesli: Combining improvements in policy optimization. In *International Conference on Machine Learning*, pages 4214–4226, 2021. 157
37. Thomas Hubert, Julian Schrittwieser, Ioannis Antonoglou, Mohammadamin Barekatain, Simon Schmitt, and David Silver. Learning and planning in complex action spaces. In *International Conference on Machine Learning*, pages 4476–4486, 2021. 156
38. Jonathan Hui. Model-based reinforcement learning https://medium.com/@jonathan_hui/rl-model-based-reinforcement-learning-3c2b6f0aa323. Medium post, 2018. 145
39. Roman Ilin, Robert Kozma, and Paul J Werbos. Efficient learning in cellular simultaneous recurrent neural networks—the case of maze navigation problem. In *2007 IEEE International Symposium on Approximate Dynamic Programming and Reinforcement Learning*, pages 324–329, 2007. 152, 155, 160
40. Michael Janner, Justin Fu, Marvin Zhang, and Sergey Levine. When to trust your model: Model-based policy optimization. In *Advances in Neural Information Processing Systems*, pages 12498–12509, 2019. 145, 148, 150, 160

41. Lukasz Kaiser, Mohammad Babaeizadeh, Piotr Milos, Blazej Osinski, Roy H. Campbell, Konrad Czechowski, Dumitru Erhan, Chelsea Finn, Piotr Kozakowski, Sergey Levine, Ryan Sepassi, George Tucker, and Henryk Michalewski. Model-based reinforcement learning for Atari. *arXiv:1903.00374*, 2019. 146, 149, 150, 160

42. Gabriel Kalweit and Joschka Boedecker. Uncertainty-driven imagination for continuous deep reinforcement learning. In *Conference on Robot Learning*, pages 195–206, 2017. 148

43. Reza Kamyar and Ehsan Taheri. Aircraft optimal terrain/threat-based trajectory planning and control. *Journal of Guidance, Control, and Dynamics*, 37(2):466–483, 2014. 149

44. Satwik Kansal and Brendan Martin. Learn data science webpage., 2018. 142

45. Maximilian Karl, Maximilian Soelch, Justin Bayer, and Patrick Van der Smagt. Deep variational Bayes filters: Unsupervised learning of state space models from raw data. *arXiv preprint arXiv:1605.06432*, 2016. 147

46. Andrej Karpathy, Justin Johnson, and Li Fei-Fei. Visualizing and understanding recurrent networks. *arXiv preprint arXiv:1506.02078*, 2015. 159

47. Henry J Kelley. Gradient theory of optimal flight paths. *American Rocket Society Journal*, 30(10):947–954, 1960. 152, 155, 160

48. Diederik P Kingma and Max Welling. Auto-encoding variational Bayes. In *International Conference on Learning Representations*, 2014. 147

49. Diederik P Kingma and Max Welling. An introduction to variational autoencoders. *Found. Trends Mach. Learn.*, 12(4):307–392, 2019. 147

50. Jens Kober, J Andrew Bagnell, and Jan Peters. Reinforcement learning in robotics: A survey. *The International Journal of Robotics Research*, 32(11):1238–1274, 2013. 160, 160

51. Basil Kouvaritakis and Mark Cannon. *Model Predictive Control*. Springer, 2016. 160

52. Thanard Kurutach, Ignasi Clavera, Yan Duan, Aviv Tamar, and Pieter Abbeel. Model-ensemble trust-region policy optimization. In *International Conference on Learning Representations*, 2018. 145, 145, 160

53. W Hi Kwon, AM Bruckstein, and T Kailath. Stabilizing state-feedback design via the moving horizon method. *International Journal of Control*, 37(3):631–643, 1983. 148

54. Balaji Lakshminarayanan, Alexander Pritzel, and Charles Blundell. Simple and scalable predictive uncertainty estimation using deep ensembles. In *Advances in Neural Information Processing Systems*, pages 6402–6413, 2017. 149

55. Matteo Leonetti, Luca Iocchi, and Peter Stone. A synthesis of automated planning and reinforcement learning for efficient, robust decision-making. *Artificial Intelligence*, 241:103–130, 2016. 148

56. Sergey Levine and Pieter Abbeel. Learning neural network policies with guided policy search under unknown dynamics. In *Advances in Neural Information Processing Systems*, pages 1071–1079, 2014. 150, 160

57. Sergey Levine and Vladlen Koltun. Guided policy search. In *International Conference on Machine Learning*, pages 1–9, 2013. 144, 160

58. Hao Li, Zheng Xu, Gavin Taylor, Christoph Studer, and Tom Goldstein. Visualizing the loss landscape of neural nets. In *Advances in Neural Information Processing Systems*, pages 6391–6401, 2018. 159

59. Timothy P Lillicrap, Jonathan J Hunt, Alexander Pritzel, Nicolas Heess, Tom Erez, Yuval Tassa, David Silver, and Daan Wierstra. Continuous control with deep reinforcement learning. In *International Conference on Learning Representations*, 2016. 156

60. David Q Mayne, James B Rawlings, Christopher V Rao, and Pierre OM Scokaert. Constrained model predictive control: Stability and optimality. *Automatica*, 36(6):789–814, 2000. 160

61. Thomas M Moerland, Joost Broekens, and Catholijn M Jonker. A framework for reinforcement learning and planning. *arXiv preprint arXiv:2006.15009*, 2020. 143

62. Thomas M Moerland, Joost Broekens, and Catholijn M Jonker. Model-based reinforcement learning: A survey. *arXiv preprint arXiv:2006.16712*, 2020. 160

63. Anusha Nagabandi, Gregory Kahn, Ronald S Fearing, and Sergey Levine. Neural network dynamics for model-based deep reinforcement learning with model-free fine-tuning. In *2018 IEEE International Conference on Robotics and Automation (ICRA)*, pages 7559–7566, 2018. 149, 151, 160, 162

64. Nantas Nardelli, Gabriel Synnaeve, Zeming Lin, Pushmeet Kohli, Philip HS Torr, and Nicolas Usunier. Value propagation networks. In *7th International Conference on Learning Representations, ICLR 2019, New Orleans, LA, USA, May 6-9, 2019*, 2018. 150, 160

65. Sufeng Niu, Siheng Chen, Hanyu Guo, Colin Targonski, Melissa C Smith, and Jelena Kovačević. Generalized value iteration networks: Life beyond lattices. In *Thirty-Second AAAI Conference on Artificial Intelligence*, 2018. 151, 155

66. Junhyuk Oh, Xiaoxiao Guo, Honglak Lee, Richard L Lewis, and Satinder Singh. Action-conditional video prediction using deep networks in Atari games. In *Advances in Neural Information Processing Systems*, pages 2863–2871, 2015. 150, 160

67. Junhyuk Oh, Satinder Singh, and Honglak Lee. Value prediction network. In *Advances in Neural Information Processing Systems*, pages 6118–6128, 2017. 146, 146, 146, 147, 147, 150, 152, 156, 160

68. David Opitz and Richard Maclin. Popular ensemble methods: An empirical study. *Journal of Artificial Intelligence Research*, 11:169–198, 1999. 145, 160

69. Aske Plaat, Walter Kosters, and Mike Preuss. High-accuracy model-based reinforcement learning, a survey. *arXiv preprint arXiv:2107.08241*, 2021. 144, 150, 154, 160

70. Sébastien Racanière, Theophane Weber, David P. Reichert, Lars Buesing, Arthur Guez, Danilo Jimenez Rezende, Adrià Puigdomènech Badia, Oriol Vinyals, Nicolas Heess, Yujia Li, Razvan Pascanu, Peter W. Battaglia, Demis Hassabis, David Silver, and Daan Wierstra. Imagination-augmented agents for deep reinforcement learning. In *Advances in Neural Information Processing Systems*, pages 5690–5701, 2017. 150, 155

71. Daniel Schleich, Tobias Klamt, and Sven Behnke. Value iteration networks on multiple levels of abstraction. In *Robotics: Science and Systems XV, University of Freiburg, Freiburg im Breisgau, Germany*, 2019. 152

72. Jürgen Schmidhuber. Making the world differentiable: On using self-supervised fully recurrent neural networks for dynamic reinforcement learning and planning in non-stationary environments. Technical report, Inst. für Informatik, 1990. 147

73. Jürgen Schmidhuber. An on-line algorithm for dynamic reinforcement learning and planning in reactive environments. In *1990 IJCNN International Joint Conference on Neural Networks*, pages 253–258. IEEE, 1990. 152, 155, 160

74. Julian Schrittwieser, Ioannis Antonoglou, Thomas Hubert, Karen Simonyan, Laurent Sifre, Simon Schmitt, Arthur Guez, Edward Lockhart, Demis Hassabis, Thore Graepel, Timothy Lillicrap, and David Silver. Mastering Atari, go, chess and shogi by planning with a learned model. *Nature*, 588(7839):604–609, 2020. 150, 156, 156, 160, 160

75. Julian Schrittwieser, Thomas Hubert, Amol Mandhane, Mohammadamin Barekatain, Ioannis Antonoglou, and David Silver. Online and offline reinforcement learning by planning with a learned model. *arXiv preprint arXiv:2104.06294*, 2021. 157

76. Ramanan Sekar, Oleh Rybkin, Kostas Daniilidis, Pieter Abbeel, Danijar Hafner, and Deepak Pathak. Planning to explore via self-supervised world models. In *International Conference on Machine Learning*, 2020. 146, 150, 156, 160

77. David Silver, Thomas Hubert, Julian Schrittwieser, Ioannis Antonoglou, Matthew Lai, Arthur Guez, Marc Lanctot, Laurent Sifre, Dharshan Kumaran, Thore Graepel, Timothy Lillicrap, Karen Simonyan, and Demis Hassabis. A general reinforcement learning algorithm that masters chess, shogi, and Go through self-play. *Science*, 362(6419):1140–1144, 2018. 160

78. David Silver, Guy Lever, Nicolas Heess, Thomas Degris, Daan Wierstra, and Martin Riedmiller. Deterministic policy gradient algorithms. In *International Conference on Machine Learning*, pages 387–395, 2014. 148

79. David Silver, Julian Schrittwieser, Karen Simonyan, Ioannis Antonoglou, Aja Huang, Arthur Guez, Thomas Hubert, Lucas Baker, Matthew Lai, Adrian Bolton, Yutian Chen, Timothy Lillicrap, Fan Hui, Laurent Sifre, George van den Driessche, Thore Graepel, and Demis Hassabis. Mastering the game of Go without human knowledge. *Nature*, 550(7676):354, 2017. 160

80. David Silver, Richard S Sutton, and Martin Müller. Temporal-difference search in computer Go. *Machine Learning*, 87(2):183–219, 2012. 147

81. David Silver, Hado van Hasselt, Matteo Hessel, Tom Schaul, Arthur Guez, Tim Harley, Gabriel Dulac-Arnold, David Reichert, Neil Rabinowitz, Andre Barreto, and Thomas Degris. The Predictron: End-to-end learning and planning. In *Proceedings of the 34th International Conference on Machine Learning*, pages 3191–3199, 2017. 146, 150, 152, 153, 160, 160

82. Aravind Srinivas, Allan Jabri, Pieter Abbeel, Sergey Levine, and Chelsea Finn. Universal planning networks. In *International Conference on Machine Learning*, pages 4739–4748, 2018. 152

83. Richard S Sutton. Integrated architectures for learning, planning, and reacting based on approximating dynamic programming. In *Machine Learning Proceedings 1990*, pages 216–224. Elsevier, 1990. 138, 139, 141, 141, 160

84. Richard S Sutton. Dyna, an integrated architecture for learning, planning, and reacting. *ACM Sigart Bulletin*, 2(4):160–163, 1991. 160

85. Richard S Sutton and Andrew G Barto. *Reinforcement learning, An Introduction, Second Edition*. MIT Press, 2018. 138, 141, 160

86. Aviv Tamar, Yi Wu, Garrett Thomas, Sergey Levine, and Pieter Abbeel. Value iteration networks. In *Advances in Neural Information Processing Systems*, pages 2154–2162, 2016. 150, 151, 152, 155, 160, 160

87. Yuval Tassa, Tom Erez, and Emanuel Todorov. Synthesis and stabilization of complex behaviors through online trajectory optimization. In *2012 IEEE/RSJ International Conference on Intelligent Robots and Systems*, pages 4906–4913, 2012. 150, 160

88. Tingwu Wang, Xuchan Bao, Ignasi Clavera, Jerrick Hoang, Yeming Wen, Eric Langlois, Shunshi Zhang, Guodong Zhang, Pieter Abbeel, and Jimmy Ba. Benchmarking model-based reinforcement learning. *arXiv:1907.02057*, 2019. 139, 155, 160

89. Weirui Ye, Shaohuai Liu, Thanard Kurutach, Pieter Abbeel, and Yang Gao. Mastering Atari games with limited data. *arXiv preprint arXiv:2111.00210*, 2021. 157

90. Lunjun Zhang, Ge Yang, and Bradly C Stadie. World model as a graph: Learning latent landmarks for planning. In *International Conference on Machine Learning*, pages 12611–12620. PMLR, 2021. 148, 150

91. Marvin Zhang, Sharad Vikram, Laura Smith, Pieter Abbeel, Matthew J Johnson, and Sergey Levine. Solar: Deep structured representations for model-based reinforcement learning. In *International Conference on Machine Learning*, pages 7444–7453, 2019. 160

Chapter 6
Two-Agent Self-Play

Previous chapters were concerned with how a single agent can learn optimal behavior for its environment. This chapter is different. We turn to problems where two agents operate whose behavior will both be modeled (and, in the next chapter, more than two).

Two-agent problems are interesting for two reasons. First, the world around us is full of active entities that interact, and modeling two agents and their interaction is a step closer to understanding the real world than modeling a single agent. Second, in two-agent problems, exceptional results were achieved—reinforcement learning agents teaching themselves to become stronger than human world champions—and by studying these methods, we may find a way to achieve similar results in other problems.

The kind of interaction that we model in this chapter is zero-sum: my win is your loss and vice versa. These two-agent zero-sum dynamics are fundamentally different from single-agent dynamics. In single-agent problems, the environment lets you probe it, lets you learn how it works, and lets you find good actions. Although the environment may not be your friend, it is also not working against you. In two-agent zero-sum problems, the environment does try to win from you, and it actively changes its replies to minimize your reward, based on what it learns from your behavior. When learning our optimal policy, we should take all possible counteractions into account.

A popular way to do so is to implement the environment's actions with self-play: we replace the environment by a copy of ourselves. In this way we let ourselves play against an opponent that has all the knowledge that we currently have, and agents learn from each other.

We start with a short review of two-agent problems, after which we dive into self-learning. We look at the situation when both agents know the transition function perfectly, so that model accuracy is no longer a problem. This is the case, for example, in games such as chess and Go, where the rules of the game determine how we can go from one state to another.

In self-learning, the environment is used to generate training examples for the agent to train a better policy, after which the better agent policy is used in this environment to train the agent, and again and again, creating a virtuous cycle of self-learning and mutual improvement. It is possible for an agent to teach itself to play a game without any prior knowledge at all, so-called *tabula rasa* learning, learning from a blank slate.

The self-play systems that we describe in this chapter use model-based methods and combine planning and learning approaches. There is a planning algorithm that we have mentioned a few times but have not yet explained in detail. In this chapter we will discuss Monte Carlo Tree Search, or MCTS, a highly popular planning algorithm. MCTS can be used in single-agent and in two-agent situations and is the core of many successful applications, including MuZero and the self-learning AlphaZero series of programs. We will explain how self-learning and self-play work in AlphaGo Zero and why they work so well. We will then discuss the concept of curriculum learning, which is behind the success of self-learning.

The chapter is concluded with exercises, a summary, and pointers to further reading.

Core Concepts

- Self-play
- Curriculum learning

Core Problem

- Use a given transition model for self-play, in order to become stronger than the current best players

Core Algorithms

- Minimax (Listing 6.2)
- Monte Carlo Tree Search (Listing 6.3)
- AlphaZero tabula rasa learning (Listing 6.1)

Self-Play in Games

We have seen in Chap. 5 that when the agent has a transition model of the environment, it can achieve greater performance, especially when the model has high accuracy. What if the accuracy of our model were perfect, if the agent's transition function is the same as the environment's, how far would that bring us? And what if we could improve our environment as part of our learning process, can we then transcend our teacher, can the sorcerer's apprentice outsmart the wizard?

To set the scene for this chapter, let us describe the first game where this has happened: backgammon.

Learning to Play Backgammon

In Sect. 3.2.3, we briefly discussed research into backgammon. Already in the early 1990s, the program TD-Gammon achieved stable reinforcement learning with a shallow network. This work was started at the end of the 1980s by Gerald Tesauro, a researcher at IBM laboratories. Tesauro was faced with the problem of getting a program to learn beyond the capabilities of any existing entity. (In Fig. 6.1, we see Tesauro in front of his program; image by IBM Watson Media.)

In the 1980s, computing was different. Computers were slow, datasets were small, and neural networks were shallow. Against this background, the success of Tesauro is quite remarkable.

His programs were based on neural networks that learned good patterns of play. His first program, Neurogammon, was trained using supervised learning, based on games of human experts. In supervised learning, the model cannot become stronger than the human games it is trained on. Neurogammon achieved an intermediate level of play [115]. His second program, TD-Gammon, was based on reinforcement

Fig. 6.1 Backgammon and Tesauro

learning, using temporal difference learning and self-play. Combined with hand-crafted heuristics and some planning, in 1992, it played at human championship level, becoming the first computer program to do so in a game of skill [118].

TD-Gammon is named after temporal difference learning because it updates its neural net after each move, reducing the difference between the evaluation of previous and current positions. The neural network used a single hidden layer with up to 80 units. TD-Gammon initially learned from a state of zero knowledge, *tabula rasa*. Tesauro describes TD-Gammon's self-play as follows: *the move that is selected is the move with maximum expected outcome for the side making the move. In other words, the neural network is learning from the results of playing against itself. This self-play training paradigm is used even at the start of learning, when the network's weights are random, and hence its initial strategy is a random strategy* [117].

TD-Gammon performed tabula rasa learning, its neural network weights initialized to small random numbers. It reached world champion level purely by playing against itself, learning the game as it played along.

Such autonomous self-learning is one of the main goals of artificial intelligence. TD-Gammon's success inspired many researchers to try neural networks and self-play approaches, culminating eventually, many years later, in high-profile results in Atari [75] and AlphaGo [107, 109], which we will describe in this chapter.[1]

In Sect. 6.1, two-agent zero-sum environments will be described. Next, in Sect. 6.2, the tabula rasa self-play method is described in detail. In Sect. 6.3, we focus on the achievements of the self-play methods. Let us now start with two-agent zero-sum problems.

6.1 Two-Agent Zero-Sum Problems

Before we look into self-play algorithms, let us look for a moment at the two-agent games that have fascinated artificial intelligence researchers for such a long time.

Games come in many shapes and sizes. Some are easy, and some are hard. The characteristics of games are described in a fairly standard taxonomy. Important characteristics of games are the number of players, whether the game is zero-sum or non-zero-sum, whether it is perfect or imperfect information, what the complexity of taking decisions is, and what the state space complexity is. We will look at these characteristics in more detail.

- *Number of Players.* One of the most important elements of a game is the number of players. One-player games are normally called puzzles and are modeled as a standard MDP. The goal of a puzzle is to find a solution. Two-player games

---

[1] A modern reimplementation of TD-Gammon in TensorFlow is available on GitHub at TD-Gammon https://github.com/fomorians/td-gammon.

are "real" games. Quite a number of two-player games exist that provide a nice balance between being too easy and being too hard for players (and for computer programmers) [34]. Examples of two-player games that are popular in AI are chess, checkers, Go, Othello, and shogi.

Multiplayer games are played by three or more players. Well-known examples of multiplayer games are the card games bridge and poker and strategy games such as Risk, Diplomacy, and StarCraft.

- *Zero Sum versus Non-Zero Sum.* An important aspect of a game is whether it is competitive or cooperative. Most two-player games are competitive: the win (+1) of player A is the loss (−1) of player B. These games are called *zero sum* because the sum of the wins for the players remains a constant zero. Competition is an important element in the real world, and these games provide a useful model for the study of conflict and strategic behavior.

  In contrast, in cooperative games, the players win if they can find win/win situations. Examples of cooperative games are Hanabi, bridge, Diplomacy [37, 63], poker, and Risk. The next chapter will discuss multi-agent and cooperative games.

- *Perfect Versus Imperfect Information.* In perfect information games, all relevant information is known to all players. This is the case in typical board games such as chess and checkers. In imperfect information games, some information may be hidden from some players. This is the case in card games such as bridge and poker, where not all cards are known to all players. Imperfect information games can be modeled as partially observable Markov processes, POMDP [82, 104]. A special form of (im)perfect information games is games of chance, such as backgammon and Monopoly, in which dice play an important role. There is no hidden information in these games, and these games are sometimes considered to be perfect information games, despite the uncertainty present at move time. Stochasticity is not the same as imperfect information.

- *Decision Complexity.* The difficulty of playing a game depends on the complexity of the game. The decision complexity is the number of end positions that define the value (win, draw, or loss) of the initial game position (also known as the critical tree or proof tree [60]). The larger the number of actions in a position, the larger the decision complexity. Games with small board sizes such as tic tac toe (3 × 3) have a smaller complexity than games with larger boards, such as Gomoku (19 × 19). When the action space is very large, it can often be treated as a continuous action space. In poker, for example, the monetary bets can be of any size, defining an action size that is practically continuous.

- *State Space Complexity.* The state space complexity of a game is the number of legal positions reachable from the initial position of a game. State space and decision complexity are normally positively correlated, since games with high decision complexity typically have high state space complexity. Determining the exact state space complexity of a game is a nontrivial task, since positions may

Table 6.1 Characteristics of games

Name	Board	State space	Zero-sum	Information
Chess	8 × 8	10^{47}	Zero sum	Perfect
Checkers	8 × 8	10^{18}	Zero sum	Perfect
Othello	8 × 8	10^{28}	Zero sum	Perfect
Backgammon	24	10^{20}	Zero sum	Chance
Go	19 × 19	10^{170}	Zero sum	Perfect
Shogi	9 × 9	10^{71}	Zero sum	Perfect
Poker	Card	10^{161}	Non-zero	Imperfect

be illegal or unreachable.[2] For many games, approximations of the state space have been calculated. In general, games with a larger state space complexity are harder to play ("require more intelligence") for humans and computers. Note that the dimensionality of the states may not correlate with the size of the state space, for example, the rules of some of the simpler Atari games limit the number of reachable states, although the states themselves are high dimensional (they consist of many video pixels).

Zero-Sum Perfect Information Games

Two-person zero-sum games of perfect information, such as chess, checkers, and Go, are among the oldest applications of artificial intelligence. Turing and Shannon published the first ideas on how to write a program to play chess more than 70 years ago [105, 125]. To study strategic reasoning in artificial intelligence, these games are frequently used. Strategies, or policies, determine the outcome. Table 6.1 summarizes some of the games that have played an important role in artificial intelligence research.

6.1.1 The Difficulty of Playing Go

After the 1997 defeat of chess world champion Garry Kasparov by IBM's Deep Blue computer (Fig. 6.2; image by Chessbase), the game of Go (Fig. 1.4) became the next benchmark game, the *Drosophila*[3] of AI, and research activity in Go intensified significantly.

The game of Go is more difficult than chess. It is played on a larger board (19 × 19 vs. 8 × 8), the action space is larger (around 250 moves available in

---

[2] For example, the maximal state space of tic-tac-toe is $3^9 = 19683$ positions (9 squares of "X," "O," or blank), where only 765 positions remain if we remove symmetrical and illegal positions [96].

[3] *Drosophila Melanogaster* is also known as the fruitfly, a favorite species of genetics researchers to test their theories, because experiments produce quick and clear answers.

Fig. 6.2 Deep Blue and Garry Kasparov in May 1997 in New York

a position versus some 25 in chess), the game takes longer (typically 300 moves versus 70) and the state space complexity is much larger: 10^{170} for Go versus 10^{47} for chess. Furthermore, rewards in Go are sparse. Only at the end of a long game, after many moves have been played, is the outcome (win/loss) known. Captures are not so frequent in Go, and no good efficiently computable heuristic has been found. In chess, in contrast, the material balance in chess can be calculated efficiently and gives a good indication of how far ahead we are. For the computer, much of the playing in Go happens in the dark. In contrast, for humans, it can be argued that the visual patterns of Go may be somewhat easier to interpret than the deep combinatorial lines of chess.

For reinforcement learning, credit assignment in Go is challenging. Rewards only occur after a long sequence of moves, and it is unclear which moves contributed the most to such an outcome or whether all moves contributed equally. Many games will have to be played to acquire enough outcomes. In conclusion, Go is more difficult to master with a computer than chess (Fig. 6.3).

Traditionally, computer Go programs followed the conventional chess design of a minimax search with a heuristic evaluation function, which, in the case of Go, was based on the influence of stones (see Sect. 6.2.1 and Fig. 6.4) [71]. This chess approach, however, did not work for Go, or at least not well enough. The level of play was stuck at mid-amateur level for many years.

The main problems were the large branching factor and the absence of an efficient and good evaluation function.

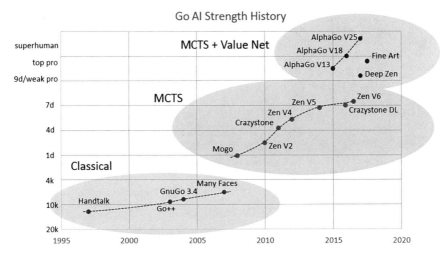

Fig. 6.3 Go playing strength of top programs over the years [3]

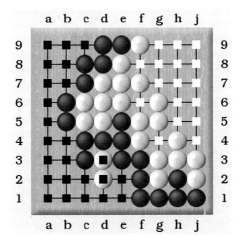

Fig. 6.4 Influence in the game of Go. Empty intersections are marked as being part of Black's or White's Territory

Subsequently, Monte Carlo Tree Search was developed, in 2006. MCTS is a variable-depth adaptive search algorithm, which did not need a heuristic function, but instead used random playouts to estimate board strength. MCTS programs caused the level of play to improve from 10 kyu to 2–3 dan and even stronger on the small 9 × 9 board.[4] However, again, at that point, performance stagnated,

---

[4] Absolute beginners in Go start at 30 kyu, progressing to 10 kyu and advancing to 1 kyu (30k–1k). Stronger amateur players then achieve 1 dan, progressing to 7 dan, the highest amateur rating for Go (1d–7d). Professional Go players have a rating from 1 dan to 9 dan, written as 1p–9p.

and researchers expected that world champion level play was still many years into the future. Neural networks had been tried but were slow and did not improve performance much.

Playing Strength in Go

Let us compare the three programming paradigms of a few different Go programs that have been written over the years (Fig. 6.3). The programs fall into three categories. First are the programs that use heuristic planning, the minimax-style programs. GNU Go is a well-known example of this group of programs. The heuristics in these programs are hand-coded. The level of play of these programs was at medium amateur level. Next come the MCTS-based programs. They reached strong amateur level. Finally come the AlphaGo programs, in which MCTS is combined with deep self-play. These reached super-human performance. The figure also shows other programs that follow a related approach.

Thus, Go provided a large and sparse state space, providing a highly challenging test, to see how far self-play with a perfect transition function can come. Let us have a closer look at the achievements of AlphaGo.

6.1.2 AlphaGo Achievements

In 2016, after decades of research, the effort in Go paid off. In the years 2015–2017, the DeepMind AlphaGo team played three matches in which it beat all human champions that it played, Fan Hui, Lee Sedol, and Ke Jie. The breakthrough performance of AlphaGo came as a surprise. Experts in computer games had expected grandmaster level play to be at least 10 years away.

The techniques used in AlphaGo are the result of many years of research and cover a wide range of topics. The game of Go worked very well as *Drosophila*. Important new algorithms were developed, most notably Monte Carlo Tree Search (MCTS), as well as major progress was made in deep reinforcement learning. We will provide a high-level overview of the research that culminated in AlphaGo (which beats the champions) and its successor AlphaGo Zero (which learns Go tabula rasa). First we will describe the Go matches.

The games against Fan Hui were played in October 2015 in London as part of the development effort of AlphaGo. Fan Hui is the 2013, 2014, and 2015 European Go Champion, then rated at 2p dan. The games against Lee Sedol were played in May 2016 in Seoul and were widely covered by the media (see Fig. 6.5; image by DeepMind). Although there is no official worldwide ranking in international Go, in

Fig. 6.5   AlphaGo versus Lee Sedol in 2016 in Seoul

2016 Lee Sedol was widely considered one of the four best players in the world. A year later another match was played, this time in China, against the Chinese champion Ke Jie, who was ranked number one in the Korean, Japanese, and Chinese ranking systems at the time of the match. All three matches were won convincingly by AlphaGo. Beating the best Go players appeared on the cover of the journal Nature, see Fig. 6.6.

The AlphaGo series of programs actually consists of three programs: AlphaGo, AlphaGo Zero, and AlphaZero. AlphaGo is the program that beats the human Go champions. It consists of a combination of supervised learning from grandmaster games and from self-play games. The second program, AlphaGo Zero, is a full re-design, based solely on self-play. It performs tabula rasa learning of Go and plays stronger than AlphaGo. AlphaZero is a generalization of this program that also plays chess and shogi. Section 6.3 will describe the programs in more detail.

Let us now have an in-depth look at the self-play algorithms as featured in AlphaGo Zero and AlphaZero.

Fig. 6.6 AlphaGo on the cover of Nature

6.2 Tabula Rasa Self-Play Agents

Model-based reinforcement learning showed us that by learning a local transition model, good sample efficiency can be achieved when the accuracy of the model is sufficient. When we have perfect knowledge of the transitions, as we have in this chapter, then we can plan far into the future, without error.

In regular agent–environment reinforcement learning, the complexity of the environment does not change as the agent learns, and as a consequence, the intelligence of the agent's policy is limited by the complexity of the environment. However, in self-play, a cycle of mutual improvement occurs; the intelligence of the environment improves *because* the agent is learning. With self-play, we can create a system that can transcend the original environment, and keep growing, and growing, in a virtuous cycle of mutual learning; it is as if intelligence emerges out of nothing. This is the kind of system that is needed when we wish to beat the best known entity in a certain domain, since copying from a teacher will not help us to transcend it.

Studying how such a high level of play is achieved is interesting, for three reasons: (1) it is exciting to follow an AI success story, (2) it is interesting to see which techniques were used and how it is possible to achieve beyond human intelligence, and (3) it is interesting to see if we can learn a few techniques that can be used in other domains, beyond two-agent zero-sum games, to see if we can achieve superintelligence there as well.

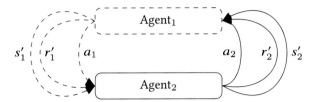

Fig. 6.7 Agent–Agent world

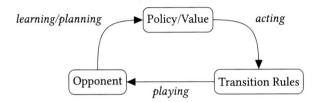

Fig. 6.8 Playing with known transition rules

Let us have a closer look at the self-learning agent architecture that is used by AlphaGo Zero. We will see that two-agent self-play actually consists of three levels of self-play: move-level self-play, example-level self-play, and tournament-level self-play.

First, we will discuss the general architecture and how it creates a cycle of virtuous improvement. Next, we will describe the levels in detail.

Cycle of Virtuous Improvement

In contrast to the agent/environment model, we now have two agents (Fig. 6.7). In comparison with the model-based world of Chap. 5 (Fig. 6.8), our learned model has been replaced by perfect knowledge of the transition rules, and the environment is now called opponent: the negative version of the same agent playing the role of agent$_2$.

The goal in this chapter is to reach the highest possible performance in terms of level of play, without using any hand-coded domain knowledge. In applications such as chess and Go, a perfect transition model is present. Together with a learned reward function and a learned policy function, we can create a self-learning system in which a virtuous cycle of ever improving performance occurs. Figure 6.9 illustrates such a system: (1) the searcher uses the evaluation network to estimate reward values and policy actions, and the search results are used in games against the opponent in self-play, (2) the game results are then collected in a buffer,

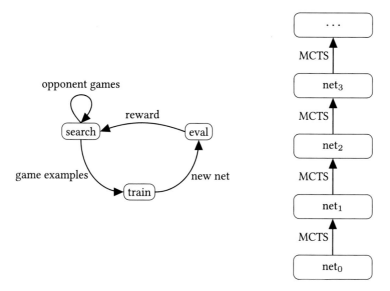

Fig. 6.9 Self-play loop improving quality of net

which is used to train the evaluation network in self-learning, and (3) by playing a tournament against a copy of ourselves, a virtuous cycle of ever-increasing function improvement is created.

AlphaGo Zero Self-Play in Detail

Let us look in more detail at how self-learning works in AlphaGo Zero. AlphaGo Zero uses a model-based actor critic approach with a planner that improves a single value/policy network. For policy improvement, it uses MCTS, for learning a single deep residual network with a policy head and a value head (Sect. B.2.6), see Fig. 6.9. MCTS improves the quality of the training examples in each iteration (left panel), and the net is trained with these better examples, improving its quality (right panel).

The output of MCTS is used to train the evaluation network, whose output is then used as evaluation function in that same MCTS. A loop is wrapped around the search-eval functions to keep training the network with the game results, creating a learning curriculum. Let us put these ideas into pseudocode.

The Cycle in Pseudocode

Conceptually self-play is as ingenious as it is elegant: a double training loop around an MCTS player with a neural network as evaluation and policy function that help MCTS. Figure 6.10 and Listing 6.1 show the self-play loop in detail. The numbers in the figure correspond to the line numbers in the pseudocode.

Let us perform an outside-in walk-through of this system. Line 1 is the main self-play loop. It controls how long the execution of the curriculum of self-play tournaments will continue. Line 2 executes the training episodes, the tournaments of self-play games after which the network is retrained. Line 4 plays such a game to

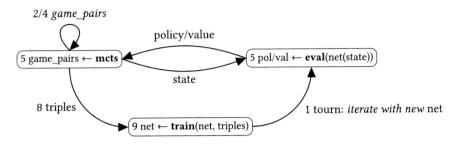

Fig. 6.10 A diagram of self-play with line numbers

```
1   for tourn in range (1, max_tourns):      # curric. of tournaments
2       for game in range(1, max_games):     # play a tourn. of games
3           trim(triples)        # if buffer full: replace old entries
4           while not game_over(): # generate the states of one game
5               move = mcts(board, eval(net))  # move is (s,a) pair
6               game_pairs += move
7               make_move_and_switch_side(board, move)
8           triples += add(game_pairs, game_outcome(game_pairs))
                # add to buf
9       net = train(net, triples)   # retrain with (s,a,outc) triples
```

Listing 6.1 Self-play pseudocode

create (*state, action*)-pairs for each move and the outcome of the game. Line 5 calls
MCTS to generate an action in each state. MCTS performs the simulations where
it uses the policy head of the net in P-UCT selection, and the value head of the net
at the MCTS leaves. Line 6 appends the state/action pair to the list of game moves.
Line 7 performs the move on the board and switches color to the other player, for
the next move in the while loop. At line 8, a full game has ended, and the outcome
is known. Line 8 adds the outcome of each game to the (*state, action*)-pairs, to
make the (*state, action, outcome*)-triples for the network to train on. Note that since
the network is a two-headed policy/value net, both an action and an outcome are
needed for network training. On the last line, this triples-buffer is then used to train
the network. The newly trained network is used in the next self-play iteration as
the evaluation function by the searcher. With this net, another tournament is played,
using the searcher's lookahead to generate a next batch of higher-quality examples,
resulting in a sequence of stronger and stronger networks (Fig. 6.9, right panel).

In the pseudocode, we see the three self-play loops where the principle of playing
against a copy of yourself is used:

1. *Move-level*: in the MCTS playouts, our opponent is actually a copy of ourselves
   (line 5)—hence, self-play at the level of *game moves*.
2. *Example-level*: the input for self-training the approximator for the policy and the
   reward functions is generated by our own games (line 2)—hence, self-play at the
   level of the *value/policy network*.

3. *Tournament-level*: the self-play loop creates a training curriculum that starts tabula rasa and ends at world champion level. The system trains at the level of the *player* against itself (line 1)—hence, self-play of the third kind.

All three of these levels use their own kind of self-play, of which we will describe the details in the following sections. We start with move-level self-play.

6.2.1 Move-Level Self-Play

At the innermost level, we use the agent to play against itself, as its own opponent. Whenever it is my opponent's turn to move, I play its move, trying to find the best move for my opponent (which will be the worst possible move for me). This scheme uses the same knowledge for player and opponent. This is different from the real world, where the agents are different, with different brains, different reasoning skills, and different experience. Our scheme is symmetrical: when we assume that our agent plays a strong game, then the opponent is also assumed to play strongly, and we can hope to learn from the strong counterplay. (We thus assume that our agent plays with the same knowledge as we have; we are not trying to consciously exploit opponent weaknesses.)[5]

6.2.1.1 Minimax

This principle of generating the counterplay by playing yourself while switching perspectives has been used since the start of artificial intelligence. It is known as minimax.

The games of chess, checkers, and Go are challenging games. The architecture that has been used to program chess and checkers players has been the same since the earliest paper designs of Turing [125]: a search routine based on minimax which searches to a certain depth and an evaluation function to estimate the score of board positions using heuristic rules of thumb when this depth is reached. In chess and checkers, for example, the number of pieces on the board of a player is a crude but effective approximation of the strength of a state for that player. Figure 6.11 shows a diagram of this classic search-eval architecture.[6]

---

[5] There is also research into opponent modeling, where we try to exploit our opponent's weaknesses [14, 47, 54]. Here, we assume an identical opponent, which often works best in chess and Go.

[6] Because the agent knows the transition function T, it can calculate the new state s' for each action a. The reward r is calculated at terminal states, where it is equal to the value v. Hence, in this diagram, the search function provides the state to the eval function. See [87, 125] for an explanation of the search-eval architecture.

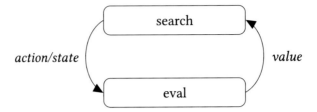

Fig. 6.11 Search-Eval architecture of games

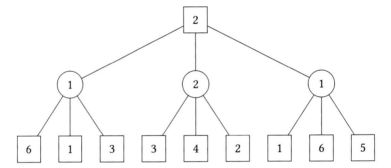

Fig. 6.12 Minimax tree

Based on this principle, many successful search algorithms have been developed, of which alpha–beta is the best known [60, 86]. Since the size of the state space is exponential in the depth of lookahead; however, many enhancements had to be developed to manage the size of the state space and to allow deep lookahead to occur [87].

The word *minimax* is a contraction of maximizing/minimizing (and then reversed for easy pronunciation). It means that in zero-sum games, the two players alternate making moves, and that on even moves, when player A is to choose a move, the best move is the one that maximizes the score for player A, while on odd moves the best move for player B is the move that minimizes the score for player A.

Figure 6.12 depicts this situation in a tree. The score values in the nodes are chosen to show how minimax works. At the top is the root of the tree, level 0, a square node where player A is to move.

Since we assume that all players rationally choose the best move, the value of the root node is determined by the value of the best move, the maximum of its children. Each child, at level 1, is a circle node where player B chooses its best move, in order to minimize the score for player A. The leaves of this tree, at level 2, are again max squares (even though there is no child to choose from anymore). Note how for each circle node, the value is the minimum of its children, and for the square node, the value is the maximum of the tree circle nodes.

Python pseudocode for a recursive minimax procedure is shown in Listing 6.2. Note the extra hyperparameter d. This is the search depth counting upward from

```
1    INF = 99999 # a value assumed to be larger than eval ever returns
2
3    def minimax(n, d):
4        if d <= 0:
5            return heuristic_eval(n)
6        elif n['type'] == 'MAX':
7            g = -INF
8            for c in n['children']:
9                g = max(g, minimax(c, d-1))
10       elif n['type'] == 'MIN':
11           g = INF
12           for c in n['children']:
13               g = min(g, minimax(c, d-1))
14       return g
15
16   print("Minimax_value:_", minimax(root, 2))
```

Listing 6.2 Minimax code [87]

the leaves. At depth 0 are the leaves, where the heuristic evaluation function is called to score the board.[7] Also note that the code for making moves on the board—transitioning actions into the new states—is not shown in the listing. It is assumed to happen inside the children dictionary. We frivolously mix actions and states in these sections, since an action fully determines which state will follow. (At the end of this chapter, the exercises provide more detail about move making and unmaking.)

AlphaGo Zero uses MCTS, a more advanced search algorithm than minimax, which we will discuss shortly.

Beyond Heuristics

Minimax-based procedures traverse the state space by recursively following all actions in each state that they visit [125]. Minimax works just like a standard depth-first search procedure, such as we have been taught in our undergraduate algorithms and data structure courses. It is a straightforward, rigid approach, which searches all branches of the node to the same search depth.

To focus the search effort on promising parts of the tree, the researchers have subsequently introduced many algorithmic enhancements, such as alpha–beta cutoffs, iterative deepening, transposition tables, null windows, and null moves [39, 60, 62, 88, 111].

In the early 1990s, experiments with a different approach started, based on random playouts of a single line of play [2, 15, 18] (Figs. 6.13 and 6.14). In

---

[7] The heuristic evaluation function is originally a linear combination of hand-crafted heuristic rules, such as material balance (which side has more pieces) or center control. At first, the linear combinations (coefficients) were not only hand-coded but also hand-tuned. Later they were trained by supervised learning [10, 46, 91, 120]. More recently, NNUE was introduced as a nonlinear neural network to use as evaluation function in an alpha–beta framework [81].

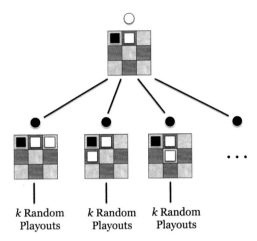

Fig. 6.13 Three lines of play [12]

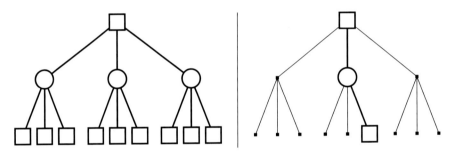

Fig. 6.14 Searching a tree versus searching a path

Fig. 6.14, this different approach is illustrated. We see a search of a single line of play versus a search of a full subtree. It turned out that averaging many such playouts could also be used to approximate the value of the root, in addition to the classic recursive tree search approach. In 2006, a tree version of this approach was introduced that proved successful in Go. This algorithm was called Monte Carlo Tree Search [17, 30]. Also in that year, Kocsis and Szepesvári created a selection rule for the exploration/exploitation trade-off that performed well and converged to the minimax value [61]. Their rule is called UCT, for upper confidence bounds applied to trees.

6.2.1.2 Monte Carlo Tree Search

Monte Carlo Tree Search has two main advantages over minimax and alpha–beta. First, MCTS is based on averaging single lines of play, instead of recursively traversing subtrees. The computational complexity of a path from the root to a

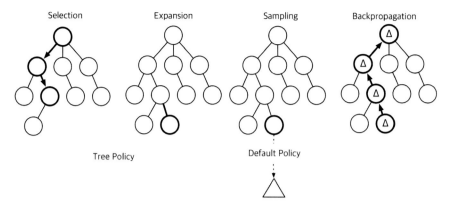

Fig. 6.15 Monte Carlo tree search [17]

leaf is polynomial in the search depth. The computational complexity of a tree is exponential in the search depth. Especially in applications with many actions per state, it is much easier to manage the search time with an algorithm that expands one path at a time.[8]

Second, MCTS does not need a heuristic evaluation function. It plays out a line of play in the game from the root to an end position. In end positions, the score of the game, a win or a loss, is known. By averaging many of these playouts, the value of the root is approximated. Minimax has to cope with an exponential search tree, which it cuts off after a certain search depth, at which point it uses the heuristic to estimate the scores at the leaves. There are, however, games where no efficient heuristic evaluation function can be found. In this case, MCTS has a clear advantage, since it works without a heuristic score function.

MCTS has proven to be successful in many different applications. Since its introduction in 2006, MCTS has transformed the field of heuristic search. Let us see in more detail how it works.

Monte Carlo Tree Search consists of four operations: select, expand, playout, and backpropagate (Fig. 6.15). The third operation (playout) is also called *rollout*, *simulation*, and *sampling*. Backpropagation is sometimes called *backup*. Select is the downward policy action trial part, and backup is the upward error/learning part of the algorithm. We will discuss the operations in more detail in a short while.

MCTS is a successful planning-based reinforcement learning algorithm, with an advanced exploration/exploitation selection rule. MCTS starts from the initial state s_0, using the transition function to generate successor states. In MCTS, the state

---

[8] Compare chess and Go: in chess, the typical number of moves in a position is 25, and for Go this number is 250. A chess-tree of depth 5 has $25^5 = 9765625$ leaves. A Go-tree of depth 5 has $250^5 = 976562500000$ leaves. A depth-5 minimax search in Go would take prohibitively long; an MCTS search of 1000 expansions expands the same number of paths from root to leaf in both games.

space is traversed iteratively, and the tree data structure is built in a step-by-step fashion, node by node, and playout by playout. A typical size of an MCTS search is to do 1000–10,000 iterations. In MCTS, each iteration starts at the root s_0, traversing a path in the tree down to the leaves using a selection rule, expanding a new node, and performing a random playout. The result of the playout is then propagated back to the root. During the backpropagation, statistics at all internal nodes are updated. These statistics are then used in future iterations by the selection rule to go to the currently most interesting part of the tree.

The statistics consist of two counters: the win count w and the visit count v. During backpropagation, the visit count v at all nodes that are on the path back from the leaf to the root is incremented. When the result of the playout was a win, then the win count w of those nodes is also incremented. If the result was a loss, then the win count is left unchanged.

The selection rule uses the win rate w/v and the visit count v to decide whether to exploit high-win-rate parts of the tree or to explore low-visit-count parts. An often-used selection rule is UCT (Sect. 6.2.1.2). It is this selection rule that governs the exploration/exploitation trade-off in MCTS.

The Four MCTS Operations

Let us look in more detail at the four operations. Please refer to Listing 6.3 and Fig. 6.15 [17]. As we see in the figure and the listing, the main steps are repeated as long as there is time left. Per step, the activities are as follows:

1. *Select.* In the selection step, the tree is traversed from the root node down until a leaf of the MCTS search tree is reached where a new child is selected that is not part of the tree yet. At each internal state, the selection rule is followed to determine which action to take and thus which state to go to next. The UCT rule works well in many applications [61].
   The selections at these states are part of the policy $\pi(s)$ of actions of the state.
2. *Expand.* Then, in the expansion step, a child is added to the tree. In most cases, only one child is added. In some MCTS versions, all successors of a leaf are added to the tree [17].
3. *Playout.* Subsequently, during the playout step, random moves are played in a form of self-play until the end of the game is reached. (These nodes are not added to the MCTS tree, but their search result is, in the backpropagation step.) The reward r of this simulated game is $+1$ in case of a win for the first player, 0 in case of a draw, and -1 in case of a win for the opponent.[9]
4. *Backpropagation.* In the backpropagation step, reward r is propagated back upward in the tree, through the nodes that were traversed down previously. Two counts are updated: the visit count, for all nodes, and the win count, depending

---

[9] Originally, playouts were random (the Monte Carlo part in the name of MCTS) following Brügmann's [18] and Bouzy and Helmstetter's [15] original approach. In practice, most Go playing programs improve on the random playouts by using databases of small 3×3 patterns with best replies and other fast heuristics [24, 31, 33, 50, 106]. Small amounts of domain knowledge are used after all, albeit not in the form of a heuristic evaluation function.

```
1   def monte_carlo_tree_search(root):
2       while resources_left(time, computational power):
3           leaf = select(root) # leaf = unvisited node
4           simulation_result = rollout(leaf)
5           backpropagate(leaf, simulation_result)
6       return best_child(root) # or: child with highest visit count
7
8   def select(node):
9       while fully_expanded(node):
10          node = best_child(node)      # traverse down path of best
                UCT nodes
11      return expand(node.children) or node # no children/node is
            terminal
12
13  def rollout(node):
14      while non_terminal(node):
15          node = rollout_policy(node)
16      return result(node)
17
18  def rollout_policy(node):
19      return pick_random(node.children)
20
21  def backpropagate(node, result):
22      if is_root(node) return
23      node.stats = update_stats(node, result)
24      backpropagate(node.parent)
25
26  def best_child(node, c_param=1.0):
27      choices_weights = [
28          (c.q / c.n) + c_param * np.sqrt((np.log(node.n) / c.n))
                    # UCT
29          for c in node.children
30      ]
31      return node.children[np.argmax(choices_weights)]
```

Listing 6.3 MCTS pseudo-Python [17, 35]

on the reward value. Note that in a two-agent game, nodes in the MCTS tree alternate color. If white has won, then only white-to-play nodes are incremented, and if black has won, then only the black-to-play nodes are incremented.

MCTS is on policy: the values that are backed up are those of the nodes that were selected.

Pseudocode

Many websites contain useful resources on MCTS, including example code (see Listing 6.3).[10] The pseudocode in the listing is from an example program for

---

[10] https://int8.io/monte-carlo-tree-search-beginners-guide/.

game play. The MCTS algorithm can be coded in many different ways. For implementation details, see [35] and the comprehensive survey [17].

MCTS is a popular algorithm. An easy way to use it in Python is by installing it from a pip package (`pip install mcts`).

Policies

At the end of the search, after the predetermined iterations have been performed, or when time is up, MCTS returns the value and the action with the highest visit count. An alternative would be to return the action with the highest win rate. However, the visit count takes into account the win rate (through UCT) and the number of simulations on which it is based. A high win rate may be based on a low number of simulations and can thus be high variance. High visit counts will be low variance. Due to selection rule, high visit count implies high win rate with high confidence, while high win rate may be low confidence [17]. The action of this initial state s_0 constitutes the deterministic policy $\pi(s_0)$.

UCT Selection

The adaptive exploration/exploitation behavior of MCTS is governed by the selection rule, for which often UCT is chosen. UCT is an adaptive exploration/exploitation rule that achieves high performance in many different domains.

UCT was introduced in 2006 by Kocsis and Szepesvári [61]. The paper provides a theoretical guarantee of eventual convergence to the minimax value. The selection rule was named UCT, for upper confidence bounds for multi-armed bandits applied to trees. (Bandit theory was also mentioned in Sect. 2.2.4.7.)

The selection rule determines the way in which the current values of the children influence which part of the tree will be explored more. The UCT formula is

$$\text{UCT}(a) = \frac{w_a}{n_a} + C_p \sqrt{\frac{\ln n}{n_a}}, \tag{6.1}$$

where w_a is the number of wins in child a, n_a is the number of times child a has been visited, n is the number of times the parent node has been visited, and $C_p \geq 0$ is a constant, the tunable exploration/exploitation hyperparameter. The first term in the UCT equation, the win rate $\frac{w_a}{n_a}$, is the exploitation term. A child with a high win rate receives through this term an exploitation bonus. The second term $\sqrt{\frac{\ln n}{n_a}}$ is for exploration. A child that has been visited infrequently has a higher exploration term. The level of exploration can be adjusted by the C_p constant. A low C_p does little exploration, and a high C_p has more exploration. The selection rule is then to select the child with the highest UCT sum (the familiar arg max function of value-based methods).

The UCT formula balances *win rate* $\frac{w_a}{n_a}$ and *"newness"* $\sqrt{\frac{\ln n}{n_a}}$ in the selection of nodes to expand.[11] Alternative selection rules have been proposed, such as Auer's UCB1 [6–8] and P-UCT [70, 93].

P-UCT

We should note that the MCTS that is used in the AlphaGo Zero program is a little different. MCTS is used inside the training loop, as an integral part of the self-generation of training examples, to enhance the quality of the examples for every self-play iteration, using both value and policy inputs to guide the search.

Also, in the AlphaGo Zero program, MCTS backups rely fully on the value function approximator; no playout is performed anymore. The MC part in the name of MCTS, which stands for the Monte Carlo playouts, really has become a misnomer for this neural network-guided tree searcher.

Furthermore, selection in self-play MCTS is different. UCT-based node selection now also uses the input from the policy head of the trained function approximators, in addition to the win rate and newness. What remains is that through the UCT mechanism MCTS can focus its search effort greedily on the part with the highest win rate, while at the same time balancing exploration of parts of the tree that are underexplored.

The formula that is used to incorporate input from the policy head of the deep network is a variant of P-UCT [70, 77, 93, 109] (for *predictor*-UCT). Let us compare P-UCT with UCT. The P-UCT formula adds the policy head $\pi(a|s)$ to Eq. 6.1

$$\text{P-UCT}(a) = \frac{w_a}{n_a} + C_p \pi(a|s) \frac{\sqrt{n}}{1 + n_a}.$$

P-UCT adds the $\pi(a|s)$ term specifying the probability of the action a to the exploration part of the UCT formula.[12]

Exploration/Exploitation

The search process of MCTS is guided by the statistics values in the tree. MCTS discovers during the search where the promising parts of the tree are. The tree expansion of MCTS is inherently variable depth and variable width (in contrast to minimax-based algorithms such as alpha–beta, which are inherently fixed-depth and fixed-width). In Fig. 6.16, we see a snapshot of the search tree of an actual MCTS optimization. Some parts of the tree are searched more deeply than others [128].

---

[11] The square root term is a measure of the variance (uncertainty) of the action value. The use of the natural logarithm ensures that, since increases get smaller over time, old actions are selected less frequently. However, since logarithm values are unbounded, eventually all actions will be selected [114].

[12] Note further the small differences under the square root (no logarithm, and the 1 in the denominator) also change the UCT function profile somewhat, ensuring correct behavior at unvisited actions [77].

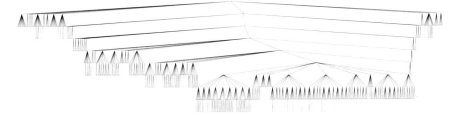

Fig. 6.16 Adaptive MCTS tree [65]

An important element of MCTS is the exploration/exploitation trade-off, which can be tuned with the C_p hyperparameter. The effectiveness of MCTS in different applications depends on the value of this hyperparameter. Typical initial choices for Go programs are $C_p = 1$ or $C_p = 0.1$ [17], although in AlphaGo we see highly explorative choices such as $C_p = 5$. In general, experience has learned that when compute power is low, C_p should be low, and when more compute power is available, more exploration (higher C_p) is advisable [17, 65].

Applications
MCTS was introduced in 2006 [30–32] in the context of computer Go programs, following work by Chang et al. [23], Auer et al. [7], and Cazenave and Helmstetter [22]. The introduction of MCTS improved performance of Go programs considerably, from medium amateur to strong amateur. Where the heuristics-based GNU Go program played around 10 kyu, Monte Carlo programs progressed to 2–3 dan in a few years' time.

Eventually, on the small 9×9 board, Go programs achieved very strong play. On the large 19×19 board, performance did not improve much beyond the 4–5 dan level, despite much effort by the researchers. It was thought that perhaps the large action space of the 19×19 board was too hard for MCTS. Many enhancements were considered, for the playout phase and for the selection. As the AlphaGo results show, a crucial enhancement was the introduction of deep function approximation.

After its introduction, MCTS quickly proved successful in other applications, both two agent and single agent: for video games [25], for single player applications [17], and for many other games. Beyond games, MCTS revolutionized the world of heuristic search [17]. Previously, in order to achieve best-first search, one had to find a domain-specific heuristic to guide the search in a smart way. With MCTS, this is no longer necessary. Now a general method exists that finds the promising parts of the search without a domain-specific heuristic, just by using statistics of the search itself.

There is a deeper relation between UCT and reinforcement learning. Grill et al. [52] showed how the second term of P-UCT acts as a regularizer on modelfree policy optimization [1]. In particular, Jacob et al. [57] showed how MCTS can be used to achieve human-like play in chess, Go, and Diplomacy, by regularizing reinforcement learning with supervised learning on human games.

MCTS in AlphaGo Zero

For policy improvement, AlphaGo Zero uses a version of on-policy MCTS that does not use random playouts anymore. To increase exploration, Dirichlet noise is added to the P-UCT value at the root node, to ensure that all moves may be tried. The C_p value of MCTS in AlphaGo is 5, heavily favoring exploration. In AlphaGo Zero, the value depends on the stage in the learning; it grows during self-play. In each self-play iteration, 25,000 games are played. For each move, MCTS performs 1600 simulations. In total, over a 3-day course of training, 4.9 million games were played, after which AlphaGo Zero outperformed the previous version, AlphaGo [109].

Conclusion

We have taken a look into the planning part of AlphaGo Zero's self-play architecture. MCTS consists of a move selection and a statistics backup phase, which corresponds to the behavior (trial) and learning (error) from reinforcement learning. MCTS is an important algorithm in reinforcement learning, and we have taken a detailed look at the algorithm.

Move-level self-play is our first self-play procedure; it is a procedure that plays itself to generate its counter moves. The move-level planning is only one part of the self-play picture. Just as important is the learning part. Let us have a look at how AlphaGo Zero achieves its function approximation. For this, we move to the second level of self-play: the example-level.

6.2.2 Example-Level Self-Play

Move-level self-play creates an environment for us that can play our countermoves. Now we need a mechanism to learn from these actions. AlphaGo Zero follows the actor critic principle to approximate both value and policy functions. It approximates these functions using a single deep residual neural network with a value head and a policy head (Sect. B.2.6). The policy and the value approximations are incorporated in MCTS in the selection and backup step.

In order to learn, reinforcement learning needs training examples. The training examples are generated at the self-play move level. Whenever a move is played, the $\langle s, a \rangle$ state–action pair is recorded, and whenever a full game has been played to the end, the outcome z is known, and the outcome is added to all pairs of game moves, to create $\langle s, a, z \rangle$ triples. The triples are stored in the replay buffer and sampled randomly to train the value/policy net. The actual implementation in AlphaGo Zero contains many more elements to improve the learning stability.

The player is designed to become stronger than the opponent, and this occurs at the example-level. Here it uses MCTS to improve the current policy, improving it with moves that are winning against the opponent's moves.

Example-level self-play is our second self-play procedure, and the examples are generated in the self-play games and are used to train the network that is used to play the moves by the two players.

6.2.2.1 Policy and Value Network

The first AlphaGo program uses three separate neural networks: for the rollout policy, for the value, and for the selection policy [107]. AlphaGo Zero uses a single network, which is tightly integrated in MCTS. Let us have a closer look at this single network.

The network is trained on the example triples $\langle s, a, z \rangle$ from the replay buffer. These triples contain search results of the board states of the game and the two loss function targets: a for the actions that MCTS predicts for each board states and z for the outcome of the game (win or loss) when it came to an end. The action a is the policy loss, and the outcome z is the value loss. All triples for a game consist of the same outcome z and the different actions that were played at each state.

AlphaGo Zero uses a dual-headed residual network (a convolutional network with extra skip links between layers, to improve regularization, see Sect. B.2.6 [20, 55]). Policy and value loss contribute equally to the loss function [129]. The network is trained by stochastic gradient descent. L2 regularization is used to reduce overfitting. The network has 19 hidden layers, and an input layer and two output layers, for policy and value. The size of the mini-batch for updates is 2048. This batch is distributed over 64 GPU workers, each with 32 data entries. The mini-batch is sampled uniformly over the last 500,000 self-play games (replay buffer). The learning rate started at 0.01 and went down to 0.0001 during self-play. More details of the AlphaGo Zero network are described in [109].

Please note the size of the replay buffer and the long training time. Go is a complex game, with sparse rewards. Only at the end of a long game, the win or loss is known, and attributing this sparse reward to the many individual moves of a game is difficult, requiring many games to even out errors.

MCTS is an on-policy algorithm that makes use of guidance in two places: in the downward action selection and in the upward value backup. In AlphaGo Zero, the function approximator returns both elements: a policy for the action selection and a value for the backup [109].

For tournament-level self-play to succeed, the training process must (1) cover enough of the state space, (2) be stable, and (3) converge. Training targets must be sufficiently challenging to learn and sufficiently diverse. The purpose of MCTS is to act as a policy improver in the actor critic setting, to generate learning targets of sufficient quality and diversity for the agent to learn.

Let us have a closer look at these aspects, to get a broader perspective on why it was so difficult to get self-play to work in Go.

6.2.2.2 Stability and Exploration

Self-play has a long history in artificial intelligence, going back to TD-Gammon, 30 years ago. Let us look at the challenges in achieving a strong level of play.

Since in AlphaGo Zero all learning is by reinforcement, the training process must now be even more stable than in AlphaGo, which also used supervised learning from grandmaster games. The slightest problem in overfitting or correlation between states can throw off the coverage, correlation, and convergence. AlphaGo Zero uses various forms of exploration to achieve stable reinforcement learning. Let us summarize how stability is achieved.

- *Coverage* of the sparse state space is improved by playing a large number of diverse games. The quality of the states is further improved by MCTS lookahead. MCTS searches for good training samples, improving the quality and diversity of the covered states. The exploration part of MCTS should make sure that enough new and unexplored parts of the state space are covered. Dirichlet noise is added at the root node, and the C_p parameter in the P-UCT formula, which controls the level of exploration, has been set quite high, around 5 (see also Sect. 6.2.1.2).
- *Correlation* between subsequent states is reduced through the use of an experience replay buffer, as in DQN and Rainbow algorithms. The replay buffer breaks correlation between subsequent training examples. Furthermore, the MCTS search also breaks correlation, by searching deep in the tree to find better states.
- *Convergence* of the training is improved by using on-policy MCTS and by taking small training steps. Since the learning rate is small, training target stability is higher, and the risk of divergence is reduced. A disadvantage is that convergence is quite slow and requires many training games.

By using these measures together, stable generalization and convergence are achieved. Although self-play is conceptually simple, achieving stable and high-quality self-play in a game as complex and sparse as Go required slow training with a large number of games and quite some hyperparameter tuning. There are many hyperparameters whose values must be set correctly, for the full list, see [109]. Although the values are published [109], the reasoning behind the values is not always clear. Reproducing the AlphaGo Zero results is not easy, and much time is spent in tuning and experimenting to reproduce the AlphaGo Zero results [89, 113, 119, 121, 135].

Two Views

At this point, it is useful to step back and reflect on the self-play architecture.

There are two different views. The one view, planning-centric, which we have followed so far, is of a searcher that is helped by a learned evaluation function (which trains on examples from games played against itself). In addition, there is move-level self-play (opponent's moves are generated with an inverted replica of itself) and there is tournament-level self-play (by the value learner).

The alternative view, learning-centric, is that a policy is learned by generating game examples from self-play. In order for these examples to be of high quality, the policy learning is helped by a policy improver, a planning function that performs lookahead to create better learning targets (and the planning is performed by making moves by a copy of the player). In addition, there is tournament-level self-play (by the policy learner) and there is move-level self-play (by the policy improver).

The difference in viewpoint is who comes first: the planning viewpoint favors the searcher, and the learner is there to help the planner; the reinforcement learning viewpoint favors the policy learner, and the planner is there to help improve the policy. Both viewpoints are equally valid, and both viewpoints are equally valuable. Knowing of the other viewpoint deepens our understanding of how these complex self-play algorithms work.

This concludes our discussion of the second type of self-play, the example-level, and we move on to the third type: tournament-level self-play.

6.2.3 Tournament-Level Self-Play

At the top level, a tournament of self-play games is played between the two (identical) players. The player is designed to increase in strength, learning from the examples at the second level, so that the player can achieve a higher level of play. In tabula rasa self-play, the players start from scratch. By becoming progressively stronger, they also become stronger opponents for each other, and their mutual level of play can increase. A virtuous cycle of ever-increasing intelligence will emerge.

For this ideal of artificial intelligence to become reality, many stars have to line up. After TD-Gammon, many researchers have tried to achieve this goal in other games but were unsuccessful.

Tournament-level self-play is only possible when move-level self-play and example-level self-play work. For move-level self-play to work, both players need to have access to the transition function, which must be completely accurate. For example-level self-play to work, the player architecture must be such that it is able to learn a stable policy of high quality (MCTS and the network have to mutually improve each other).

Tournament-level self-play is the third self-play procedure, where a tournament is created with games starting from easy learning tasks, changing to harder tasks, increasing all the way to world champion level training. This third procedure allows reinforcement learning to transcend the level of play ("intelligence") of previous teachers.

Curriculum Learning
As mentioned before, the AlphaGo effort consists of three programs: AlphaGo, AlphaGo Zero, and AlphaZero. The first AlphaGo program used supervised learning based on grandmaster games, followed by reinforcement learning on self-play games. The second program, AlphaGo Zero, used reinforcement learning only, in a self-play architecture that starts from zero knowledge. The first program trained many weeks, yet the second program needed only a few days to become stronger than the first [107, 109].

Why did the self-play approach of AlphaGo Zero learn faster than the original AlphaGo that could benefit from all the knowledge of Grandmaster games? Why is self-play faster than a combination of supervised and reinforcement learning? The reason is a phenomenon called curriculum learning: self-play is faster because it creates a sequence of learning tasks that are ordered from easy to hard. Training such an ordered sequence of small tasks is quicker than one large unordered task.

Curriculum learning starts the training process with easy concepts before the hard concepts are learned; this is, of course, the way in which humans learn. Before we learn to run, we learn to walk, and before we learn about multiplication, we learn about addition. In curriculum learning, the examples are ordered in batches from easy to hard. Learning such an ordered sequence of batches goes better since understanding the easy concepts helps understanding of the harder concepts; learning everything all at once typically takes longer and may result in lower accuracy.[13]

6.2.3.1 Self-Play Curriculum Learning

In ordinary deep reinforcement learning, the network tries to solve a fixed problem in one large step, using environment samples that are not sorted from easy to hard. With examples that are not sorted, the program has to achieve the optimization step from no knowledge to human-level play in one big, unsorted leap, by optimizing many times over challenging samples where the error function is large. Overcoming such a large training step (from beginner to advanced) costs much training time.

In contrast, in AlphaGo Zero, the network is trained in many small steps, starting against a very weak opponent, just as a human child learns to play the game by playing against a teacher that plays simple moves. As our level of play increases, so does the difficulty of the moves that our teacher proposes to us. Subsequently, harder problems are generated and trained for, refining the network that has already been pretrained with the easier examples.

Self-play naturally generates a curriculum with examples from easy to hard. The learning network is always in lock step with the training target—errors are low throughout the training. As a consequence, training times go down and the playing strength goes up.

6.2.3.2 Supervised and Reinforcement Curriculum Learning

Curriculum learning has been studied before in psychology and education science. Selfridge et al. [100] first connected curriculum learning to machine learning, where they trained the proverbial Cartpole controller. First they trained the controller on long and light poles, while gradually moving toward shorter and heavier poles. Schmidhuber [98] proposed a related concept, to improve exploration for world models by artificial curiosity. Curriculum learning was subsequently applied to

---

[13] Such a sequence of related learning tasks corresponds to a meta-learning problem. In meta-learning, the aim is to learn a new task fast, by using the knowledge learned from previous, related, tasks; see Chap. 9.

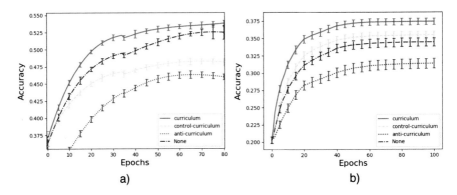

Fig. 6.17 Effectiveness of a Sorted Curriculum [132]. (**a**) Large network. (**b**) Small network

match the order of training examples to the growth in model capacity in different supervised learning settings [13, 41, 64]. Another related development is in developmental robotics, where curriculum learning can help to self-organize open-ended developmental trajectories [83], related to intrinsic motivation (see Sect. 8.3.2). The AMIGo approach uses curriculum learning to generate subgoals in hierarchical reinforcement learning [19].

To order the training examples from easy to hard, we need a measure to quantify the difficulty of the task. One idea is to use the minimal loss with respect to some of the upper layers of a high-quality pretrained model [132]. In a supervised learning experiment, Weinshall et al. compared the effectiveness of curriculum learning on a set of test images (5 images of mammals from CIFAR100). Figure 6.17 shows the accuracy of a curriculum ordering (green), no curriculum (blue), randomly ordered groups (yellow), and the labels sorted in reverse order (red). Both networks are regular networks with multiple convolutional layers followed by a fully connected layer. The large network has 1,208,101 parameters, and the small network has 4557 parameters. We can clearly see the effectiveness of ordered learning [132].

Procedural Content Generation
Finding a good way to order the sequence of examples is often difficult. A possible method to generate a sequence of tasks that are related is by using procedural content generation (PCG) [79, 103]. Procedural content generation uses randomized algorithms to generate images and other content for computer games; the difficulty of the examples can often be controlled. It is frequently used to automatically generate different levels in games, so that they do not have to be all created manually by the game designers and programmers [112, 124].[14]

---

[14] See also generative adversarial networks and deep dreaming, for a connectionist approach to content generation, Sect. B.2.6.2.

The Procgen benchmark suite has been built upon procedurally generated games [29]. Another popular benchmark is the general video game AI competition (GVGAI) [68]. Curriculum learning reduces overfitting to single tasks. Justesen et al. [59] have used GVGAI to show that a policy easily overfits to specific games and that training over a curriculum improves its generalization to levels that were designed by humans. MiniGrid is a procedurally generated world that can be used for hierarchical reinforcement learning [26, 92].

Active Learning

Curriculum learning is also related to active learning.

Active learning is a type of machine learning that is in-between supervised and reinforcement learning. Active learning is relevant when labels are in principle available (as in supervised learning) but at a cost.

Active learning performs a kind of iterative supervised learning, in which the agent can choose to query which labels to reveal during the learning process. Active learning is related to reinforcement learning and to curriculum learning and is for example of interest for studies into recommender systems, where acquiring more information may come at a cost [36, 94, 102].

Single-Agent Curriculum Learning

Curriculum learning has been studied for many years. A problem is that it is difficult to find an ordering of tasks from easy to hard in most learning situations. In two-player self-play, the ordering comes naturally, and the successes have inspired recent work on single-agent curriculum learning. For example, Laterre et al. introduce the Ranked Reward method for solving bin packing problems [66] and Wang et al. presented a method for Morpion Solitaire [130]. Feng et al. use an AlphaZero-based approach to solve hard Sokoban instances [44]. Their model is an 8-block standard residual network, with MCTS as planner. They create a curriculum by constructing simpler subproblems from hard instances, using the fact that Sokoban problems have a natural hierarchical structure. This approach was able to solve harder Sokoban instances than had been solved before. Florensa et al. [45] study the generation of goals for curriculum learning using a generator network (GAN).

Conclusion

Although curriculum learning has been studied in artificial intelligence and in psychology for some time, it has not been a popular method, since it is difficult to find well-sorted training curricula [73, 74, 131]. Due to the self-play results, curriculum learning is now attracting more interest, see [80, 133]. Work is reported in single-agent problems [38, 44, 66, 80] and in multi-agent games, as we will see in Chap. 7.

After this detailed look at self-play algorithms, it is time to look in more detail at the environments and benchmarks for self-play.

6.3 Self-Play Environments

Progress in reinforcement learning is determined to a large extent by the application domains that provide the learning challenge. The domains of checkers, backgammon, and especially chess and Go, have provided highly challenging domains, for which substantial progress was achieved, inspiring many researchers.

The previous section provided an overview of the planning and learning algorithms. In this section we will have a closer look at the environments and systems that are used to benchmark these algorithms.

Table 6.2 lists the AlphaGo and self-play approaches that we discuss in this chapter. First we will discuss the three AlphaGo programs, and then we will list open self-play frameworks. We will start with the first program, AlphaGo.

6.3.1 How to Design a World Class Go Program?

Figure 6.18 shows the playing strength of traditional programs (right panel, in red) and different versions of the AlphaGo programs, in blue. We see how much stronger the 2015, 2016, and 2017 versions of AlphaGo are than the earlier heuristic minimax program GnuGo and two MCTS-only programs Pachi and Crazy Stone.

How did the AlphaGo authors design such a strong Go program? Before AlphaGo, the strongest programs used the Monte Carlo Tree Search planning algo-

Table 6.2 Self-Play approaches

Name	Approach	Ref
TD-Gammon	Tabula rasa self-play, shallow network, small alpha–beta search	[116]
AlphaGo	Supervised, self-play, 3×CNN, MCTS	[107]
AlphaGo Zero	Tabula rasa self-play, dual-head-ResNet, MCTS	[109]
AlphaZero	Tabula rasa self-play, dual-head-ResNet, MCTS on Go, chess, shogi	[108]

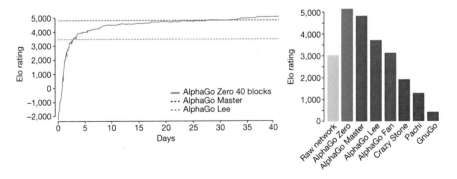

Fig. 6.18 Performance of AlphaGo Zero [109]

rithm, without neural networks. For some time, neural networks were considered to be too slow for use as value function in MCTS, and random playouts were used, often improved with small pattern-based heuristics [25, 32, 43, 49, 50]. Around 2015, a few researchers tried to improve performance of MCTS by using deep learning evaluation functions [4, 28, 48]. These efforts were strengthened by the strong results in Atari [76].

The AlphaGo team also tried to use neural networks. Except for backgammon, pure self-play approaches had not been shown to work well, and the AlphaGo team did the sensible thing to pretrain the network with the games of human grandmasters, using supervised learning. Next, a large number of self-play games were used to further train the networks. In total, no less than three neural networks were used: one for the MCTS playouts, one for the policy function, and one for the value function [107].

Thus, the original AlphaGo program consisted of three neural networks and used both supervised learning and reinforcement learning. The diagram in Fig. 6.19 illustrates the AlphaGo architecture. Although this design made sense at the time given the state-of-the-art of the field, and although it did convincingly beat the three strongest human Go players, managing and tuning such a complicated piece of software is quite difficult. The authors of AlphaGo tried to improve performance further by simplifying their design. Could TD-gammon's elegant self-play-only design be replicated in Go after all?

Indeed, a year later a reinforcement learning-only version was ready that learned to play Go from zero knowledge, tabula rasa: no grandmaster games, only self-play, and just a single neural network [109]. Surprisingly, this simpler version played stronger and learned faster. The program was called AlphaGo Zero, since it learned

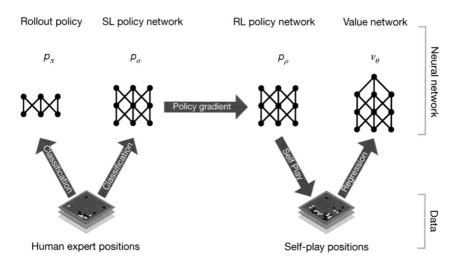

Fig. 6.19 All AlphaGo networks [107]

from zero knowledge; not a single grandmaster game was learned from, nor was there any heuristic domain knowledge hand coded into the program.

This new result, tabula rasa learning in Go with a pure self-play design, inspired much further research in self-play reinforcement learning.

6.3.2 AlphaGo Zero Performance

In their paper, Silver et al. [109] describe that learning progressed smoothly throughout the training. AlphaGo Zero outperformed the original AlphaGo after just 36 hours. The training time for the version of AlphaGo that played Lee Sedol was several months. Furthermore, AlphaGo Zero used a single machine with 4 tensor processing units, whereas AlphaGo Lee was distributed over many machines and used 48 TPUs.[15] Figure 6.18 shows the performance of AlphaGo Zero. Also shown is the performance of the raw network, without MCTS search. The importance of MCTS is large, around 2000 Elo points.[16]

AlphaGo Zero's reinforcement learning is truly learning Go knowledge from scratch, and, as the development team discovered, it did so in a way similar to how humans are discovering the intricacies of the game. In their paper [109], they published a picture of how this knowledge acquisition progressed (Fig. 6.20).

Joseki are standard corner openings that all Go players become familiar with as they learn to play the game. There are beginner's and advanced joseki. Over the course of its learning, AlphaGo Zero did learn joseki, and it learned them from beginner to advanced. It is interesting to see how it did so, as it reveals the progression of AlphaGo Zero's Go intelligence. Figure 6.20 shows sequences from games played by the program. Not to anthropomorphize too much,[17] but you can see the little program getting smarter.

The top row shows five joseki that AlphaGo Zero discovered. The first joseki is one of the standard beginner's openings in Go theory. As we move to the right, more difficult joseki are learned, with stones being played in looser configurations. The bottom row shows five joseki favored at different stages of the self-play training. It starts with a preference for a weak corner move. After 10 more hours of training, a better 3-3 corner sequence is favored. More training reveals more, and better, variations.

AlphaGo Zero discovered a remarkable level of Go knowledge during its self-play training process. This knowledge included not only fundamental elements of

---

[15] TPU stands for tensor processing unit, a low-precision design specifically developed for fast neural network processing.

[16] The basis of the Elo rating is pairwise comparison [42]. Elo is often used to compare playing strength in board games.

[17] Treat as if human.

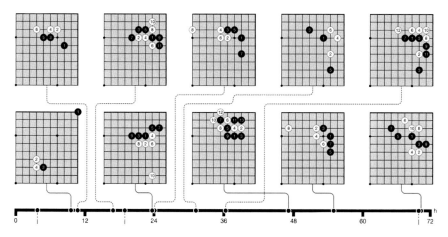

Fig. 6.20 AlphaGo Zero is Learning Joseki in a Curriculum from Easy to Hard [109]

human Go knowledge but also nonstandard strategies beyond the scope of traditional Go knowledge.

For a human Go player, it is remarkable to see this kind of progression in computer play, reminding them of the time when they discovered these joseki themselves. With such evidence of the computer's learning, it is hard not to anthropomorphize AlphaGo Zero.

6.3.3 AlphaZero

The AlphaGo story does not end with AlphaGo Zero. A year after AlphaGo Zero, a version was created with different input and output layers that learned to play chess and shogi (also known as Japanese chess, Fig. 6.21). AlphaZero uses the same MCTS and deep reinforcement learning architecture as for learning to play Go (the only differences are the input and output layers) [108]. This new program, AlphaZero, beats the strongest chess and shogi programs, Stockfish and Elmo. Both these programs followed a conventional heuristic minimax design, optimized by hand and machine learning, and improved with many heuristics for decades. AlphaZero used zero knowledge, zero grandmaster games, and zero hand-crafted heuristics, yet it played stronger. The AlphaZero architecture allows not only very strong play but is also a general architecture, suitable for three different games.[18]

---

[18] Although an AlphaZero version that has learned to play Go cannot play chess, it has to re-learn chess from scratch, with different input and output layers.

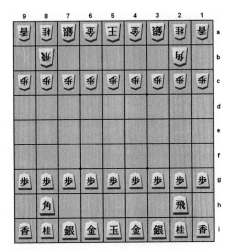

Fig. 6.21 A Shogi Board

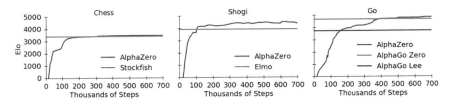

Fig. 6.22 Elo rating of AlphaZero in Chess, Shogi, and Go [108]

The Elo rating of AlphaZero in chess, shogi, and Go is shown in Fig. 6.22, [107, 109]. AlphaZero is stronger than the other programs. In chess, the difference is the smallest. In this field, the program has benefited from a large community of researchers that have worked intensely on improving performance of the heuristic alpha–beta approach. For shogi, the difference is larger.

General Game Architecture

AlphaZero can play three different games with the same architecture. The three games are quite different. Go is a static game of strategy. Stones do not move and are rarely captured. Stones, once played, are of strategic importance. In chess, the pieces move. Chess is a dynamic game where tactics are important. Chess also features sudden death a checkmate can occur in the middle of the game by capturing the king. Shogi is even more dynamic since captured pieces can be returned to the game, creating even more complex game dynamics.

Table 6.3 Self-learning environments

Name	Type	URL	Ref.
AlphaZero General	AlphaZero in Python	https://github.com/suragnair/alpha-zero-general	[119]
ELF	Game framework	https://github.com/pytorch/ELF	[121]
Leela	AlphaZero for Chess, Go	https://github.com/LeelaChessZero/lczero	[85]
PhoenixGo	AlphaZero-based Go prog.	https://github.com/Tencent/PhoenixGo	[135]
PolyGames	Env. for Zero learning	https://github.com/facebookincubator/Polygames	[21]

It is testament to the generality of AlphaZero's architecture, that games that differ so much in tactics and strategy can be learned so successfully. Conventional programs must be purposely developed for each game, with different search hyperparameters and different heuristics. Yet the MCTS/ResNet self-play architecture is able to learn all three from scratch.

6.3.4 Open Self-Play Frameworks

Tabula rasa learning for the game of Go is a remarkable achievement that inspired many researchers. The code of AlphaGo Zero and AlphaZero, however, is not public. Fortunately, the scientific publications [108, 109] provide many details, allowing other researchers to reproduce similar results.

Table 6.3 summarizes some of the self-learning environments, which we will briefly discuss.

- *A0G: AlphaZero General.* Thakoor et al. [119] created a self-play system called AlphaZero General (A0G).[19] It is implemented in Python for TensorFlow, Keras, and PyTorch and suitably scaled down for smaller computational resources. It has implementations for 6×6 Othello, tic tac toe, gobang, and connect4, all small games of significantly less complexity than Go. Its main network architecture is a four-layer CNN followed by two fully connected layers. The code is easy to understand in an afternoon of study and is well suited for educational purposes. The project write-up provides some documentation [119].
- *Facebook ELF.* ELF stands for Extensible Lightweight Framework. It is a framework for game research in C++ and Python [121]. Originally developed for real-time strategy games by Facebook, it includes the Arcade Learning Envi-

---

[19] https://github.com/suragnair/alpha-zero-general.

ronment and the Darkforest[20] Go program [123]. ELF can be found on GitHub.[21] ELF also contains the self-play program OpenGo [122], a reimplementation of AlphaGo Zero (in C++).

- *Leela.* Another reimplementation of AlphaZero is Leela. Both a chess and a Go version of Leela exist. The chess version is based on chess engine Sjeng. The Go[22] version is based on Go engine Leela. Leela does not come with trained weights of the network. Part of Leela is a community effort to compute these weights.
- *PhoenixGo.* PhoenixGo is a strong self-play Go program by Tencent [135]. It is based on the AlphaGo Zero architecture.[23] A trained network is available as well.
- *PolyGames.* PolyGames [21] is an environment for zero-based learning (MCTS with deep reinforcement learning) inspired by AlphaGo Zero. Relevant learning methods are implemented, and bots for Hex, Othello, and Havannah have been implemented. PolyGames can be found on GitHub.[24] A library of games is provided, as well as a checkpoint zoo of neural network models.

6.3.5 Hands On: Hex in PolyGames Example

Let us get some hands-on experience with MCTS-based self-play. We will implement the game of Hex with the PolyGames suite. Hex is a simple and fun board game invented independently by Piet Hein and John Nash in the 1940s. Its simplicity makes it easy to learn and play and also a popular choice for mathematical analysis. The game is played on a hexagonal board, player A wins if its moves connect the right to the left side, and player B wins if top connects to bottom (see Fig. 6.23; image by Wikimedia). A simple page with resources is here;[25] extensive strategy and background books have been written about Hex [16, 53]. We use Hex because it is simpler than Go, to get you up to speed quickly with self-play learning; we will also use PolyGames.

Click on the link[26] and start by reading the introduction to PolyGames on GitHub. Download the paper [21] and familiarize yourself with the concepts behind PolyGames. Clone the repository and build it by following the instructions. PolyGames uses PyTorch,[27] so install that too (follow the instructions on the PolyGames page).

---

[20] https://github.com/facebookresearch/darkforestGo.

[21] https://github.com/pytorch/ELF.

[22] https://github.com/gcp/leela-zero.

[23] https://github.com/Tencent/PhoenixGo.

[24] https://github.com/facebookincubator/Polygames.

[25] https://www.maths.ed.ac.uk/~csangwin/hex/index.html.

[26] https://github.com/facebookincubator/Polygames.

[27] https://pytorch.org.

Fig. 6.23 A Hex win for blue

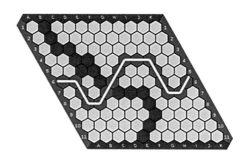

PolyGames is interfaced via the pypolygames Python package. The games, such as Hex, can be found in `src/games` and are coded in C++ for speed. The command

```
pypolygames train
```

is used to train a game and a model.
The command

```
pypolygames eval
```

is used to use a previously trained model.
The command

```
pypolygames human
```

allows a human to play against a trained model.
Type

```
python -m pypolygames {train,eval,traineval,human} --help
```

for help with each of the commands `train`, `eval`, `traineval`, or `human`.
A command to start training a Hex model with the default options is

```
python -m pypolygames train --game_name="Hex"
```

Try loading a pre-trained model from the zoo. Experiment with different training options, and try playing against the model that you just trained. When everything works, you can also try training different games. Note that more complex games may take (very) long to train.

Summary and Further Reading

We will now summarize the chapter and provide pointers for further reading.

Summary

For two-agent zero-sum games, when the transition function is given by the rules of the game, a special kind of reinforcement learning becomes possible. Since the agent can perfectly simulate the moves of the opponent, accurate planning far into the future becomes possible, allowing strong policies to be learned. Typically, the second agent becomes the environment. Previously, environments were static, but they will now evolve as the agent is learning, creating a virtuous cycle of increasing (artificial) intelligence in agent (and environment). The promise of this self-play setup is to achieve high levels of intelligence in a specific field. The challenges to overcome instability, however, are large, since this kind of self-play combines different kinds of unstable learning methods. Both TD-Gammon and AlphaGo Zero have overcome these challenges, and we have described their approach in quite some detail.

Self-play is a combination of planning, learning, and a self-play loop. The self-play loop in AlphaGo Zero uses MCTS to generate high-quality examples, which are used to train the neural net. This new neural net is then used in a further self-play iteration to generate more difficult games and refine the network further (and again, and again, and again). Alpha(Go) Zero thus learns starting at zero knowledge, *tabula rasa*.

Self-play makes use of many reinforcement learning techniques. In order to ensure stable learning, exploration is important. MCTS is used for deep planning. The exploration parameter in MCTS is set high, and convergent training is achieved by a low learning rate α. Because of these hyperparameter settings and because of the sparse rewards in Go, many games have to be played. The computational demands of stable self-play are large.

AlphaGo Zero uses function approximation of two functions: value and policy. Policy is used to help guide action selection in the P-UCT selection operation in MCTS, and value is used instead of random playouts to provide the value function at the leaves of the MCTS tree. MCTS has been changed significantly to work in the self-play setting. Gone are the random playouts that gave MCTS the name

Monte Carlo, and much of the performance is due to a high-quality policy and value approximation residual network.

Originally, the AlphaGo program (not AlphaGo Zero) used grandmaster games in supervised learning in addition to using reinforcement learning; it started from the knowledge of grandmaster games. Next came AlphaGo Zero, which does not use grandmaster games or any other domain-specific knowledge. All learning is based on reinforcement learning, playing itself to build up the knowledge of the game from zero knowledge. A third experiment has been published, called AlphaZero (without the "Go"). In this paper, the same network architecture and MCTS design (and the same learning hyperparameters) were used to learn three games: chess, shogi, and Go. This presented the AlphaZero architecture as a general learning architecture, stronger than the best alpha–beta-based chess and shogi programs.

Interestingly, the all-reinforcement learning AlphaGo Zero architecture was not only stronger than the supervised/reinforcement hybrid AlphaGo but also faster: it learned world champion level play in days, not weeks. Self-play learns quickly because of curriculum learning. It is more efficient to learn a large problem in many small steps, starting with easy problems, ending with hard ones, then in one large step. Curriculum learning works both for humans and for artificial neural networks.

Further Reading

One of the main interests of artificial intelligence is the study of how intelligence emerges out of simple, basic interactions. In self-learning systems, this is happening [67].

The work on AlphaGo is a landmark achievement in artificial intelligence. The primary sources of information for AlphaGo are the three AlphaGo/AlphaZero papers by Silver et al. [107–109]. The systems are complex and so are the papers and their supplemental methods sections. Many blogs have been written about AlphaGo that are more accessible. A movie has been made about AlphaGo.[28] There are also explanations on YouTube.[29]

A large literature on minimax and minimax enhancements for games exists; an overview is in [87]. A book devoted to building your own state-of-the-art self-learning Go bot is *Deep Learning and the Game of Go* by Pumperla and Ferguson [90], which came out before PolyGames [21].

MCTS has been a landmark algorithm by itself in artificial intelligence [17, 30, 61]. In the contexts of MCTS, many researchers worked on combining MCTS with learned patterns, especially to improve the random rollouts of MCTS. Other developments, such as asymmetrical and continuous MCTS, are [77, 78] or parallelizations such as [72].

---

[28] https://www.alphagomovie.com.
[29] https://www.youtube.com/watch?v=MgowR4pq3e8.

Supervised learning on grandmaster games was used to improve playouts and also to improve UCT selection. Gelly and Silver published notable works in this area [49, 50, 110]. Graf et al. [51] describe experiments with adaptive playouts in MCTS with deep learning. Convolutional neural nets were also used in Go by Clark and Storkey [27, 28], who had used a CNN for supervised learning from a database of human professional games, showing that it outperformed GNU Go and scored wins against Fuego, a strong open-source Go program [43] based on MCTS without deep learning.

Tesauro's success inspired many others to try temporal difference learning. Wiering et al. [134] and Van der Ree [126] report on self-play and TD learning in Othello and Backgammon. The program Knightcap [9, 10] and Beal et al. [11] also use temporal difference learning on evaluation function features. Arenz [5] applied MCTS to chess. Heinz reported on self-play experiments in chess [56].

Since the AlphaGo results, many other applications of machine learning have been shown to be successful. There is interest in theoretical physics [84, 95], chemistry [58], and pharmacology, specifically for retrosynthetic molecular design [99] and drug design [127]. High-profile results have been achieved by AlphaFold, a program that can predict protein structures [58, 101].

To learn more about curriculum learning, see [13, 45, 69, 133]. Wang et al. [129] study the optimization target of a dual-headed self-play network in AlphaZero-General. The success of self-play has led to interest in curriculum learning in single-agent problems [38, 40, 44, 66]. The relation between classical single-agent and two-agent search is studied by Schaeffer et al. [97].

Exercises

To review your knowledge of self-play, here are some questions and exercises. We start with questions to check your understanding of this chapter. Each question is a closed question where a simple, one sentence answer is possible.

Questions

1. What are the differences between AlphaGo, AlphaGo Zero, and AlphaZero?
2. What is MCTS?
3. What are the four steps of MCTS?
4. What does UCT do?
5. Give the UCT formula. How is P-UCT different?
6. Describe the function of each of the four operations of MCTS.
7. How does UCT achieve trading-off exploration and exploitation? Which inputs does it use?
8. When C_p is small, does MCTS explore more or exploit more?

9. For small numbers of node expansions, would you prefer more exploration or more exploitation?
10. What is a double-headed network? How is it different from regular actor critic?
11. Which three elements make up the self-play loop? (You may draw a picture.)
12. What is tabula rasa learning?
13. How can tabula rasa learning be faster than reinforcement learning on top of supervised learning of grandmaster games?
14. What is curriculum learning?

Implementation: New or Make/Undo

You may have noticed that the minimax and MCTS pseudocode in the figures lacks implementation details for performing actions, to arrive at successor states. Such board manipulation and move making details are important for creating a working program.

Game playing programs typically call the search routine with the current board state, often indicated with parameter n for the new node. This board can be created and allocated anew in each search node, in a value-passing style (local variable). Another option is to pass a reference to the board and to apply a makemove operation on the board, placing the stone on the board before the recursive call, and an undomove operation removing the stone from the board when it returns back out of the recursion (global variable). This reference-passing style may be quicker if allocating the memory for a new board is an expensive operation. It may also be more difficult to implement correctly, since the makemove/undomove protocol must be followed strictly on all relevant places in the code. If capture moves cause many changes to the board, then these must be remembered for the subsequent undo.

For parallel implementations in a shared memory, at least all parallel threads must have their own copy of a value-passing style board. (On a distributed memory cluster, the separate machines will have their own copy of the board by virtue of the distributed memory.)

Exercises

For the programming exercises, we use PolyGames. See the previous section on how to install PolyGames. If training takes a long time, consider using the GPU support of Polygames and Pytorch.

1. *Hex* Install PolyGames and train a Hex player with self-play. Experiment with different board sizes. Keep the training time constant, and draw a graph where you contrast playing strength against board size. Do you see a clear correlation that the program is stronger on smaller boards?

2. *Visualize* Install the visualization support `torchviz`. Visualize the training process using the `draw_model` script.
3. *Hyperparameters* Use different models, and try different hyperparameters, as specified in the PolyGames documentation.
4. *MCTS* Run an evaluation tournament from the trained Hex model against a pure MCTS player. See below for tips on make and undo of moves. How many nodes can MCTS search in a reasonable search time? Compare the MCTS Hex player against the self-play player. How many games do you need to play to have statistically significant results? Is the random seed randomized or fixed? Which is stronger: MCTS or trained Hex?

References

1. Abbas Abdolmaleki, Jost Tobias Springenberg, Yuval Tassa, Remi Munos, Nicolas Heess, and Martin Riedmiller. Maximum a posteriori policy optimisation. *arXiv preprint arXiv:1806.06920*, 2018. 192
2. Bruce Abramson. Expected-outcome: A general model of static evaluation. *IEEE Transactions on Pattern Analysis and Machine Intelligence*, 12(2):182–193, 1990. 185
3. Anonymous. Go AI strength vs. time. *Reddit post*, 2017. 176
4. Thomas Anthony, Zheng Tian, and David Barber. Thinking fast and slow with deep learning and tree search. In *Advances in Neural Information Processing Systems*, pages 5360–5370, 2017. 201
5. Oleg Arenz. Monte Carlo Chess. Master's thesis, Universität Darmstadt, 2012. 210
6. Peter Auer. Using confidence bounds for exploitation-exploration trade-offs. *Journal of Machine Learning Research*, 3(Nov):397–422, 2002. 191
7. Peter Auer, Nicolo Cesa-Bianchi, and Paul Fischer. Finite-time analysis of the multiarmed bandit problem. *Machine Learning*, 47(2–3):235–256, 2002. 192
8. Peter Auer and Ronald Ortner. UCB revisited: Improved regret bounds for the stochastic multi-armed bandit problem. *Periodica Mathematica Hungarica*, 61(1–2):55–65, 2010. 191
9. Jonathan Baxter, Andrew Tridgell, and Lex Weaver. Knightcap: a chess program that learns by combining TD (λ) with game-tree search. *arXiv preprint cs/9901002*, 1999. 210
10. Jonathan Baxter, Andrew Tridgell, and Lex Weaver. Learning to play chess using temporal differences. *Machine Learning*, 40(3):243–263, 2000. 185, 210
11. Don Beal and Martin C. Smith. Temporal difference learning for heuristic search and game playing. *Information Sciences*, 122(1):3–21, 2000. 210
12. Laurens Beljaards. AI agents for the abstract strategy game Tak. Master's thesis, Leiden University, 2017. 186
13. Yoshua Bengio, Jérôme Louradour, Ronan Collobert, and Jason Weston. Curriculum learning. In *Proceedings of the 26th Annual International Conference on Machine Learning*, pages 41–48, 2009. 198, 210
14. Darse Billings, Denis Papp, Jonathan Schaeffer, and Duane Szafron. Opponent modeling in poker. *AAAI/IAAI*, 493:499, 1998. 183
15. Bruno Bouzy and Bernard Helmstetter. Monte Carlo Go developments. In *Advances in Computer Games*, pages 159–174. Springer, 2004. 185, 188
16. Cameron Browne. *Hex Strategy*. AK Peters/CRC Press, 2000. 206

17. Cameron B Browne, Edward Powley, Daniel Whitehouse, Simon M Lucas, Peter I Cowling, Philipp Rohlfshagen, Stephen Tavener, Diego Perez, Spyridon Samothrakis, and Simon Colton. A survey of Monte Carlo Tree Search methods. *IEEE Transactions on Computational Intelligence and AI in Games*, 4(1):1–43, 2012. 186, 187, 188, 188, 189, 190, 190, 192, 192, 192, 192, 209

18. Bernd Brügmann. Monte Carlo Go. Technical report, Syracuse University, 1993. 185, 188

19. Andres Campero, Roberta Raileanu, Heinrich Küttler, Joshua B Tenenbaum, Tim Rocktäschel, and Edward Grefenstette. Learning with AMIGo: Adversarially motivated intrinsic goals. In *International Conference on Learning Representations*, 2020. 198

20. Tristan Cazenave. Residual networks for computer Go. *IEEE Transactions on Games*, 10(1):107–110, 2018. 194

21. Tristan Cazenave, Yen-Chi Chen, Guan-Wei Chen, Shi-Yu Chen, Xian-Dong Chiu, Julien Dehos, Maria Elsa, Qucheng Gong, Hengyuan Hu, Vasil Khalidov, Cheng-Ling Li, Hsin-I Lin, Yu-Jin Lin, Xavier Martinet, Vegard Mella, Jérémy Rapin, Baptiste Rozière, Gabriel Synnaeve, Fabien Teytaud, Olivier Teytaud, Shi-Cheng Ye, Yi-Jun Ye, Shi-Jim Yen, and Sergey Zagoruyko. Polygames: Improved zero learning. *arXiv preprint arXiv:2001.09832*, 2020. 205, 206, 206, 209

22. Tristan Cazenave and Bernard Helmstetter. Combining tactical search and Monte-Carlo in the game of Go. In *Proceedings of the 2005 IEEE Symposium on Computational Intelligence and Games (CIG05), Essex University*, volume 5, pages 171–175, 2005. 192

23. Hyeong Soo Chang, Michael C Fu, Jiaqiao Hu, and Steven I Marcus. An adaptive sampling algorithm for solving Markov decision processes. *Operations Research*, 53(1):126–139, 2005. 192

24. Guillaume Chaslot. *Monte-Carlo tree search*. PhD thesis, Maastricht University, 2010. 188

25. Guillaume Chaslot, Sander Bakkes, Istvan Szita, and Pieter Spronck. Monte-Carlo tree search: A new framework for game AI. In *AIIDE*, 2008. 192, 201

26. Maxime Chevalier-Boisvert, Lucas Willems, and Sumans Pal. Minimalistic gridworld environment for OpenAI Gym https://github.com/maximecb/gym-minigrid, 2018. 199

27. Christopher Clark and Amos Storkey. Teaching deep convolutional neural networks to play Go. arxiv preprint. *arXiv preprint arXiv:1412.3409*, 1, 2014. 210

28. Christopher Clark and Amos Storkey. Training deep convolutional neural networks to play Go. In *International Conference on Machine Learning*, pages 1766–1774, 2015. 201, 210

29. Karl Cobbe, Chris Hesse, Jacob Hilton, and John Schulman. Leveraging procedural generation to benchmark reinforcement learning. In *International Conference on Machine Learning*, pages 2048–2056. PMLR, 2020. 199

30. Rémi Coulom. Efficient selectivity and backup operators in Monte-Carlo Tree Search. In *International Conference on Computers and Games*, pages 72–83. Springer, 2006. 186, 192, 209

31. Rémi Coulom. Monte-Carlo tree search in Crazy Stone. In *Proceedings Game Programming Workshop, Tokyo, Japan*, pages 74–75, 2007. 188

32. Rémi Coulom. The Monte-Carlo revolution in Go. In *The Japanese-French Frontiers of Science Symposium (JFFoS 2008), Roscoff, France*, 2009. 192, 201

33. Joseph C Culberson and Jonathan Schaeffer. Pattern databases. *Computational Intelligence*, 14(3):318–334, 1998. 188

34. Wojciech Marian Czarnecki, Gauthier Gidel, Brendan Tracey, Karl Tuyls, Shayegan Omidshafiei, David Balduzzi, and Max Jaderberg. Real world games look like spinning tops. In *Advances in Neural Information Processing Systems*, 2020. 173

35. Kamil Czarnogórski. Monte Carlo Tree Search beginners guide https://int8.io/monte-carlo-tree-search-beginners-guide/, 2018. 189, 190

36. Shubhomoy Das, Weng-Keen Wong, Thomas Dietterich, Alan Fern, and Andrew Emmott. Incorporating expert feedback into active anomaly discovery. In *2016 IEEE 16th International Conference on Data Mining (ICDM)*, pages 853–858. IEEE, 2016. 199

37. Dave De Jonge, Tim Baarslag, Reyhan Aydoğan, Catholijn Jonker, Katsuhide Fujita, and Takayuki Ito. The challenge of negotiation in the game of diplomacy. In *International Conference on Agreement Technologies*, pages 100–114. Springer, 2018. 173

38. Thang Doan, Joao Monteiro, Isabela Albuquerque, Bogdan Mazoure, Audrey Durand, Joelle Pineau, and R Devon Hjelm. On-line adaptive curriculum learning for GANs. In *Proceedings of the AAAI Conference on Artificial Intelligence*, volume 33, pages 3470–3477, 2019. 199, 210

39. Christian Donninger. Null move and deep search. *ICGA Journal*, 16(3):137–143, 1993. 185

40. Yan Duan, John Schulman, Xi Chen, Peter L Bartlett, Ilya Sutskever, and Pieter Abbeel. RL^2: Fast reinforcement learning via slow reinforcement learning. *arXiv preprint arXiv:1611.02779*, 2016. 210

41. Jeffrey L Elman. Learning and development in neural networks: The importance of starting small. *Cognition*, 48(1):71–99, 1993. 198

42. Arpad E Elo. *The Rating of Chessplayers, Past and Present*. Arco Pub., 1978. 202

43. Markus Enzenberger, Martin Muller, Broderick Arneson, and Richard Segal. Fuego—an open-source framework for board games and Go engine based on Monte Carlo tree search. *IEEE Transactions on Computational Intelligence and AI in Games*, 2(4):259–270, 2010. 201, 210

44. Dieqiao Feng, Carla P Gomes, and Bart Selman. Solving hard AI planning instances using curriculum-driven deep reinforcement learning. *arXiv preprint arXiv:2006.02689*, 2020. 199, 199, 210

45. Carlos Florensa, David Held, Xinyang Geng, and Pieter Abbeel. Automatic goal generation for reinforcement learning agents. In *International Conference on Machine Learning*, pages 1515–1528. PMLR, 2018. 199, 210

46. David B Fogel, Timothy J Hays, Sarah L Hahn, and James Quon. Further evolution of a self-learning chess program. In *Computational Intelligence in Games*, 2005. 185

47. Sam Ganzfried and Tuomas Sandholm. Game theory-based opponent modeling in large imperfect-information games. In *The 10th International Conference on Autonomous Agents and Multiagent Systems*, volume 2, pages 533–540, 2011. 183

48. Sylvain Gelly, Levente Kocsis, Marc Schoenauer, Michele Sebag, David Silver, Csaba Szepesvári, and Olivier Teytaud. The grand challenge of computer Go: Monte Carlo tree search and extensions. *Communications of the ACM*, 55(3):106–113, 2012. 201

49. Sylvain Gelly and David Silver. Achieving master level play in 9×9 computer Go. In *AAAI*, volume 8, pages 1537–1540, 2008. 201, 210

50. Sylvain Gelly, Yizao Wang, and Olivier Teytaud. Modification of UCT with patterns in Monte-Carlo Go. Technical Report RR-6062, INRIA, 2006. 188, 201, 210

51. Tobias Graf and Marco Platzner. Adaptive playouts in Monte-Carlo tree search with policy-gradient reinforcement learning. In *Advances in Computer Games*, pages 1–11. Springer, 2015. 210

52. Jean-Bastien Grill, Florent Altché, Yunhao Tang, Thomas Hubert, Michal Valko, Ioannis Antonoglou, and Rémi Munos. Monte-Carlo tree search as regularized policy optimization. In *International Conference on Machine Learning*, pages 3769–3778. PMLR, 2020. 192

53. Ryan B Hayward and Bjarne Toft. *Hex: The Full Story*. CRC Press, 2019. 206

54. He He, Jordan Boyd-Graber, Kevin Kwok, and Hal Daumé III. Opponent modeling in deep reinforcement learning. In *International Conference on Machine Learning*, pages 1804–1813. PMLR, 2016. 183

55. Kaiming He, Xiangyu Zhang, Shaoqing Ren, and Jian Sun. Deep residual learning for image recognition. In *Proceedings of the IEEE Conference on Computer Vision and Pattern Recognition*, pages 770–778, 2016. 194

56. Ernst A Heinz. New self-play results in computer chess. In *International Conference on Computers and Games*, pages 262–276. Springer, 2000. 210

57. Athul Paul Jacob, David J Wu, Gabriele Farina, Adam Lerer, Anton Bakhtin, Jacob Andreas, and Noam Brown. Modeling strong and human-like gameplay with KL-regularized search. *arXiv preprint arXiv:2112.07544*, 2021. 192

58. John Jumper, Richard Evans, Alexander Pritzel, Tim Green, Michael Figurnov, Olaf Ronneberger, Kathryn Tunyasuvunakool, Russ Bates, Augustin Žídek, Anna Potapenko, Alex Bridgland, Clemens Meyer, Simon A. A. Kohl, Andrew J. Ballard, Andrew Cowie, Bernardino Romera-Paredes, Stanislav Nikolov, Rishub Jain, Jonas Adler, Trevor Back, Stig Petersen, David Reiman, Ellen Clancy, Michal Zielinski, Martin Steinegger, Michalina Pacholska, Tamas Berghammer, Sebastian Bodenstein, David Silver, Oriol Vinyals, Andrew W. Senior, Koray Kavukcuoglu, Pushmeet Kohli, and Demis Hassabis. Highly accurate protein structure prediction with AlphaFold. *Nature*, 596(7873):583–589, 2021. 210, 210

59. Niels Justesen, Ruben Rodriguez Torrado, Philip Bontrager, Ahmed Khalifa, Julian Togelius, and Sebastian Risi. Illuminating generalization in deep reinforcement learning through procedural level generation. *arXiv preprint arXiv:1806.10729*, 2018. 199

60. Donald E Knuth and Ronald W Moore. An analysis of alpha-beta pruning. *Artificial Intelligence*, 6(4):293–326, 1975. 173, 184, 185

61. Levente Kocsis and Csaba Szepesvári. Bandit based Monte-Carlo planning. In *European Conference on Machine Learning*, pages 282–293. Springer, 2006. 186, 188, 190, 209

62. Richard E Korf. Depth-first iterative-deepening: An optimal admissible tree search. *Artificial intelligence*, 27(1):97–109, 1985. 185

63. Sarit Kraus, Eithan Ephrati, and Daniel Lehmann. Negotiation in a non-cooperative environment. *Journal of Experimental & Theoretical Artificial Intelligence*, 3(4):255–281, 1994. 173

64. Kai A Krueger and Peter Dayan. Flexible shaping: How learning in small steps helps. *Cognition*, 110(3):380–394, 2009. 198

65. Jan Kuipers, Aske Plaat, Jos AM Vermaseren, and H Jaap van den Herik. Improving multivariate Horner schemes with Monte Carlo tree search. *Computer Physics Communications*, 184(11):2391–2395, 2013. 192, 192

66. Alexandre Laterre, Yunguan Fu, Mohamed Khalil Jabri, Alain-Sam Cohen, David Kas, Karl Hajjar, Torbjorn S Dahl, Amine Kerkeni, and Karim Beguir. Ranked reward: Enabling self-play reinforcement learning for combinatorial optimization. *arXiv preprint arXiv:1807.01672*, 2018. 199, 199, 210

67. Joel Z Leibo, Edward Hughes, Marc Lanctot, and Thore Graepel. Autocurricula and the emergence of innovation from social interaction: A manifesto for multi-agent intelligence research. *arXiv preprint arXiv:1903.00742*, 2019. 209

68. Diego Pérez Liébana, Simon M Lucas, Raluca D Gaina, Julian Togelius, Ahmed Khalifa, and Jialin Liu. General video game artificial intelligence. *Synthesis Lectures on Games and Computational Intelligence*, 3(2):1–191, 2019. 199

69. Tambet Matiisen, Avital Oliver, Taco Cohen, and John Schulman. Teacher-student curriculum learning. *IEEE Trans. Neural Networks Learn. Syst.*, 31(9):3732–3740, 2020. 210

70. Kiminori Matsuzaki. Empirical analysis of PUCT algorithm with evaluation functions of different quality. In *2018 Conference on Technologies and Applications of Artificial Intelligence (TAAI)*, pages 142–147. IEEE, 2018. 191, 191

71. Jonathan K Millen. Programming the game of Go. *Byte Magazine*, 1981. 175

72. S Ali Mirsoleimani, Aske Plaat, Jaap Van Den Herik, and Jos Vermaseren. Scaling Monte Carlo tree search on Intel Xeon Phi. In *Parallel and Distributed Systems (ICPADS), 2015 IEEE 21st International Conference on*, pages 666–673. IEEE, 2015. 209

73. Tom M Mitchell. The need for biases in learning generalizations. Technical Report CBM-TR-117, Department of Computer Science, Rutgers University, 1980. 199

74. Tom M Mitchell. The discipline of machine learning. Technical Report CMU-ML-06-108, Carnegie Mellon University, School of Computer Science, Machine Learning, 2006. 199

75. Volodymyr Mnih, Koray Kavukcuoglu, David Silver, Alex Graves, Ioannis Antonoglou, Daan Wierstra, and Martin Riedmiller. Playing Atari with deep reinforcement learning. *arXiv preprint arXiv:1312.5602*, 2013. 172

76. Volodymyr Mnih, Koray Kavukcuoglu, David Silver, Andrei A. Rusu, Joel Veness, Marc G. Bellemare, Alex Graves, Martin A. Riedmiller, Andreas Fidjeland, Georg Ostrovski, Stig Petersen, Charles Beattie, Amir Sadik, Ioannis Antonoglou, Helen King, Dharshan Kumaran,

Daan Wierstra, Shane Legg, and Demis Hassabis. Human-level control through deep reinforcement learning. *Nature*, 518(7540):529–533, 2015. 201

77. Thomas M Moerland, Joost Broekens, Aske Plaat, and Catholijn M Jonker. A0C: Alpha zero in continuous action space. *arXiv preprint arXiv:1805.09613*, 2018. 191, 191, 209

78. Thomas M Moerland, Joost Broekens, Aske Plaat, and Catholijn M Jonker. Monte Carlo tree search for asymmetric trees. *arXiv preprint arXiv:1805.09218*, 2018. 209

79. Matthias Müller-Brockhausen, Mike Preuss, and Aske Plaat. Procedural content generation: Better benchmarks for transfer reinforcement learning. In *Conference on Games*, 2021. 198

80. Sanmit Narvekar, Bei Peng, Matteo Leonetti, Jivko Sinapov, Matthew E Taylor, and Peter Stone. Curriculum learning for reinforcement learning domains: A framework and survey. *Journal Machine Learning Research*, 2020. 199, 199

81. Yu Nasu. Efficiently updatable neural-network-based evaluation functions for computer shogi. *The 28th World Computer Shogi Championship Appeal Document*, 2018. 185

82. Frans A Oliehoek and Christopher Amato. *A Concise Introduction to Decentralized POMDPs*. Springer, 2016. 173

83. Pierre-Yves Oudeyer, Frederic Kaplan, and Verena V Hafner. Intrinsic motivation systems for autonomous mental development. *IEEE Transactions on Evolutionary Computation*, 11(2):265–286, 2007. 198

84. Giuseppe Davide Paparo, Vedran Dunjko, Adi Makmal, Miguel Angel Martin-Delgado, and Hans J Briegel. Quantum speedup for active learning agents. *Physical Review X*, 4(3):031002, 2014. 210

85. Gian-Carlo Pascutto. Leela zero. https://github.com/leela-zero/leela-zero, 2017. 205

86. Judea Pearl. *Heuristics: Intelligent Search Strategies for Computer Problem Solving*. Addison-Wesley, Reading, MA, 1984. 184

87. Aske Plaat. *Learning to Play: Reinforcement Learning and Games*. Springer Verlag, Heidelberg, https://learningtoplay.net, 2020. 183, 184, 185, 209

88. Aske Plaat, Jonathan Schaeffer, Wim Pijls, and Arie De Bruin. Best-first fixed-depth minimax algorithms. *Artificial Intelligence*, 87(1-2):255–293, 1996. 185

89. Aditya Prasad. Lessons from implementing AlphaZero https://medium.com/oracledevs/lessons-from-implementing-alphazero-7e36e9054191, 2018. 195

90. Max Pumperla and Kevin Ferguson. *Deep Learning and the Game of Go*. Manning, 2019. 209

91. J Ross Quinlan. Learning efficient classification procedures and their application to chess end games. In *Machine Learning*, pages 463–482. Springer, 1983. 185

92. Roberta Raileanu and Tim Rocktäschel. RIDE: rewarding impact-driven exploration for procedurally-generated environments. In *International Conference on Learning Representations*, 2020. 199

93. Christopher D Rosin. Multi-armed bandits with episode context. *Annals of Mathematics and Artificial Intelligence*, 61(3):203–230, 2011. 191, 191

94. Neil Rubens, Mehdi Elahi, Masashi Sugiyama, and Dain Kaplan. Active learning in recommender systems. In *Recommender Systems Handbook*, pages 809–846. Springer, 2015. 199

95. Ben Ruijl, Jos Vermaseren, Aske Plaat, and Jaap van den Herik. HEPGAME and the simplification of expressions. *arXiv preprint arXiv:1405.6369*, 2014. 210

96. Steve Schaefer. Mathematical recreations. http://www.mathrec.org/old/2002jan/solutions.html, 2002. 174

97. Jonathan Schaeffer, Aske Plaat, and Andreas Junghanns. Unifying single-agent and two-player search. *Information Sciences*, 135(3-4):151–175, 2001. 210

98. Jürgen Schmidhuber. Curious model-building control systems. In *Proceedings International Joint Conference on Neural Networks*, pages 1458–1463, 1991. 197

99. Marwin HS Segler, Mike Preuss, and Mark P Waller. Planning chemical syntheses with deep neural networks and symbolic AI. *Nature*, 555(7698):604, 2018. 210

100. Oliver G Selfridge, Richard S Sutton, and Andrew G Barto. Training and tracking in robotics. In *International Joint Conference on Artificial Intelligence*, pages 670–672, 1985. 197

101. Andrew W. Senior, Richard Evans, John Jumper, James Kirkpatrick, Laurent Sifre, Tim Green, Chongli Qin, Augustin Zídek, Alexander W. R. Nelson, Alex Bridgland, Hugo Penedones, Stig Petersen, Karen Simonyan, Steve Crossan, Pushmeet Kohli, David T. Jones, David Silver, Koray Kavukcuoglu, and Demis Hassabis. Improved protein structure prediction using potentials from deep learning. *Nature*, 577(7792):706–710, 2020. 210

102. Burr Settles. Active learning literature survey. Technical report, University of Wisconsin-Madison Department of Computer Sciences, 2009. 199

103. Noor Shaker, Julian Togelius, and Mark J Nelson. *Procedural Content Generation in Games*. Springer, 2016. 198

104. Guy Shani, Joelle Pineau, and Robert Kaplow. A survey of point-based POMDP solvers. *Autonomous Agents and Multi-Agent Systems*, 27(1):1–51, 2013. 173

105. Claude E Shannon. Programming a computer for playing chess. In *Computer Chess Compendium*, pages 2–13. Springer, 1988. 174

106. David Silver. *Reinforcement learning and simulation based search in the game of Go*. PhD thesis, University of Alberta, 2009. 188

107. David Silver, Aja Huang, Chris J. Maddison, Arthur Guez, Laurent Sifre, George van den Driessche, Julian Schrittwieser, Ioannis Antonoglou, Veda Panneershelvam, Marc Lanctot, Sander Dieleman, Dominik Grewe, John Nham, Nal Kalchbrenner, Ilya Sutskever, Timothy Lillicrap, Madeleine Leach, Koray Kavukcuoglu, Thore Graepel, and Demis Hassabis. Mastering the game of Go with deep neural networks and tree search. *Nature*, 529(7587):484, 2016. 172, 194, 196, 200, 201, 201, 204, 209

108. David Silver, Thomas Hubert, Julian Schrittwieser, Ioannis Antonoglou, Matthew Lai, Arthur Guez, Marc Lanctot, Laurent Sifre, Dharshan Kumaran, Thore Graepel, Timothy Lillicrap, Karen Simonyan, and Demis Hassabis. A general reinforcement learning algorithm that masters chess, shogi, and Go through self-play. *Science*, 362(6419):1140–1144, 2018. 200, 203, 204, 205

109. David Silver, Julian Schrittwieser, Karen Simonyan, Ioannis Antonoglou, Aja Huang, Arthur Guez, Thomas Hubert, Lucas Baker, Matthew Lai, Adrian Bolton, Yutian Chen, Timothy Lillicrap, Fan Hui, Laurent Sifre, George van den Driessche, Thore Graepel, and Demis Hassabis. Mastering the game of Go without human knowledge. *Nature*, 550(7676):354, 2017. 172, 191, 193, 194, 194, 195, 195, 196, 200, 200, 201, 202, 202, 203, 204, 205, 209

110. David Silver, Richard S Sutton, and Martin Müller. Reinforcement learning of local shape in the game of Go. In *International Joint Conference on Artificial Intelligence*, volume 7, pages 1053–1058, 2007. 210

111. David J Slate and Lawrence R Atkin. Chess 4.5—the northwestern university chess program. In *Chess skill in Man and Machine*, pages 82–118. Springer, 1983. 185

112. Gillian Smith. An analog history of procedural content generation. In *Foundations of Digital Games*, 2015. 198

113. Darin Straus. Alphazero implementation and tutorial. https://towardsdatascience.com/alphazero-implementation-and-tutorial-f4324d65fdfc, 2018. 195

114. Richard S Sutton and Andrew G Barto. *Reinforcement learning, An Introduction, Second Edition*. MIT Press, 2018. 191

115. Gerald Tesauro. Neurogammon wins Computer Olympiad. *Neural Computation*, 1(3):321–323, 1989. 171

116. Gerald Tesauro. TD-gammon: A self-teaching backgammon program. In *Applications of Neural Networks*, pages 267–285. Springer, 1995. 200

117. Gerald Tesauro. Temporal difference learning and TD-Gammon. *Communications of the ACM*, 38(3):58–68, 1995. 172

118. Gerald Tesauro. Programming backgammon using self-teaching neural nets. *Artificial Intelligence*, 134(1-2):181–199, 2002. 172

119. Shantanu Thakoor, Surag Nair, and Megha Jhunjhunwala. Learning to play Othello without human knowledge. Stanford University CS238 Final Project Report, 2017. 195, 205, 205, 205

120. Sebastian Thrun. Learning to play the game of chess. In *Advances in Neural Information Processing Systems*, pages 1069–1076, 1995. 185
121. Yuandong Tian, Qucheng Gong, Wenling Shang, Yuxin Wu, and C Lawrence Zitnick. ELF: An extensive, lightweight and flexible research platform for real-time strategy games. In *Advances in Neural Information Processing Systems*, pages 2659–2669, 2017. 195, 205, 205
122. Yuandong Tian, Jerry Ma, Qucheng Gong, Shubho Sengupta, Zhuoyuan Chen, and C. Lawrence Zitnick. ELF OpenGo. https://github.com/pytorch/ELF, 2018. 206
123. Yuandong Tian and Yan Zhu. Better computer Go player with neural network and long-term prediction. In *International Conference on Learning Representations*, 2016. 206
124. Julian Togelius, Alex J Champandard, Pier Luca Lanzi, Michael Mateas, Ana Paiva, Mike Preuss, and Kenneth O Stanley. Procedural content generation: Goals, challenges and actionable steps. In *Artificial and Computational Intelligence in Games*. Schloss Dagstuhl-Leibniz-Zentrum für Informatik, 2013. 198
125. Alan M Turing. *Digital Computers Applied to Games*. Pitman & Sons, 1953. 174, 183, 183, 185
126. Michiel Van Der Ree and Marco Wiering. Reinforcement learning in the game of Othello: learning against a fixed opponent and learning from self-play. In *IEEE Adaptive Dynamic Programming and Reinforcement Learning*, pages 108–115. IEEE, 2013. 210
127. Gerard JP Van Westen, Jörg K Wegner, Peggy Geluykens, Leen Kwanten, Inge Vereycken, Anik Peeters, Adriaan P IJzerman, Herman WT van Vlijmen, and Andreas Bender. Which compound to select in lead optimization? Prospectively validated proteochemometric models guide preclinical development. *PloS One*, 6(11):e27518, 2011. 210
128. Jos AM Vermaseren. New features of form. *arXiv preprint math-ph/0010025*, 2000. 191
129. Hui Wang, Michael Emmerich, Mike Preuss, and Aske Plaat. Alternative loss functions in AlphaZero-like self-play. In *2019 IEEE Symposium Series on Computational Intelligence (SSCI)*, pages 155–162, 2019. 194, 210
130. Hui Wang, Mike Preuss, Michael Emmerich, and Aske Plaat. Tackling Morpion Solitaire with AlphaZero-like Ranked Reward reinforcement learning. In *22nd International Symposium on Symbolic and Numeric Algorithms for Scientific Computing, SYNASC 2020, Timisoara, Romania*, 2020. 199
131. Panqu Wang and Garrison W Cottrell. Basic level categorization facilitates visual object recognition. *arXiv preprint arXiv:1511.04103*, 2015. 199
132. Daphna Weinshall, Gad Cohen, and Dan Amir. Curriculum learning by transfer learning: Theory and experiments with deep networks. In *International Conference on Machine Learning*, pages 5235–5243, 2018. 198, 198, 198
133. Lilian Weng. Curriculum for reinforcement learning https://lilianweng.github.io/lil-log/2020/01/29/curriculum-for-reinforcement-learning.html. Lil'Log, January 2020. 199, 210
134. Marco A Wiering. Self-play and using an expert to learn to play backgammon with temporal difference learning. *JILSA*, 2(2):57–68, 2010. 210
135. Qinsong Zeng, Jianchang Zhang, Zhanpeng Zeng, Yongsheng Li, Ming Chen, and Sifan Liu. PhoenixGo. https://github.com/Tencent/PhoenixGo, 2018. 195, 205, 206

Chapter 7
Multi-Agent Reinforcement Learning

On this planet, in our societies, millions of people live and work together. Each individual has their own individual set of goals and performs their actions accordingly. Some of these goals are shared. When we want to achieve shared goals, we organize ourselves in teams, groups, companies, organizations, and societies. In many intelligent species—humans, mammals, birds, insects—impressive displays of collective intelligence emerge from such organization [75, 135, 177]. We are learning as individuals, and we are learning as groups. This setting is studied in multi-agent learning: through their independent actions, agents lcarn to interact and to compete and cooperate with other agents and form groups.

Most research in reinforcement learning has focused on single-agent problems. Progress has been made in many topics, such as path finding, robot locomotion, and video games. In addition, research has been performed in two-agent problems, such as competitive two-person board games. Both single-agent and two-agent problems are questions of optimization. The goal is to find the policy with the highest reward, the shortest path, and the best moves and countermoves. The basic setting is one of reward optimization in the face of natural adversity or competitors, modeled by the environment.

As we move closer toward modeling real-world problems, we encounter another category of sequential decision making problems, and that is the category of multi-agent decision making problems. Multi-agent decision making is a difficult problem.

Agents that share the same goal might collaborate, and finding a policy with the highest reward for oneself or for the group may include achieving win/win solutions with other agents. Coalition forming and collusion are an integral part of the field of multi-agent reinforcement learning.

In real-world decision making, both competition and cooperation are important. If we want our agents to behave realistically in settings with multiple agents, then they should understand cooperation in order to perform well.

From a computational perspective, studying the emergence of group behavior and collective intelligence is challenging: the environment for the agents consists of

A. Plaat, *Deep Reinforcement Learning*,
https://doi.org/10.1007/978-981-19-0638-1_7

many other agents; goals may move, many interactions have to be modeled in order to be understood, and the world to be optimized against is constantly changing. A lack of compute power has held back experimental research in multi-agent reinforcement learning for some time. The recent growth in compute power and advances in deep reinforcement learning methods are making progress increasingly possible.

Multi-agent decision making problems that have been studied experimentally include the games of bridge, poker, Diplomacy, StarCraft, Hide and Seek, and Capture the Flag. To a large extent, the algorithms for multi-agent reinforcement learning are still being developed. This chapter is full of challenges and development.

We will start by reviewing the theoretical framework and defining multi-agent problems. We will look deeper at cooperation and competition and introduce stochastic games, extensive-form games, and the Nash equilibrium and Pareto optimality. We will discuss population-based methods and curriculum learning in multiplayer teams. Next, we will discuss environments on which multi-agent reinforcement learning methods can be tested, such as poker and StarCraft. In these games, often team structure plays an important role. The next chapter discusses hierarchical methods, which can be used to model team structures.

This chapter is concluded, as usual, with exercises, a summary, and pointers to further reading.

Core Concepts

- Competition
- Cooperation
- Team learning

Core Problem

- Find efficient methods for large competitive and cooperative multi-agent problems

Core Algorithms

- Counterfactual regret minimization (Sect. 7.2.1)
- Population-based algorithms (Sect. 7.2.3.1)

Self-driving Car

To illustrate the key components of multi-agent decision making, let us assume a self-driving car is on its way to the supermarket and is approaching an intersection. What action should it take? What is the best outcome for the car? For other agents? For society?

The goal of the car is to exit the intersection safely and reach the destination. Possible decisions are going straight or turning left or right into another lane. All these actions consist of sub-decisions. At each time step, the car can move by steering, accelerating, and braking. The car must be able to detect objects, such as traffic lights, lane markings, and other cars. Furthermore, we aim to find a policy that can control the car to make a sequence of maneuvers to achieve the goal. In a decision making setting such as this, two additional challenges arise.

The first challenge is that we should be able to anticipate the actions of other agents. During the decision making process, at each time step, the robot car should not only consider the immediate value of its own current action but also adapt to consequences of actions by other agents. (For example, it would not be good to choose a certain direction that appears safe early on, stick to it, and not adapt the policy as new information comes in, such as a car heading our way.)

The second challenge is that these other agents, in turn, may anticipate the actions of our agent in choosing their actions. Our agent needs to take their anticipatory behavior into account in its own policy—recursion at the policy level. Human drivers, for example, often predict likely movements of other cars and then take action in response (such as giving way or accelerating while merging into another lane).[1]

Multi-agent reinforcement learning addresses the sequential decision making problem of multiple autonomous agents that operate in a common stochastic environment, each of which aims to maximize their own long-term reward by interacting with the environment and with other agents. Multi-agent reinforcement learning combines the fields of multi-agent systems [137, 176] and reinforcement learning.

7.1 Multi-Agent Problems

We have seen impressive results of single-agent reinforcement learning in recent years. In addition, important results have been achieved in games such as Go and poker and in autonomous driving. These application domains involve the

---

[1] Human drivers have a *theory of mind* of other drivers. Theory of mind and the related concept of mirror neurons [56] are a psychological theory of empathy and understanding, which allows a limited amount of prediction of future behavior. Theory of mind studies how individuals simulate in their minds the actions of others, including their simulation of our actions (and of our simulations, etc.) [18, 71].

participation of more than one agent. The study of interactions between more than one agent has a long history in fields such as economics and social choice.

Multi-agent problems introduce many new kinds of possibilities, such as cooperation and simultaneous actions. These phenomena invalidate some of the assumptions of the theory behind single-agent reinforcement learning, and new theoretical concepts have been developed.

To begin our study of multi-agent problems, we will start with game theory.

Game Theory

Game theory is the study of strategic interaction among rational decision making agents. Game theory originally addressed only the theory of two-person zero-sum games and was later extended to cooperative games of multiple rational players, such as simultaneous action, non-zero sum, and imperfect information games. A classic work in game theory is John von Neumann and Oskar Morgenstern's *Theory of Games and Economic Behavior* [167, 168], published first in 1944. This book has laid the foundation for the mathematical study of economic behavior and social choice. Game theory has been instrumental in developing the theory of common goods and the role of government in a society of independent self-interested rational agents. In mathematics and artificial intelligence, game theory has given a formal basis for the computation of strategies in multi-agent systems.

The Markov decision process that we have used to formalize single-agent reinforcement learning assumes perfect information. Most multi-agent reinforcement learning problems, however, are imperfect information problems, where some of the information is private, or moves are simultaneous. We will have to extend our MDP model appropriately to be able to model imperfect information.

Stochastic Games and Extensive-Form Games

One direct generalization of MDP that captures the interaction of multiple agents is the *Markov game*, also known as the stochastic game [136]. Described by Littman [97], the framework of Markov games has long been used to express multi-agent reinforcement learning algorithms [92, 157].

This multi-agent version of an MDP is defined as follows [180]. At time t, each agent $i \in N$ executes an action a_t^i, for the state s_t of the system. The system then transitions to state s_{t+1} and rewards each agent i with reward $R_{a_t}^i(s_t, s_{t+1})$. The goal of agent i is to optimize its own long-term reward, by finding the policy $\pi^i : S \to \mathbb{E}(A^i)$ as a mapping from the state space to a distribution over the action space, such that $a_t^i \sim \pi^i(\cdot|s_t)$. Then, the value function $V^i : S \to R$ of agent i becomes a function of the joint policy $\pi : S \to \mathbb{E}(A)$ defined as $\pi(a|s) = \Pi_{i \in N}\pi^i(a^i|s)$.

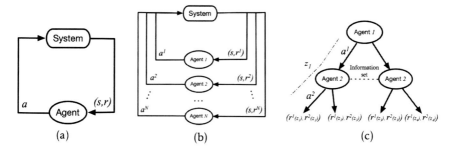

Fig. 7.1 Multi-agent models [180]. (**a**) MDP. (**b**) Markov game. (**c**) Extensive-form game

To visualize imperfect information in multiple agents, new diagrams are needed. Figure 7.1 shows three schematic diagrams. First, in (a), the familiar agent/environment diagram is shown that we used in earlier chapters. Next, in (b), the multi-agent version of this diagram for Markov games is shown. Finally, in (c), an extensive-form game tree is shown [180]. Extensive-form game trees are introduced to model imperfect information. Choices by the agent are shown as solid links, and the hidden private information of the other agent is shown dashed, as the information set of possible situations.

In (a), the agent observes the state s, performs action a, and receives reward r from the environment. In (b), in the Markov game, all agents choose their actions a^i simultaneously and receive their individual reward r^i. In (c), in a two-player extensive-form game, the agents make decisions on choosing actions a^i. They receive their individual reward $r^i(z)$ at the end of the game, where z is the score of a terminal node for the branch that resulted from the information set. The information set is indicated with a dotted line, to signal stochastic behavior, and the environment (or the other agent) chooses an unknown outcome among the dotted actions. The extensive-form notation is designed for expressing imperfect information games, where all possible (unknown) outcomes are represented in the information set. In order to calculate the value function, the agent has to regard all possible different choices in the information sets; the unknown choices of the opponents create a large state space.

Competitive, Cooperative, and Mixed Strategies

Multi-agent reinforcement learning problems fall into three groups: problems with competitive behavior, with cooperative behavior, and with mixed behavior. In the competitive setting, the reward of the agents sums up to zero. A win for one agent is a loss for another. In the cooperative setting, agents collaborate to optimize a common long-term reward. A win for one agent is a win for all agents (example: an embankment being built around an area prone to flooding; the embankment benefits all agents in the area). The mixed setting involves both cooperative and competitive agents, with so-called *general-sum* rewards; some action may lead to win/win and some other to win/loss.

These three behaviors are a useful guide to navigate the landscape of multi-agent algorithms, and we will do so in this chapter. Let us have a closer look at each type of behavior.

7.1.1 Competitive Behavior

One of the hallmarks of the field of game theory is a result by John Nash, who defined the conditions for a stable (and in a certain sense optimal) solution among multiple rational non-cooperative agents. The Nash equilibrium is defined as a situation in which no agent has anything to gain by changing its own strategy. For two competitive agents, the Nash equilibrium is the minimax strategy.

In single-agent reinforcement learning, the goal is to find the policy that maximizes the cumulative future reward of the agent. In multi-agent reinforcement learning, the goal is to find a combined policy of all agents that simultaneously achieves that goal: the multi-policy that for each agent maximizes their cumulative future reward. If you have a set of competitive (near-)zero exploitability strategies, this is called a Nash equilibrium.

The Nash equilibrium characterizes an equilibrium point $\pi^\star$, from which none of the agents has an incentive to deviate. In other words, for any agent $i \in N$, the policy $\pi^{i,\star}$ is the best response to $\pi^{-i,\star}$, where $-i$ are all agents except i [71].

An agent that follows a Nash strategy is guaranteed to do no worse than tie, against any other opponent strategy. For games of imperfect information or chance, such as many card games, this is an *expected* outcome. Since the cards are randomly dealt, there is no theoretical guarantee that a Nash strategy will win or draw every single hand, although on average, it cannot do worse than tie against the other agents.

If the opponents also play a Nash strategy, then all will tie. If the opponents make mistakes, however, then they can lose some hands, allowing the Nash equilibrium strategy to win. Such a mistake by the opponents would be a deviation from the Nash strategy, following a hunch or other non-rational reason, despite having no theoretical incentive to do so. A Nash equilibrium plays perfect defense. It does not try to exploit the opponent strategy's flaws and instead just wins when the opponent makes mistakes [26, 91].

The Nash strategy gives the best possible outcome we can achieve against our adversaries when they work against us. In the sense that a Nash strategy is on average unbeatable, it is considered to be an optimal strategy, and solving a game is equivalent to computing a Nash equilibrium.

In a few moments, we will introduce a method to calculate Nash strategies, called counterfactual regret minimization, but we will first look at cooperation.

7.1.2 Cooperative Behavior

In single-agent reinforcement learning, reward functions return scalar values: an action results in a win or a loss or a single numeric value. In multi-agent reinforcement learning, the reward functions may still return a scalar value, but the functions may be different for each agent; the overall reward function is a vector of the individual rewards. In fully cooperative stochastic games, the individual rewards are the same, and all agents have the same goal: to maximize the common return.

When choices in our problem are made in a decentralized manner by a set of individual decision makers, the problem can be modeled naturally as a decentralized partially observable Markov decision process (dec-POMDP) [117]. Large dec-POMDPs are hard to solve; in general, dec-POMDPs are known to be NEXP-complete. These problems are not solvable with polynomial-time algorithms and searching directly for an optimal solution in the policy space is intractable [21].

A central concept in cooperative problems is Pareto efficiency, after the work of Vilfredo Pareto, who studied economic efficiency and income distributions. The Pareto efficient solution is the situation where no cooperative agent can be better off without making at least one other agent worse off. All non-Pareto efficient combinations are dominated by the Pareto efficient solution. The Pareto frontier is the set of all Pareto efficient combinations, usually drawn as a curve. Figure 7.2 shows a situation where we can choose to produce different quantities of two separate goods, item 1 and item 2 (image by Wikimedia). The gray items close to the origin are choices for which better choices exist. The red symbols all represent production combinations that are deemed more favorable by consumers [139].

Fig. 7.2 Pareto frontier of production possibilities

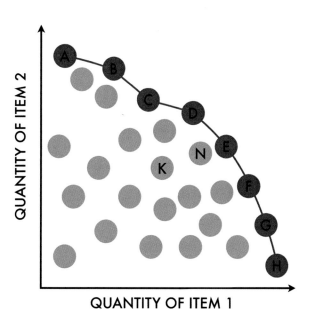

Table 7.1 Prisoner's
dilemma

	Confess	Silent
	Defect	Cooperate
Confess	$(-5, -5)$	$(0, -10)$
Defect	*Nash*	
Silent	$(-10, 0)$	$(-2, -2)$
Cooperate		*Pareto*

In multi-objective reinforcement learning, different agents can have different preferences. A policy is Pareto optimal when deviating from the policy will make at least one agent worse off. We will compare Nash and Pareto soon by looking at how they are related in the prisoner's dilemma (Table 7.1).

The Pareto optimum is the best possible outcome for us where we do not hurt others, and others do not hurt us. It is a cooperative strategy. Pareto equilibria assume communication and trust. In some sense, it is the opposite of the non-cooperative Nash strategy. Pareto calculates the situation for a cooperative world and Nash for a competitive world.

7.1.2.1   Multi-Objective Reinforcement Learning

Many real-world problems involve the optimization of multiple, possibly conflicting objectives. In a fully cooperative setting, all agents share such a common reward function. When the reward function consists of different values for each agent, the problem is said to be multi-objective: different agents may have different preferences or objectives. Note that in principle single-agent problems can also have multiple objectives, although in this book all single-agent problems have had single, scalar rewards. Multi-objective problems become especially relevant in multi-agent problems, where different agents may have different reward preferences.

In decentralized decision processes, the agents must communicate their preferences to each other. Such heterogeneity necessitates the incorporation of communication protocols into multi-agent reinforcement learning, and the analysis of communication-efficient algorithms [67, 128].

Multi-objective reinforcement learning [63, 72, 124, 127] is a generalization of standard reinforcement learning where the scalar reward signal is extended to multiple feedback signals, such as one for each agent. A multi-objective reinforcement learning algorithm optimizes multiple objectives at the same time, and algorithms such as Pareto Q-learning have been introduced for this purpose [126, 164].

7.1.3   *Mixed Behavior*

Starting as a theory for non-cooperative games, game theory has progressed to include cooperative games, where it analyzes optimal strategies for groups of individuals, assuming that they can communicate or enforce agreements between them about proper strategies. The prisoner's dilemma is a well-known example of this type of problem [89].[2] The prisoner's dilemma is thought to be invented by John Nash, although in a more neutral (non-cops and robbers) setting.

The prisoner's dilemma is as follows (see Table 7.1). Two burglars, called Row and Column, have been arrested for breaking into a bank. Row and Column understand that they cannot trust the other person, and hence they operate in a non-cooperative setting. The police officer is offering the following options: *if you both confess, then you will both get a light sentence of 5 years. If you both keep quiet, then I will only have evidence to get you a short sentence of 2 years. However, if one of you confesses, and the other stays silent, then the confessor will walk free, and the one keeping quiet goes to prison for 10 years. Please tell me your choice tomorrow morning.*

This leaves our two criminals with a tough, but clear, choice. If the other stays silent, and I confess, then I walk free; if I also stay silent, then we get 2 years in prison, so confessing is better for me in this case. If the other confesses, and I also confess, then we both get 5 years; if I then stay silent, then I get 10 years in prison, so confessing is again better for me. Whatever the other chooses, confessing gives the lighter sentence, and since they cannot coordinate their action, they will each independently confess. Both will get 5 years in prison, even though both would have only gotten 2 years if both would have stayed silent. The police is happy, since both confess and the case is solved.

If they would have been able to communicate, or if they would have trusted each other, then both would have stayed silent, and both would have gotten off with a 2 year sentence.

The dilemma faced by the criminals is that, whatever the other does, each is better off confessing than remaining silent. However, both also know that if the other could have been trusted, or if they could have coordinated their answer, then a better outcome would have been within reach.

The dilemma illustrates individual self-interest against the interest of the group. In the literature on game theory, confessing is also known as defecting, and staying silent is the cooperative choice. The confess/confess situation (both defect) is the optimal non-cooperative strategy, the Nash equilibrium, because for each agent that stays silent a strategy exists where the sentence will be made worse (10 years) when the other agent confesses. Hence the agent will not stay silent but confess so as to limit the loss.

Silent/silent (both cooperate) is Pareto optimal at 2 years for both, since going to all other cases will make at least one agent worse off.

The Nash strategy is a non-cooperative, non-communication, non-trust strategy; while the Pareto strategy is the cooperative, communication, trust strategy.

---

[2] https://plato.stanford.edu/entries/prisoner-dilemma/.

7.1.3.1 Iterated Prisoner's Dilemma

The prisoner's dilemma is a one-time game. What would happen if we play this game repeatedly, being able to identify and remember the choices of our opponent? Could some kind of communication or trust arise, even if the setting is initially non-cooperative?

This question is answered by the iterated version of the prisoner's dilemma. Interest in the iterated prisoner's dilemma grew after a series of publications following a computer tournament [7, 8]. The tournament was organized by political scientist Robert Axelrod, around 1980. Game theorists were invited to send in computer programs to play iterated prisoner's dilemmas. The programs were organized in a tournament and played each other hundreds of times. The goal was to see which strategy would win, and if cooperation would emerge in this simplest of settings. A range of programs were entered, some using elaborate response strategies based on psychology, while others used advanced machine learning to try to predict the opponent's actions.

Surprisingly, one of the simplest strategies won. It was submitted by Anatol Rapoport, a mathematical psychologist who specialized in the modeling of social interaction. Rapoport's program played a strategy known as *tit for tat*. It would start by cooperating (staying silent) in the first round and in the next rounds if would play whatever the opponent did in the previous round. Tit for tat thus rewards cooperation with cooperation and punishes defecting with defection—hence the name.

Tit for tat wins if it is paired with a cooperative opponent and does not lose too much by stubbornly cooperating with a non-cooperative opponent. In the long run, it ends up either in the Pareto optimum or in the Nash equilibrium. Axelrod attributes the success of tit for tat to a number of properties. First, it is nice, that is, it is never the first to defect. In his tournament, the top eight programs played nice strategies. Second, tit for tat is also retaliatory, and it is difficult to exploit the strategy by the non-cooperative strategies. Third, tit for tat is forgiving, when the opponent plays nice, it rewards the play by reverting to being nice, being willing to forget past non-cooperative behavior. Finally, the rule has the advantage of being clear and predictable. Others easily learn its behavior and adapt to it, by being cooperative, leading to the mutual win/win Pareto optimum.

7.1.4 Challenges

After this brief overview of theory, let us see how well practical algorithms are able to solve multi-agent problems. With the impressive results in single-agent and two-agent reinforcement learning, the interest (and the expectations) in multi-agent problems has increased. In the next section, we will have a closer look at algorithms for multi-agent problems, but let us first look at the challenges that these algorithms face.

Multi-agent problems are studied in competitive, cooperative, and mixed settings. The main challenges faced by multi-agent reinforcement learning are threefold: (1) partial observability, (2) nonstationary environments, and (3) large state space. All three of these aspects increase the size of the state space. We will discuss these challenges in order.

7.1.4.1 Partial Observability

Most multi-agent settings are imperfect information settings, where agents have some private information that is not revealed to other agents. The private information can be in the form of hidden cards whose value is unknown, such as in poker, blackjack, or bridge. In real-time strategy games, players often do not see the entire map of the game, but parts of the world are obscured. Another reason for imperfect information can be that the rules of the game allow simultaneous moves, such as in Diplomacy, where all agents are determining their next action at the same time, and agents have to act without full knowledge of the other actions.

All these situations require that all possible states of the world have to be considered, increasing the number of possible states greatly in comparison to perfect information games. Imperfect information is best expressed as an extensive-form game. In fact, given the size of most multi-agent systems, it quickly becomes unfeasible to communicate and keep track of all the information of all agents, even if all agents would make their state and intentions public (which they rarely do).

Imperfect information increases the size of the state space, and computing the unknown outcomes quickly becomes unfeasible.

7.1.4.2 Nonstationary Environments

Moreover, as all agents are improving their policies according to their own interests concurrently, the agents are faced with a nonstationary environment. The environment's dynamics are determined by the joint action space of all agents, in which the best policies of agents depend on the best policies of the other agents. This mutual dependence creates an unstable situation.

In a single-agent setting, a single state needs to be tracked to calculate the next action. In a multi-agent setting, all states and all agents' policies need to be taken into account and mutually so. Each agent faces a moving target problem.

In multi-agent reinforcement learning, the agents learn concurrently, updating their behavior policies concurrently and often simultaneously [71]. Actions taken by one agent affect the reward of other agents and therefore of the next state. This invalidates the Markov property, which states that all information that is necessary to determine the next state is present in the current state and the agent's action. The powerful arsenal of single-agent reinforcement theory must be adapted before it can be used.

To handle nonstationarity, the agents must account for the joint action space of the other agents' actions. The size of the space increases exponentially with the number of agents. In two-agent settings, the agent has to consider all possible replies of a single opponent to each of its own moves, increasing the state space greatly. In multi-agent settings, the number of replies increases even more, and computing solutions to these problems quickly becomes quite expensive. A large number of agents complicates convergence analysis and increases the computational demands substantially.

On the other hand, the agents may learn from each other, line up their goals, and collaborate. Collaboration and group forming reduce the number of independent agents that must be tracked.

7.1.4.3 Large State Space

In addition to the problems caused by partial observability and nonstationarity, the size of the state space is significantly increased in a multi-agent setting, simply because every additional agent exponentially increases the state space. Furthermore, the action space of multi-agent reinforcement learning is a joint action space whose size increases exponentially with the number of agents. The size of the joint action space often causes scalability issues.

Solving such a large state space to optimality is infeasible for all but the smallest multi-agent problems, and much work has been done to create models (abstractions) that make simplifying, but sometimes realistic, assumptions to reduce the size of the state space. In the next section, we will look at some of these assumptions.

7.2 Multi-Agent Reinforcement Learning Agents

In the preceding section, the problem of multi-agent reinforcement learning has been introduced, as well as game theory and the links with social choice theory. We have also seen that the introduction of multiple agents has, unsurprisingly, increased the size of the state space even more than for single-agent state spaces. Recent work has introduced some approaches for solving multi-agent problems, which we will introduce here.

First, we will discuss an approach based on planning, with the name counterfactual regret minimization (CFR). This algorithm is successful for computing complex social choice problems, such as occur in the game of poker. CFR is suitable for competitive multi-agent problems.

Second, we will discuss cooperative reinforcement learning methods that are also suitable for mixed multi-agent problems.

Third, we discuss population-based approaches such as evolutionary strategies and swarm computing [11, 55, 74]. These approaches are inspired by computational behavior that occurs in nature, such as in flocks of birds and societies of insects [24,

25, 45]. Such methods are well known from single-agent optimization problems; indeed, population-based methods are successful in solving many complex and large single-agent optimization problems (including stochastic gradient descent optimization) [131]. In this section, we will see that evolutionary methods are also a natural match for mixed multi-agent problems, although they are typically used for cooperative problems with many homogeneous agents.

Finally, we will discuss an approach based on multi-play self-learning, in which evolutionary and hierarchical aspects are used. Here different groups of reinforcement learning agents are trained against each other. This approach has been highly successful in the games of Capture the Flag and StarCraft and is suitable for mixed multi-agent problems.

Let us start with counterfactual regret minimization.

7.2.1 Competitive Behavior

The first setting that we will discuss is the competitive setting. This setting is still close to single- and two-agent reinforcement learning, which are also based on competition.

7.2.1.1 Counterfactual Regret Minimization

Counterfactual regret minimization (CFR) is an iterative method for approximating the Nash equilibrium of an extensive-form game [183]. CFR is suitable for imperfect information games (such as poker) and computes a strategy that is (on average) non-exploitable and that is therefore robust in a competitive setting. Central in counterfactual regret minimization is the notion of regret. Regret is the loss in expected reward that an agent suffers for not having selected the best strategy, with respect to fixed choices by the other players. Regret can only be known in hindsight. We can, however, statistically sample expected regret, by averaging the regret that did not happen. CFR finds the Nash equilibrium by comparing two hypothetical players against each other, where the opponent chooses the action that minimizes our value.

CFR is a statistical algorithm that converges to a Nash equilibrium. Just like minimax, it is a self-play algorithm that finds the optimal strategy under the assumption of optimal play by both sides. Unlike minimax, it is suitable for imperfect information games, where information sets describe a set of possible worlds that the opponent may hold. Like MCTS, it samples, repeating this process for billions of games, improving its strategy each time. As it plays, it gets closer and closer toward an optimal strategy for the game: a strategy that can do no worse than tie against any opponent [79, 91, 148]. The quality of the strategy that it computes is measured by its *exploitability*. Exploitability is the maximum amount that a perfect counterstrategy could win (on expectation) against it.

Although the Nash equilibrium is a strategy that is theoretically proven to be not exploitable, in practice, typical human play is far from the theoretical optimum, even for top players [31]. Poker programs based on counterfactual regret minimization started beating the world's best human players in heads-up limit hold'em in 2008, even though these programs were still very much exploitable by this worst-case measure [79].

Many papers on counterfactual regret minimization are quite technical, and the codes for algorithms are too long to explain here in detail. CFR is in important algorithm that is essential for understanding progress in poker. To make the work on poker and CFR more accessible, introductory papers and blogs have been written. Trenner [156] has written a blog[3] in which CFR is used to play Kuhn poker, one of the simplest Poker variants. The CFR pseudocode and the function that calls it iteratively are shown in Listing 7.1; for the other routines, see the blog. The CFR code works as follows. First it checks for being in a terminal state and returns the payoff, just as a regular tree traversal code does. Otherwise, it retrieves the information set and the current regret-matching strategy. It uses the reach probability, which is the probability that we reach the current node according to our strategy in the current iteration. Then CFR loops over the possible actions (lines 17–24) computes the new reach probabilities for the next game state and calls itself recursively. As there are 2 players taking turns in Kuhn poker, the utility for one player is exactly −1 times the utility for the other, hence the minus sign in front of the `cfr()` call. What is computed here for each action is the counterfactual value. When the loop over all possible actions finishes, the value of the node-value of the current state is computed (line 26), with our current strategy. This value is the sum of the counterfactual values per action, weighted by the likelihood of taking this action. Then the cumulative counterfactual regrets are updated by adding the node-value times the reach probability of the opponent. Finally, the node-value is returned.

Another accessible blog post where the algorithm is explained step by step has been written by Kamil Czarnogòrski,[4] with code on GitHub.[5]

7.2.1.2 Deep Counterfactual Regret Minimization

Counterfactual regret minimization is a tabular algorithm that traverses the extensive-form game tree from root to terminal nodes, coming closer to the Nash equilibrium with each iteration. Tabular algorithms do not scale well to large problems, and the researchers often have to use domain-specific heuristic

---

[3] https://ai.plainenglish.io/building-a-poker-ai-part-6-beating-kuhn-poker-with-cfr-using-python-1b4172a6ab2d.

[4] https://int8.io/counterfactual-regret-minimization-for-poker-ai/.

[5] https://github.com/int8/counterfactual-regret-minimization/blob/master/games/algorithms.py.

```
1   def cfr(
2           self,
3           cards: List[str],
4           history: str,
5           reach_probabilities: np.array,
6           active_player: int) -> int:
7       if KuhnPoker.is_terminal(history):
8           return KuhnPoker.get_payoff(history, cards)
9
10      my_card = cards[active_player]
11      info_set = self.get_information_set(my_card + history)
12
13      strategy = info_set.get_strategy(reach_probabilities[
            active_player])
14      opponent = (active_player + 1) % 2
15      counterfactual_values = np.zeros(len(Actions))
16
17      for ix, action in enumerate(Actions):
18          action_probability = strategy[ix]
19
20          new_reach_probabilities = reach_probabilities.copy()
21          new_reach_probabilities[active_player] *= \
                action_probability
22
23          counterfactual_values[ix] = -self.cfr(
24              cards, history + action, new_reach_probabilities,
                    opponent)
25
26      node_value = counterfactual_values.dot(strategy)
27      for ix, action in enumerate(Actions):
28          counterfactual_regret[ix] = \
29              reach_probabilities[opponent] * (
                    counterfactual_values[ix] - node_value)
30          info_set.cumulative_regrets[ix] += counterfactual_regret[
                ix]
31
32      return node_value
33
34  def train(self, num_iterations: int) -> int:
35      util = 0
36      kuhn_cards = ['J', 'Q', 'K']
37      for _ in range(num_iterations):
38          cards = random.sample(kuhn_cards, 2)
39          history = ''
40          reach_probabilities = np.ones(2)
41          util += self.cfr(cards, history, reach_probabilities, 0)
42      return util
```

Listing 7.1 Counter-factual regret minimization [156]

abstraction schemes [29, 58, 134], alternate methods for regret updates [148], or sampling variants [91] to achieve acceptable performance.

For large problems, a deep learning version of the algorithm has been developed [30]. The goal of deep counterfactual regret minimization is to approximate the behavior of the tabular algorithm without calculating regrets at each individual information set. It generalizes across similar InfoSets using approximation of the value function via a deep neural network with alternating player updates.

7.2.2 Cooperative Behavior

CFR is an algorithm for the competitive setting. The Nash equilibrium defines the competitive win/lose multi-agent case—the $(-5, -5)$ situation of the prisoner's dilemma of Table 7.1.

We will now move to the cooperative setting. As we have seen, in a cooperative setting, win/win situations are possible, with higher rewards, both for society as a whole and for the individuals. The Pareto optimum for the prisoner's dilemma example is $(-2, -2)$, only achievable through norms, trust or cooperation by the agents (as close-knit criminal groups aim to achieve, for example, through a code of silence), see also Leibo et al. [95].

The achievements in single-agent reinforcement learning inspire researchers to achieve similar results in multi-agent. However, partial observability and nonstationarity create a computational challenge. Researchers have tried many different approaches, some of which we will cover, although the size of problems for which the current algorithms work is still limited. Wong et al. [175] provide a review of these approaches, and open problems in cooperative reinforcement learning are listed by Dafoe et al. [39]. First we will discuss approaches based on single-agent reinforcement learning methods. Next we will discuss approaches based on opponent modeling, communication, and psychology [175].

7.2.2.1 Centralized Training/Decentralized Execution

While the dec-POMDP model offers an appropriate framework for cooperative sequential decision making under uncertainty, solving a large dec-POMDP is intractable [21], and therefore many relaxations of the problems have been developed, where some elements, such as communication, or training, are centralized, to increase tractability [150, 175].

One of the easiest approaches to train a policy for a multi-agent problem is to train the collaborating agents with a centralized controller, effectively reducing a decentralized multi-agent computation to a centralized single-agent computation. In this approach, all agents send their observations and local policies to a central controller, which now has perfect information and decides which action to take for each agent. However, as large collaborative problems are computationally expensive, the single controller becomes overworked, and this approach does not scale.

On the other extreme, we can ignore communication and nonstationarity and let agents train separately. In this approach, the agents learn an individual action value function and view other agents as part of the environment. This approach simplifies the computational demands at the cost of gross oversimplification, ignoring multi-agent interaction.

An in-between approach is centralized training and decentralized execution [87]. Here agents can access extra information during training, such as other agents' observations, rewards, gradients, and parameters. However, they execute their policy decentrally based on their local observations. The local computation and inter-agent communication mitigate nonstationarity while still modeling partial observability and (some) interaction. This approach stabilizes the local policy learning of agents, even when other agents' policies are changing [87].

When value functions are learned centrally, how should this function then be used for decentral execution by the agents? A popular method is value function factorization. The value decomposition networks method (VDN) decomposes the central value function as a sum of individual value functions [146], who are executed greedily by the agents. QMIX and QTRAN are two methods that improve on VDN by allowing nonlinear combinations [125, 144]. Another approach is multi-agent variational exploration (MAVEN) which improves the inefficient exploration problem of QMIX using a latent space model [102].

Policy-based methods focus on actor critic approaches, with a centralized critic training decentralized actors. Counterfactual multi-agent (COMA) uses such a centralized critic to approximate the Q-function that has access to the actors that train the behavior policies [53].

Lowe et al. [100] introduce a multi-agent version of a popular off-policy single-agent deep policy gradient algorithm DDPG (Sect. 4.2.7), called MADDPG. It considers action policies of other agents and their coordination. MADDPG uses an ensemble of policies for each agent. It uses a decentralized actor, centralized critic approach, with deterministic policies. MADDPG works for both competitive and collaborative multi-agent problems. An extension for collision avoidance is presented by Cai et al. [34]. A popular on-policy single-agent method is PPO. Han et al. [65] achieve sample efficient results modeling the continuous Half-Cheetah task as a model-based multi-agent problem. Their model-based multi-agent work is inspired by MVE [52] (Sect. 5.2.2.1). Yu et al. [179] achieve good results in cooperative multi-agent games (StarCraft, Hanabi, and Particle world) with MAPPO.

Modeling cooperative behavior in reinforcement learning in a way that is computationally feasible is an active area of research. Li et al. [96] use implicit coordination graphs to model the structure of interactions. They use graph neural networks to model the coordination graphs [64] for StarCraft and traffic environments [132], allowing scaling of interaction patterns that are learned by the graph convolutional network.

7.2.2.2 Opponent Modeling

The state space of multi-agent problems is large, yet the previous approaches tried to learn this large space with adaptations of single-agent algorithms. Another approach is to reduce the size of the state space, for example, by explicitly modeling opponent behavior in the agents. These models can then be used to guide the agent's decision making, reducing the state space that it has to traverse. Albrecht and Stone have written a survey of approaches [2].

One approach to reduce the state space is to assume a set of stationary policies between which agents switch [51]. The switching agent model (SAM) [182] learns an opponent model from observed trajectories with a Bayesian network. The deep reinforcement open network (DRON) [68] uses two networks, one to learn the Q-values and the other to learn the opponent policy representation.

Opponent modeling is related to the psychological theory of mind [123]. According to this theory, people attribute mental states to others, such as beliefs, intents, and emotions. Our theory of the minds of others helps us to analyze and predict their behavior. Theory of mind also holds that we assume that the other has theory of mind; it allows for a nesting of beliefs of the form: "I believe that you believe that I believe" [161–163]. Building on these concepts, learning with opponent-learning awareness (LOLA) anticipates opponent's behavior [54]. Probabilistic recursive reasoning (PR2) models our own and our opponent's behavior as a hierarchy of perspectives [170]. Recursive reasoning has been shown to lead to faster convergence and better performance [40, 107]. Opponent modeling is also an active area of research.

7.2.2.3 Communication

Another step toward modeling the real world is taken when we explicitly model communication between agents. A fundamental question is how language between agents emerges when no predefined communication protocol exists and how syntax and meaning evolve out of interaction [175]. A basic approach to communication is with referential games: a sender sends two images and a message; the receiver then has to identify which of the images was the target [93]. Language also emerges in more complicated versions or in negotiation between agents [35, 86].

Another area where multi-agent systems are frequently used is the study of coordination, social dilemmas, emergent phenomena, and evolutionary processes, see, for example [49, 95, 149]. In the card game bridge [143], bidding strategies have been developed to signal to the other player in the team which cards a player has [143]. In the game of Diplomacy, an explicit negation phase is part of each game round [3, 43, 88, 121]. Work is ongoing to design communication-aware variants of reinforcement learning algorithms [66, 141].

7.2.2.4 Psychology

Many of the key ideas in reinforcement learning, such as operant conditioning and trial-and-error, originated in cognitive science. Faced with the large state space, multi-agent reinforcement learning methods are moving toward human-like agents. In addition to opponent modeling, studies focus on coordination, pro-social behavior, and intrinsic motivation. A large literature exists on emergence of social norms and cultural evolution in multi-agent systems [5, 28, 42, 95]. To deal with nonstationarity and large state spaces, humans use heuristics and approximation [59, 104]. However, heuristics can lead to biases and suboptimal decision making [60]. It is interesting to see how multi-agent modeling is discovering concepts from psychology. More research in this area is likely to improve the human-like behavior of artificial agents.

7.2.3 Mixed Behavior

To discuss solution methods for agents in the mixed setting, we will look at one important approach that is again inspired by biology: population-based algorithms.

Population-based methods such as evolutionary algorithms and swarm computing work by evolving (or optimizing) a large number of agents at the same time. We will look closer at evolutionary algorithms and at swarm computing, and then we will look at the role they play in multi-agent reinforcement learning.

7.2.3.1 Evolutionary Algorithms

Evolutionary algorithms are inspired by bio-genetic processes of reproduction: mutation, recombination, and selection [9]. Evolutionary algorithms work with large populations of simulated individuals, which typically makes it easy to parallelize and run them on large computation clusters.

Evolutionary algorithms often achieve good results in optimizing diverse problems. For example, in optimizing single-agent problems, an evolutionary approach would model the problem as a population of individuals, in which each individual represents a candidate solution to the problem. The candidate's quality is determined by a fitness function, and the best candidates are selected for reproduction. New candidates are created through crossover and mutation of genes, and the cycle starts again, until the quality of candidates stabilizes. In this way an evolutionary algorithm iteratively approaches the optimum. Evolutionary algorithms are randomized algorithms that can circumvent local optima.

Although they are best known for solving single-agent optimization problems, we will use them here to model multi-agent problems.

Let us look in more detail at how an evolutionary algorithm works (see Algorithm 7.1) [9]. First an initial population is generated. The fitness of each

---

Algorithm 7.1 Evolutionary framework [9]

1: Generate the initial population randomly
2: **repeat**
3:     Evaluate the fitness of each individual of the population
4:     Select the fittest individuals for reproduction
5:     Through crossover and mutation generate new individuals
6:     Replace the least fit individuals by the new individuals
7: **until** terminated

---

individual is computed, and the fittest individuals are selected for reproduction, using crossover and mutation to create a new generation of individuals. The least fit individuals of the old populations are replaced by the new individuals.

Compared to reinforcement learning, in evolutionary algorithms, the agents are typically homogeneous, in the sense that the reward (fitness) function for the individuals is the same. Individuals do have different genes and thus differ in their behavior (policy). In reinforcement learning, there is a single current behavior policy, where an evolutionary approach has many candidate policies (individuals). The fitness function can engender in principle both competitive and cooperative behavior between individuals, although a typical optimization scenario is to select a single individual with the genes for the highest fitness (survival of the fittest competitor).[6]

Changes to genes of individuals (policies) occur explicitly via crossover and (random) mutation and implicitly via selection for fitness. In reinforcement learning, the reward is used more directly as a policy goal; in evolutionary algorithms, the fitness does not directly influence the policy of an individual, only its survival.

Individuals in evolutionary algorithms are passive entities that do not communicate or act, although they do combine to create new individuals.

There are similarities and differences between evolutionary and multi-agent reinforcement learning algorithms. First of all, in both approaches, the goal is to find the optimal solution, the policy that maximizes (social) reward. In reinforcement learning, this occurs by learning a policy through interaction with an environment, in evolutionary algorithms by evolving a population through survival of the fittest. Reinforcement learning deals with a limited number of agents whose policy determines their actions, and evolutionary algorithms deals with many individuals whose genes determine their survival. Policies are improved using a reward function that assesses how good actions are, genes mutate and combine, and individuals are selected using a fitness function. Policies are improved "in place" and agents do not die, and in evolutionary computation the traits (genes) of the best individuals are selected and copied to new individuals in the next generation after which the old generation does die.

---

[6] Survival of the fittest *cooperative group* of individuals can also be achieved with an appropriate fitness function [101].

Although different at first sight, the two approaches share many traits, including the main goal: optimizing behavior. Evolutionary algorithms are inherently multi-agent and may work well in finding good solutions in large and nonstationary sequential decision problems.

7.2.3.2   Swarm Computing

Swarm computing is related to evolutionary algorithms [20, 24]. Swarm computing focuses on emerging behavior in decentralized, collective, self-organized systems. Agents are typically simple, numerous, and homogeneous and interact locally with each other and the environment. Biological examples of swarm intelligence are behavior in ant colonies, bee-hives, flocks of birds, and schools of fish, Fig. 7.3 [147], image by Wikimedia. Behavior is typically cooperative through decentralized communication mechanisms. In artificial swarm intelligence, indi-

Fig. 7.3 A flock of starlings and one predator

viduals are sometimes able to imagine the behavior of other individuals, when they have a theory of mind [18, 56]. Swarm intelligence, or collective intelligence in general, is a form of decentralized computing, as opposed to reinforcement learning, where external algorithms calculate optimal behavior in a single classical centralized algorithm [177].

Although both approaches work for the mixed setting, evolutionary algorithms tend to stress competition and survival of the fittest (Nash), where swarm computing stresses cooperation and survival of the group (Pareto).

A well-known example of artificial swarm intelligence is Dorigo's ant colony optimization (ACO) algorithm, which is a probabilistic optimization algorithm modeled after the pheromone-based communication of biological ants [44, 46, 47].

Emergent behavior in multi-agent reinforcement learning is specifically studied in [15, 69, 78, 94, 99, 106]. For decentralized algorithms related to solving multi-agent problems, see, for example [117–119, 181].

7.2.3.3 Population-Based Training

Translating traditional value or policy-based reinforcement learning algorithms to the multi-agent setting is nontrivial. It is interesting to see that, in contrast, evolutionary algorithms, first designed for single-agent optimization using a population of candidate solutions, translate so naturally to the multi-agent setting, where a population of agents is used to find a shared solution that is optimal for society.

In evolutionary algorithms, agents are typically homogeneous, although they can be heterogeneous, with different fitness functions. The fitness functions may be competitive or cooperative. In the latter case, the increase of reward for one agent can also imply an increase for other agents, possibly for the entire group, or population. Evolutionary algorithms are quite effective optimization algorithms. Salimans et al. [131] report that evolution strategies rival the performance of standard reinforcement learning techniques on modern benchmarks, while being easy to parallelize. In particular, evolutionary strategies are simpler to implement (there is no need for backpropagation), are easier to scale in a distributed setting, do not suffer in settings with sparse rewards, and have fewer hyperparameters.

Evolutionary algorithms are a form of population-based computation that can be used to compute strategies that are optimal in a game-theoretic Nash sense or that try to find strategies that outperform other agents. The evolutionary approach to optimization is efficient, robust, and easy to parallelize and thus has some advantages over the stochastic gradient approach. How does this approach relate to multi-agent reinforcement learning, and can it be used to find efficient solutions for multi-agent reinforcement learning problems?

In recent years, a number of research teams have reported successful efforts in creating reinforcement learning players for multi-agent strategy games (see Sect. 7.3). These were all large research efforts, where a range of different approaches were used, from self-play reinforcement learning, cooperative learning, hierarchical modeling, and evolutionary computing.

Algorithm 7.2 Population based training [78]

---

procedure TRAIN($\mathcal{P}$)   ▷ initial population $\mathcal{P}$
    Population $\mathcal{P}$, weights θ, hyperparameters h, model evaluation p, time step t
    for $(\theta, h, p, t) \in \mathcal{P}$ (asynchronously in parallel) **do**
        while not end of training **do**
            $\theta \leftarrow \text{step}(\theta|h)$ ▷ one step of optimisation using hyperparameters h
            $p \leftarrow \text{eval}(\theta)$ ▷ current model evaluation
            if $\text{ready}(p, t, \mathcal{P})$ **then**
                $h', \theta' \leftarrow \text{exploit}(h, \theta, p, \mathcal{P})$ ▷ use the rest of population for improvement
                if $\theta \neq \theta'$ **then**
                    $h, \theta \leftarrow \text{explore}(h', \theta', \mathcal{P})$ ▷ produce new hyperparameters h
                    $p \leftarrow \text{eval}(\theta)$ ▷ new model evaluation
                end if
            end if
            update $\mathcal{P}$ with new $(\theta, h, p, t + 1)$ ▷ update population
        end while
    end for
    return θ with the highest p in $\mathcal{P}$
end procedure

Jaderberg et al. [78] report success in Capture the Flag games with a combination of ideas from evolutionary algorithms and self-play reinforcement learning. Here a population-based approach of self-play is used where teams of diverse agents are trained in tournaments against each other. Algorithm 7.2 describes this population-based training (PBT) approach in more detail. Diversity is enhanced through mutation of policies, and performance is improved through culling of under-performing agents.

Population-based training uses two methods. The first is `exploit`, which decides whether a worker should abandon the current solution and focus on a more promising one. The second is `explore`, which, given the current solution and hyperparameters, proposes new solutions to explore the solution space. Members of the population are trained in parallel. Their weights θ are updated and `eval` measures their current performance. When a member of the population is deemed *ready* because it has reached a certain performance threshold, its weights and hyperparameters are updated by `exploit` and `explore`, to replace the current weights with the weights that have the highest recorded performance in the rest of the population and to randomly perturb the hyperparameters with noise. After `exploit` and `explore`, iterative training continues as before until convergence.

Let us have a look at this fusion approach for training leagues of players.

7.2.3.4 Self-play Leagues

Self-play league learning refers to the training of a multi-agent league of individual game playing characters. As we will see in the next section, variants have been used to play games such as StarCraft, Capture the Flag, and Hide and Seek.

League learning combines population-based training with self-play training as in AlphaZero. In league learning, the agent plays against a league of different opponents, while being part of a larger team of agents that is being trained. The team of agents is managed to have enough diversity in order to provide a stable training goal to reduce divergence or local minima. League learning employs evolutionary concepts such as mutation of behavior policies and culling of under-performing agents from the population. The team employs cooperative strategies, in addition to competing against the other teams. Agents are trained in an explicit hierarchy.

The goal of self-play league training is to find stable policies for all agents, which maximize their team reward. In mixed and cooperative settings, the teams of agents may benefit from each other's strength increase [100]. Self-play league learning also uses aspects of hierarchical reinforcement learning, a topic that will be covered in the next chapter.

In the next section, we will look deeper into how self-play league learning is implemented in specific multiplayer games.

7.3 Multi-Agent Environments

Reinforcement learning has achieved quite a few imaginative results in which it has succeeded in emulating behavior that approaches human behavior in the real world. In this chapter, we have made a step toward modeling more behavior that is even closer to the real world. Let us summarize in this section the results in four different multi-agent games, some of which have been published in prestigious scientific journals.

We will use the familiar sequence of (1) competitive, (2) cooperative, and (3) mixed environments. For each, we will sketch the problem, outline the algorithmic approach, and summarize the achievements.

Table 7.2 lists the multi-agent games and their dominant approach.

Table 7.2 Multi-agent game approaches

Environment	Behavior	Approach	Ref
Poker	Competitive	(Deep) counterfactual regret minimization	[26, 32, 105]
Hide and Seek	Cooperative	Self-play, team hierarchy	[13]
Capture the Flag	Mixed	Self-play, hierarchical, population-based	[77]
StarCraft II	Mixed	Self-play, population-based	[165]

7.3.1 *Competitive Behavior: Poker*

Poker is a popular imperfect information game. It is played competitively and human championships are organized regularly. Poker has been studied for some time in artificial intelligence, and computer poker championships have been conducted regularly [17, 22]. Poker is a competitive game. Cooperation (collusion and collaboration between two players to the detriment of a third player) is possible but often does not occur in practice; finding the Nash equilibrium therefore is a successful approach to playing the game in practice. The method of counterfactual regret minimization has been developed specifically to make progress in poker.

There are many variants of poker that are regularly played. No-limit Texas hold'em is a popular variant; the two-player version is called Heads Up and is easier to analyze because no opponent collusion can occur. Heads-up no-limit Texas hold'em (HUNL) has been the primary AI benchmark for imperfect information game play for several years.

Poker has hidden information (the face-down cards). Because of this, agents are faced with a large number of possible states; poker is a game that is far more complex than chess or checkers. The state space of the two-person HUNL version is reported to be around 10^{161} [31]. A further complication is that during the course of the game information is revealed by players through their bets; high bets indicate good cards or the wish of the player to make the opponent believe that this is the case (bluffing). Therefore, a player must choose between betting high on good cards and on doing the opposite, so that the opponent does not find out too much and can counteract.

In 2018, one of the top two-player poker programs, Libratus, defeated top human professionals in HUNL in a 20-day 120,000-hand competition featuring a $200,000 prize pool. Brown et al. [31] describe the architecture of Libratus. The program consists of three main modules, one for computing a quick CFR Nash policy using a smaller version of the game, a second module for constructing a finer-grained strategy once a later stage of the game is reached, and a third module to enhance the first policy by filling in missing branches.

In the experiment against top human players, Libratus analyzed the bet sizes that were played most often by its opponents at the end of each day of the competition. The programs would then calculate a response overnight, in order to improve as the competition proceeded.

Two-agent Libratus was originally based on heuristics, abstraction, and tabular counterfactual regret minimization—not on deep reinforcement learning. For multi-agent poker, the program Pluribus was developed. For Pluribus, the authors used *deep* counterfactual regret minimization [30], with a 7-layer neural network that followed the AlphaZero self-play approach. Pluribus defeated top players in six-player poker [32] (Fig. 7.4). Pluribus uses an AlphaZero approach of self-play in combination with search. Another top program, DeepStack, also uses randomly generated games to train a deep value function network [105].

Fig. 7.4 Pluribus on the cover of Science

7.3.2 Cooperative Behavior: Hide and Seek

In addition to competition, cooperative behavior is part of real-world behavior. Indeed, cooperation is what defines our social fabric, and much of our society consists of different ways in which we organize ourselves in families, groups, companies, parties, filter bubbles, and nations. The question of how voluntary cooperation can emerge between individuals has been studied extensively, and the work in tit for tat (Sect. 7.1.3.1) is just one of many studies in this fascinating field [5, 28, 95, 130, 177].

One study into emergent cooperation has been performed with a version of the game of Hide and Seek. Baker et al.[13] report on an experiment in which they used MuJoCo to build a new multi-agent game environment. The environment was created with the specific purpose of studying how cooperation emerged out of the combination of a few simple rules and reward maximization (see Fig. 7.5).

In the Hide and Seek experiment, the game is played on a randomly generated Grid world where props are available, such as boxes. The reward function stimulates hiders to avoid the line of sight of seekers and vice versa. There are objects scattered throughout the environment that the agents can grab and lock in place. The environment contains one to three hiders and one to three seekers, and there are

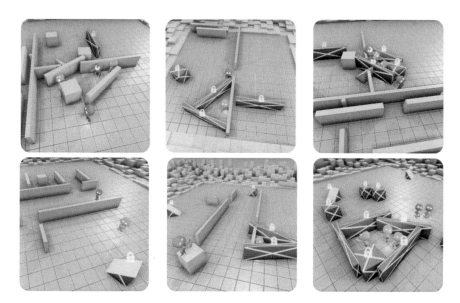

Fig. 7.5 Six strategies in hide and seek: running and chasing, fort building, ramp use, ramp defense, box surfing, and surf defense (left to right, top to bottom) [13]

three to nine movable boxes, some elongated. There are also two movable ramps. Walls and rooms are static and are randomly generated. Agents can see, move, and grab objects. A good way to understand the challenge is to view the videos[7] on the Hide and Seek blog.[8]

With only a visibility-based reward function, the agents are able to learn many different skills, including collaborative tool use. For example, hiders learn to create shelters by barricading doors or constructing multi-object forts, so that the seekers can never see them anymore, until, as a counter strategy, seekers learned to use ramps to jump into the shelter. In effect, out of the agents' interaction, a training curriculum emerges in which the agents learn tasks, many of which require sophisticated tool use and coordination.

Hide and Seek features cooperation (inside the team) and competition (between the hiders and the seekers). It uses a self-play version of PPO for policy learning. It is interesting to see how easy cooperation emerges. The game does not have explicit communication for the team to coordinate cooperation; all cooperative behavior emerges out of basic interaction between agents that are guided by their reward functions. Cooperative strategies thus emerge out of the game design: the homogeneous reward functions for each team and the environment in which blocks are present and follow laws of physics.

---

[7] https://www.youtube.com/watch?v=kopoLzvh5jY&t=10s.

[8] https://openai.com/blog/emergent-tool-use/.

During play, the agents essentially construct an autocurriculum for themselves [13, 94]. Six different behavior strategies are reported, each more advanced than the other, increasing the competitive pressure to find counter strategies for the opponent (see Fig. 7.5).

Baker et al. [13] report that initially the hiders and seekers learn the basic strategy of running away and chasing. However, after much training (25 million episodes), hiders start to use boxes to construct shelters behind which they hide. Then, after another 75 million episodes, the seekers learn to move and use ramps to jump over obstacles into the shelter. A mere 10 million episodes later, the hiders learn to defend by moving the ramps to the edge and lock them in place out of range of the shelters. Then, after a long 270 million episodes of training, the seekers learned *box-surfing*. They moved a box to the edge of the play area next to the locked ramps. One seeker then used the ramp to climb on top of the box and the other seekers push it to the shelter, where the seeker could peek over the edge and see the hiders. Finally, in response, the hiders locked all of the boxes in place before building their shelter, and they were safe from the seekers.

The Hide and Seek experiment is interesting because of the emergence of diverse behavior strategies. The strategies emerged out of a basic reward function and random exploration (see also the *reward is enough* argument [140]).

The emergence of strategies out of basic reward and exploration suggests an evolutionary process. However, Hide and Seek does not employ population-based training or evolutionary algorithms, in contrast to the work in Capture the Flag, in the next section.

7.3.3 Mixed Behavior: Capture the Flag and StarCraft

The world around us exhibits a mix of competitive and cooperative behavior. Team collaboration is an important aspect of human life, and it has been studied extensively in biology, sociology, and artificial intelligence. It emerges (evolves) out of the most basic settings—the need to achieve an ambitious goal—as we just saw. In recent years, many research groups have studied the mixed multi-agent model in real-time strategy games, such as Capture the Flag and StarCraft. We will discuss both games.

7.3.3.1 Capture the Flag

First we will discuss the game Capture the Flag, which is played in a Quake III Arena (see Fig. 7.6). Jaderberg et al. have reported on an extensive experiment with this game [77], in which the agents learn from scratch to see, act, cooperate, and compete. In this experiment, the agents are trained with population-based self-play [78], Algorithm 7.2. The agents in the population are all different (they have different genes). The population learns by playing against each other, providing

Outdoor map overview

Fig. 7.6 Capture the flag [77]

increased diversity of teammates and opponents and a more stable and faster learning process than traditional single-agent deep reinforcement learning methods. In total, 30 different bots were created and pitted against each other. Agents are part of a team, and the reward functions form a hierarchy. A two-layer optimization process optimizes the internal rewards for winning and uses reinforcement learning on the internal rewards to learn the policies.

In Capture the Flag, the bots start by acting randomly. After 450,000 games, a bot strategy was found that performed well, and they developed cooperative strategies, such as following teammates in order to outnumber opponents and loitering near the enemy base when their teammate has the flag. Again, as in Hide and Seek, cooperative strategies emerged out of the basic rules, by combining environment feedback and survival of the fittest.

The work on Capture the Flag is notable since it demonstrated that with only pixels as input an agent can learn to play competitively in a rich multi-agent environment. To do so, it used a combination of population-based training, internal reward optimization, and temporally hierarchical reinforcement learning (see the next chapter).

7.3.3.2 StarCraft

The final game that we will discuss in this chapter is StarCraft. StarCraft is a multi-player real-time strategy game of even larger complexity. The state space has been estimated to be on the order of 10^{1685} [120], a very large number. StarCraft features multi-agent decision making under uncertainty, spatial and temporal reasoning, competition, team collaboration, opponent modeling, and real-time planning. Figure 1.6 shows a picture of a StarCraft II scene.

Research on StarCraft has been ongoing for some time [120], and a special StarCraft multi-agent challenge has been introduced [132]. A team from DeepMind has created a program called AlphaStar. In a series of test matches held in December 2018, DeepMind's AlphaStar beat two top players in two-player single-map matches, using a different user interface.

AlphaStar plays the full game of StarCraft II. The neural network was initially trained by supervised learning from anonymized human games, which were then further trained by playing against other AlphaStar agents, using a population-based version of self-play reinforcement learning [78, 165].[9] These agents are used to seed a multi-agent reinforcement learning process. A continuous competitive league was created, with the agents of the league playing games in competition against each other. By branching from existing competitors, new competitors were added. Agents learn from games against other competitors. Population-based learning was taken further, creating a process that explores the very large space of StarCraft game play by pitting agents against strong opponent strategies and retaining strong early strategies.

Diversity in the league is increased by giving each agent its own learning objective, such as which competitors it should focus on and which game unit it should build. A form of prioritized league-self-play actor critic training is used, called prioritized fictitious self-play—details are in [165]. AlphaStar was trained on a custom-built scalable distributed training system using Google's tensor processing units (TPUs). The AlphaStar league was run for 14 days. In this training, each agent experienced the equivalent of 200 years of real-time StarCraft play.

In StarCraft, players can choose to play one of the three alien races: Terran, Zerg, or Protoss. AlphaStar was trained to play Protoss only, to reduce training time, although the same training pipeline could be applied to any race. AlphaStar was first tested against a human grandmaster named TLO, a top professional Zerg player and a grandmaster level Protoss player. The human player remarked: *I was surprised by how strong the agent was. AlphaStar takes well-known strategies and turns them on their head. The agent demonstrated strategies I had not thought of before, which means there may still be new ways of playing the game that we have not fully explored yet.*

7.3.4 Hands On: Hide and Seek in the Gym Example

Many of the research efforts reported in these final chapters describe significant efforts by research teams working on complicated and large games. These games

---

[9] Similar to the first approach in AlphaGo, where self-play reinforcement learning was also bootstrapped by supervised learning from human games.

represent the frontier of artificial intelligence, and research teams use all available computational and software engineering power that they can acquire to achieve the best results. Also, typically a large amount of time is spent in training and in finding the right hyperparameters for the learning to work.

Replicating results of this scale is highly challenging, and most research efforts are focused on replicating the results on a smaller, more manageable scale, with more manageable computational resources.

In this section, we will try to replicate some aspects with modest computational requirements. We will focus on Hide and Seek (Sect. 7.3.2). The original code for the Hide and Seek experiments is on GitHub.[10] Please visit and install the code. Hide and Seek uses MuJoCo and the mujoco-worldgen package. Install them and the dependencies with

```
pip install -r mujoco-worldgen/requirements.txt
pip install -e mujoco-worldgen/
pip install -e multi-agent-emergence-environments/
```

Examples of environments can be found in the mae_envs/envs folder. You can also build your own environments, starting from the Base environment, in mae_envs/envs/base.py, and then adding boxes, ramps, as well as the appropriate wrappers. Look in the other environments for how to do this.

Try out environments by using the bin/examine script; example usage:

```
bin/examine.py base
```

See further the instructions in the GitHub repository.

7.3.4.1 Multiplayer Environments

To conclude this hands-on section, we mention a multiplayer implementation of the Arcade Learning Environment. It is presented by Terry et al. [152], who also present baseline performance results for a multiplayer version of DQN, Ape-X DQN, which performed well elsewhere [16, 76]. The environment is also presented as part of PettingZoo, a multi-agent version of Gym [85, 153, 159].

Another multi-agent research environment is the Google Football Research Environment [90]. A physics-based simulator is provided, as well as three baseline implementations (DQN, IMPALA, and PPO).

---

[10] https://github.com/openai/multi-agent-emergence-environments.

Summary and Further Reading

We will now summarize the chapter and provide pointers to further reading.

Summary

Multi-agent reinforcement learning learns optimal policies for environments that consist of multiple agents. The optimal policy of the agents is influenced by the policies of the other agents, whose policy is also being optimized. This gives rise to the problem of nonstationarity, and agent behavior violates the Markov property.

Multi-agent reinforcement learning adds the element of cooperative behavior to the repertoire of reinforcement learning, which now consists of competition, cooperation, and mixed behavior. The field is closely related to game theory—the basis of the study of rational behavior in economics. A famous problem of game theory is the prisoner's dilemma. A famous result of non-cooperative game theory is the Nash equilibrium, which is defined as the joint strategy where no player has anything to gain by changing their own strategy. A famous result from cooperative game theory is the Pareto optimum, the situation where no individual can be better off without making someone else worse off.

When agents have private information, a multi-agent problem is partially observable. Multi-agent problems can be modeled by stochastic games or as extensive-form games. The behavior of agents is ultimately determined by the reward functions, which can be homogeneous or heterogeneous. When agents have different reward functions, multi-agent becomes multi-objective reinforcement learning.

The regret of an action is the amount of reward that is missed by the agent for not choosing the actions with the highest payoff. A regret minimization algorithm is the stochastic and multi-agent equivalent of minimax. Counterfactual regret minimization is an approach for finding Nash strategies in competitive multi-agent games, such as poker.

Variants of single-agent algorithms are often used for cooperative multi-agent situations. The large state space due to nonstationarity and partial observability precludes solving large problems. Other promising approaches are opponent modeling and explicit communication modeling.

Population-based methods such as evolutionary algorithms and swarm intelligence are used frequently in multi-agent systems. These approaches are suitable for homogeneous reward functions and competitive, cooperative, and mixed problems. Evolutionary methods evolve a population of agents, combining behaviors and selecting the best according to some fitness function. Evolutionary methods are a natural fit for parallel computers and are among the most popular and successful optimization algorithms. Swarm intelligence often introduces (rudimentary) forms of communication between agents, such as in ant colony optimization where agents communicate through artificial pheromones to indicate which part of the solution space they have traveled.

For some of the most complicated problems that have recently been tackled, such as StarCraft, Capture the Flag, and Hide and Seek, hierarchical and evolutionary principles are often combined in league training, where leagues of teams of agents are trained in a self-play fashion and where the fittest agents survive. Current achievements require large amounts of computational power; the future work is trying to reduce these requirements.

Further Reading

Multi-agent learning is a widely studied field. Surveys—both on early multi-agent reinforcement learning and on deep multi-agent reinforcement learning—can be found in [1, 2, 33, 63, 71, 72, 149, 158, 175, 178]. After Littman [97], Shoham et al. [138] look deeper into MDP modeling.

The classic work on game theory is Von Neumann and Morgenstern [167]. Modern introductions are [41, 57, 111]. Game theory underlies much of the theory of rational behavior in classical economics. Seminal works of John Nash are [113–115]. In 1950, he introduced the Nash equilibrium in his dissertation of 28 pages, which won him the Nobel prize in Economics in 1994. A biography and film have been made about the life of John Nash [112].

The game of rock–paper–scissors plays an important role in game theory and the study of computer poker [22, 32, 129, 169]. Prospect theory [81], introduced in 1979, studies human behavior in the face of uncertainty, a topic that evolved into the field of behavioral economics [36, 110, 174]. Gigerenzer introduced fast and frugal heuristics to explain human decision making [59].

For more intriguing works on the field of evolution of cooperation and the emergence of social norms, see, for example [4–6, 8, 28, 70, 73].

More recently, multi-*objective* reinforcement learning has been studied; a survey is [98]. In this field, the more realistic assumption is adopted that agents have different rewards functions, leading to different Pareto optima [109, 164, 172]. Oliehoek et al. have written a concise introduction to decentralized multi-agent modeling [116, 117].

Counterfactual regret minimization has been fundamental for the success in computer poker [31, 32, 79, 183]. An often-used Monte Carlo version is published in [91]. A combination with function approximation is studied in [30].

Evolutionary algorithms have delivered highly successful optimization algorithms. Some entries to this vast field are [9–12, 50]. A related field is swarm intelligence, where communication between homogeneous agents is taking place [19, 45, 48, 82]. For further research in multi-agent systems, refer to [160, 176]. For collective intelligence, see, for example [46, 56, 82, 122, 177].

Many other works report on evolutionary algorithms in a reinforcement learning setting [37, 38, 84, 108, 131, 145, 171, 173]. Most of these approaches concern single-agent approaches, although some are specifically applied to multi-agent approaches [78, 83, 85, 99, 103, 142].

Research into benchmarks is active. Among interesting approaches are procedural content generation [154], MuJoCo Soccer [99], and the Obstacle Tower Challenge [80]. There is an extensive literature on computer poker; see, for example [17, 22, 23, 23, 26, 27, 61, 61, 105, 129, 133]. StarCraft research can be found in [120, 132, 132, 165, 166]. Other game studies are [151, 155]. Approaches inspired by results in poker and Go are now also being applied with success in no-press Diplomacy [3, 14, 62, 121].

Exercises

Below are a few quick questions to check your understanding of this chapter. For each question, a simple, single sentence answer is sufficient.

Questions

1. Why is there so much interest in multi-agent reinforcement learning?
2. What is one of the main challenges of multi-agent reinforcement learning?
3. What is a Nash strategy?
4. What is a Pareto Optimum?
5. In a competitive multi-agent system, what algorithm can be used to calculate a Nash strategy?
6. What makes it difficult to calculate the solution for a game of imperfect information?
7. Describe the Prisoner's dilemma.
8. Describe the iterated Prisoner's dilemma.
9. Name two multi-agent card games of imperfect information.
10. What is the setting with a heterogeneous reward function usually called?
11. Name three kinds of strategies that can occur a multi-agent reinforcement learning.
12. Name two solution methods that are appropriate for solving mixed strategy games.
13. What AI method is named after ant colonies, bee swarms, bird flocks, or fish schools? How does it work in general terms?
14. Describe the main steps of an evolutionary algorithm.
15. Describe the general form of Hide and Seek and three strategies that emerged from the interactions of the hiders or seekers.

Exercises

Here are some programming exercises to become more familiar with the methods that we have covered in this chapter.

1. *CFR.* Implement counterfactual regret minimization for a Kuhn poker player. Play against the program, and see if you can win. Do you see possibilities to extend it to a more challenging version of poker?
2. *Hide and Seek.* Implement Hide and Seek with cooperation and competition. Add more types of objects. See if other cooperation behavior emerges.
3. *Ant Colony.* Use the DeepMind Control Suite and set up a collaborative level and a competitive level, and implement Ant Colony Optimization. Find problem instances on the web or in the original paper [46]. Can you implement more swarm algorithms?
4. *Football.* Go to the Google Football blog,[11] and implement algorithms for football agents. Consider using a population-based approach.
5. *StarCraft.* Go to the StarCraft Python interface,[12] and implement a StarCraft player (highly challenging) [132].

References

1. Stefano Albrecht and Peter Stone. Multiagent learning: foundations and recent trends. In *Tutorial at IJCAI-17 conference*, 2017. 251
2. Stefano Albrecht and Peter Stone. Autonomous agents modelling other agents: A comprehensive survey and open problems. *Artificial Intelligence*, 258:66–95, 2018. 236, 251
3. Thomas Anthony, Tom Eccles, Andrea Tacchetti, János Kramár, Ian M. Gemp, Thomas C. Hudson, Nicolas Porcel, Marc Lanctot, Julien Pérolat, Richard Everett, Satinder Singh, Thore Graepel, and Yoram Bachrach. Learning to play no-press diplomacy with best response policy iteration. In *Advances in Neural Information Processing Systems*, 2020. 236, 252
4. Robert Axelrod. An evolutionary approach to norms. *The American Political Science Review*, pages 1095–1111, 1986. 251
5. Robert Axelrod. *The complexity of cooperation: Agent-based models of competition and collaboration*, volume 3. Princeton university press, 1997. 237, 244
6. Robert Axelrod. The dissemination of culture: A model with local convergence and global polarization. *Journal of Conflict Resolution*, 41(2):203–226, 1997. 251
7. Robert Axelrod and Douglas Dion. The further evolution of cooperation. *Science*, 242(4884):1385–1390, 1988. 228
8. Robert Axelrod and William D Hamilton. The evolution of cooperation. *Science*, 211(4489):1390–1396, 1981. 228, 251
9. Thomas Bäck. *Evolutionary Algorithms in Theory and Practice: Evolutionary Strategies, Evolutionary Programming, Genetic Algorithms*. Oxford University Press, 1996. 237, 237, 238, 251

---

[11] https://ai.googleblog.com/2019/06/introducing-google-research-football.html.

[12] https://github.com/deepmind/pysc2.

10. Thomas Bäck, David B Fogel, and Zbigniew Michalewicz. Handbook of evolutionary computation. *Release*, 97(1):B1, 1997.
11. Thomas Bäck, Frank Hoffmeister, and Hans-Paul Schwefel. A survey of evolution strategies. In *Proceedings of the fourth International Conference on Genetic Algorithms*, 1991. 230
12. Thomas Bäck and Hans-Paul Schwefel. An overview of evolutionary algorithms for parameter optimization. *Evolutionary Computation*, 1(1):1–23, 1993. 251
13. Bowen Baker, Ingmar Kanitscheider, Todor Markov, Yi Wu, Glenn Powell, Bob McGrew, and Igor Mordatch. Emergent tool use from multi-agent autocurricula. *arXiv preprint arXiv:1909.07528*, 2019. 242, 244, 245, 246, 246
14. Anton Bakhtin, David Wu, Adam Lerer, and Noam Brown. No-press diplomacy from scratch. *Advances in Neural Information Processing Systems*, 34, 2021. 252
15. Trapit Bansal, Jakub Pachocki, Szymon Sidor, Ilya Sutskever, and Igor Mordatch. Emergent complexity via multi-agent competition. *arXiv preprint arXiv:1710.03748*, 2017. 240
16. Nolan Bard, Jakob N. Foerster, Sarath Chandar, Neil Burch, Marc Lanctot, H. Francis Song, Emilio Parisotto, Vincent Dumoulin, Subhodeep Moitra, Edward Hughes, Iain Dunning, Shibl Mourad, Hugo Larochelle, Marc G. Bellemare, and Michael Bowling. The Hanabi challenge: A new frontier for AI research. *Artificial Intelligence*, 280:103216, 2020. 249
17. Nolan Bard, John Hawkin, Jonathan Rubin, and Martin Zinkevich. The annual computer poker competition. *AI Magazine*, 34(2):112, 2013. 243, 252
18. Simon Baron-Cohen, Alan M Leslie, and Uta Frith. Does the autistic child have a "theory of mind"? *Cognition*, 21(1):37–46, 1985. 221, 240
19. Gerardo Beni. Swarm intelligence. *Complex Social and Behavioral Systems: Game Theory and Agent-Based Models*, pages 791–818, 2020. 251
20. Gerardo Beni and Jing Wang. Swarm intelligence in cellular robotic systems. In *Robots and Biological Systems: Towards a New Bionics?*, pages 703–712. Springer, 1993. 239
21. Daniel S Bernstein, Robert Givan, Neil Immerman, and Shlomo Zilberstein. The complexity of decentralized control of Markov decision processes. *Mathematics of Operations Research*, 27(4):819–840, 2002. 225, 234
22. Darse Billings, Aaron Davidson, Jonathan Schaeffer, and Duane Szafron. The challenge of poker. *Artificial Intelligence*, 134(1-2):201–240, 2002. 243, 251, 252
23. Darse Billings, Aaron Davidson, Terence Schauenberg, Neil Burch, Michael Bowling, Robert Holte, Jonathan Schaeffer, and Duane Szafron. Game-tree search with adaptation in stochastic imperfect-information games. In *International Conference on Computers and Games*, pages 21–34. Springer, 2004. 252, 252
24. Christian Blum and Daniel Merkle. *Swarm Intelligence: Introduction and Applications*. Springer Science & Business Media, 2008. 230, 239
25. Eric Bonabeau, Marco Dorigo, and Guy Theraulaz. *Swarm Intelligence: From Natural to Artificial Systems*. Oxford University Press, 1999. 231
26. Michael Bowling, Neil Burch, Michael Johanson, and Oskari Tammelin. Heads-up Limit Hold'em poker is solved. *Science*, 347(6218):145–149, 2015. 224, 242, 252
27. Michael H. Bowling, Nicholas Abou Risk, Nolan Bard, Darse Billings, Neil Burch, Joshua Davidson, John Alexander Hawkin, Robert Holte, Michael Johanson, Morgan Kan, Bryce Paradis, Jonathan Schaeffer, David Schnizlein, Duane Szafron, Kevin Waugh, and Martin Zinkevich. A demonstration of the polaris poker system. In *Proceedings of The 8th International Conference on Autonomous Agents and Multiagent Systems*, volume 2, pages 1391–1392, 2009. 252
28. Robert Boyd and Peter J Richerson. *Culture and the Evolutionary Process*. University of Chicago press, 1988. 237, 244, 251
29. Noam Brown, Sam Ganzfried, and Tuomas Sandholm. Hierarchical abstraction, distributed equilibrium computation, and post-processing, with application to a champion No-Limit Texas Hold'em agent. In *AAAI Workshop: Computer Poker and Imperfect Information*, 2015. 234

30. Noam Brown, Adam Lerer, Sam Gross, and Tuomas Sandholm. Deep counterfactual regret minimization. In *International Conference on Machine Learning*, pages 793–802. PMLR, 2019. 234, 243, 251

31. Noam Brown and Tuomas Sandholm. Superhuman AI for Heads-up No-limit poker: Libratus beats top professionals. *Science*, 359(6374):418–424, 2018. 232, 243, 243, 251

32. Noam Brown and Tuomas Sandholm. Superhuman AI for multiplayer poker. *Science*, 365(6456):885–890, 2019. 242, 243, 251, 251

33. Lucian Busoniu, Robert Babuska, and Bart De Schutter. A comprehensive survey of multi-agent reinforcement learning. *IEEE Transactions on Systems, Man, and Cybernetics, Part C (Applications and Reviews)*, 38(2):156–172, 2008. 251

34. Zhiyuan Cai, Huanhui Cao, Wenjie Lu, Lin Zhang, and Hao Xiong. Safe multi-agent reinforcement learning through decentralized multiple control barrier functions. *arXiv preprint arXiv:2103.12553*, 2021. 235

35. Kris Cao, Angeliki Lazaridou, Marc Lanctot, Joel Z Leibo, Karl Tuyls, and Stephen Clark. Emergent communication through negotiation. In *International Conference on Learning Representations*, 2018. 236

36. Edward Cartwright. *Behavioral Economics*. Routledge, 2018. 251

37. Patryk Chrabaszcz, Ilya Loshchilov, and Frank Hutter. Back to basics: Benchmarking canonical evolution strategies for playing Atari. In *Proceedings of the Twenty-Seventh International Joint Conference on Artificial Intelligence, IJCAI 2018, July 13-19, 2018, Stockholm*, pages 1419–1426, 2018. 251

38. Edoardo Conti, Vashisht Madhavan, Felipe Petroski Such, Joel Lehman, Kenneth O Stanley, and Jeff Clune. Improving exploration in evolution strategies for deep reinforcement learning via a population of novelty-seeking agents. In *Advances in Neural Information Processing Systems*, pages 5032–5043, 2018. 251

39. Allan Dafoe, Edward Hughes, Yoram Bachrach, Tantum Collins, Kevin R McKee, Joel Z Leibo, Kate Larson, and Thore Graepel. Open problems in cooperative AI. *arXiv preprint arXiv:2012.08630*, 2020. 234

40. Zhongxiang Dai, Yizhou Chen, Bryan Kian Hsiang Low, Patrick Jaillet, and Teck-Hua Ho. R2-B2: recursive reasoning-based Bayesian optimization for no-regret learning in games. In *International Conference on Machine Learning*, pages 2291–2301. PMLR, 2020. 236

41. Morton D Davis. *Game Theory: a Nontechnical Introduction*. Courier Corporation, 2012. 251

42. Richard Dawkins and Nicola Davis. *The Selfish Gene*. Macat Library, 2017. 237

43. Dave De Jonge, Tim Baarslag, Reyhan Aydoğan, Catholijn Jonker, Katsuhide Fujita, and Takayuki Ito. The challenge of negotiation in the game of diplomacy. In *International Conference on Agreement Technologies*, pages 100–114. Springer, 2018. 236

44. Marco Dorigo. Optimization, learning and natural algorithms. *PhD Thesis, Politecnico di Milano*, 1992. 240

45. Marco Dorigo and Mauro Birattari. Swarm intelligence. *Scholarpedia*, 2(9):1462, 2007. 231, 251

46. Marco Dorigo, Mauro Birattari, and Thomas Stutzle. Ant colony optimization. *IEEE Computational Intelligence Magazine*, 1(4):28–39, 2006. 240, 251, 253

47. Marco Dorigo and Luca Maria Gambardella. Ant colony system: a cooperative learning approach to the traveling salesman problem. *IEEE Transactions on Evolutionary Computation*, 1(1):53–66, 1997. 240

48. Russell C Eberhart, Yuhui Shi, and James Kennedy. *Swarm Intelligence*. Elsevier, 2001. 251

49. Tom Eccles, Edward Hughes, János Kramár, Steven Wheelwright, and Joel Z Leibo. Learning reciprocity in complex sequential social dilemmas. *arXiv preprint arXiv:1903.08082*, 2019. 236

50. Agoston E Eiben and Jim E Smith. What is an evolutionary algorithm? In *Introduction to Evolutionary Computing*, pages 25–48. Springer, 2015. 251

51. Richard Everett and Stephen Roberts. Learning against non-stationary agents with opponent modelling and deep reinforcement learning. In *2018 AAAI Spring Symposium Series*, 2018. 236

52. Vladimir Feinberg, Alvin Wan, Ion Stoica, Michael I Jordan, Joseph E Gonzalez, and Sergey Levine. Model-based value estimation for efficient model-free reinforcement learning. *arXiv preprint arXiv:1803.00101*, 2018. 235

53. Jakob Foerster, Gregory Farquhar, Triantafyllos Afouras, Nantas Nardelli, and Shimon Whiteson. Counterfactual multi-agent policy gradients. In *Proceedings of the AAAI Conference on Artificial Intelligence*, volume 32, 2018. 235

54. Jakob N Foerster, Richard Y Chen, Maruan Al-Shedivat, Shimon Whiteson, Pieter Abbeel, and Igor Mordatch. Learning with opponent-learning awareness. *arXiv preprint arXiv:1709.04326*, 2017. 236

55. David B Fogel. An introduction to simulated evolutionary optimization. *IEEE Transactions on Neural Networks*, 5(1):3–14, 1994. 230

56. Vittorio Gallese and Alvin Goldman. Mirror neurons and the simulation theory of mind-reading. *Trends in Cognitive Sciences*, 2(12):493–501, 1998. 221, 240, 251

57. Sam Ganzfried and Tuomas Sandholm. Game theory-based opponent modeling in large imperfect-information games. In *The 10th International Conference on Autonomous Agents and Multiagent Systems*, volume 2, pages 533–540, 2011. 251

58. Sam Ganzfried and Tuomas Sandholm. Endgame solving in large imperfect-information games. In *Proceedings of the 2015 International Conference on Autonomous Agents and Multiagent Systems*, pages 37–45, 2015. 234

59. Gerd Gigerenzer and Daniel G Goldstein. Reasoning the fast and frugal way: models of bounded rationality. *Psychological review*, 103(4):650, 1996. 237, 251

60. Thomas Gilovich, Dale Griffin, and Daniel Kahneman. *Heuristics and Biases: The Psychology of Intuitive Judgment*. Cambridge university press, 2002. 237

61. Andrew Gilpin and Tuomas Sandholm. A competitive Texas Hold'em poker player via automated abstraction and real-time equilibrium computation. In *Proceedings of the National Conference on Artificial Intelligence*, volume 21, page 1007, 2006. 252, 252

62. Jonathan Gray, Adam Lerer, Anton Bakhtin, and Noam Brown. Human-level performance in no-press diplomacy via equilibrium search. *arXiv preprint arXiv:2010.02923*, 2020. 252

63. Sven Gronauer and Klaus Diepold. Multi-agent deep reinforcement learning: a survey. *Artificial Intelligence Review*, pages 1–49, 2021. 226, 251

64. Carlos Guestrin, Daphne Koller, and Ronald Parr. Multiagent planning with factored MDPs. In *Advances in Neural Information Processing Systems*, volume 1, pages 1523–1530, 2001. 235

65. Dongge Han, Chris Xiaoxuan Lu, Tomasz Michalak, and Michael Wooldridge. Multiagent model-based credit assignment for continuous control, 2021. 235

66. Matthew John Hausknecht. *Cooperation and Communication in Multiagent Deep Reinforcement Learning*. PhD thesis, University of Texas at Austin, 2016. 236

67. Conor F. Hayes, Roxana Radulescu, Eugenio Bargiacchi, Johan Källström, Matthew Macfarlane, Mathieu Reymond, Timothy Verstraeten, Luisa M. Zintgraf, Richard Dazeley, Fredrik Heintz, Enda Howley, Athirai A. Irissappane, Patrick Mannion, Ann Nowé, Gabriel de Oliveira Ramos, Marcello Restelli, Peter Vamplew, and Diederik M. Roijers. A practical guide to multi-objective reinforcement learning and planning. *arXiv preprint arXiv:2103.09568*, 2021. 226

68. He He, Jordan Boyd-Graber, Kevin Kwok, and Hal Daumé III. Opponent modeling in deep reinforcement learning. In *International Conference on Machine Learning*, pages 1804–1813. PMLR, 2016. 236

69. Nicolas Heess, Dhruva TB, Srinivasan Sriram, Jay Lemmon, Josh Merel, Greg Wayne, Yuval Tassa, Tom Erez, Ziyu Wang, SM Eslami, Martin Riedmiller, and David Silver. Emergence of locomotion behaviours in rich environments. *arXiv preprint arXiv:1707.02286*, 2017. 240

70. Joseph Henrich, Robert Boyd, and Peter J Richerson. Five misunderstandings about cultural evolution. *Human Nature*, 19(2):119–137, 2008. 251

71. Pablo Hernandez-Leal, Michael Kaisers, Tim Baarslag, and Enrique Munoz de Cote. A survey of learning in multiagent environments: Dealing with non-stationarity. *arXiv preprint arXiv:1707.09183*, 2017. 221, 224, 229, 251

72. Pablo Hernandez-Leal, Bilal Kartal, and Matthew E Taylor. A survey and critique of multiagent deep reinforcement learning. *Autonomous Agents and Multi-Agent Systems*, 33(6):750–797, 2019. 226, 251

73. Francis Heylighen. *What makes a Meme Successful? Selection Criteria for Cultural Evolution.* Association Internationale de Cybernetique, 1998. 251

74. John Holland. Adaptation in natural and artificial systems: an introductory analysis with application to biology. *Control and Artificial Intelligence*, 1975. 230

75. Bert Hölldobler and Edward O Wilson. *The Superorganism: the Beauty, Elegance, and Strangeness of Insect Societies.* WW Norton & Company, 2009. 219

76. Dan Horgan, John Quan, David Budden, Gabriel Barth-Maron, Matteo Hessel, Hado Van Hasselt, and David Silver. Distributed prioritized experience replay. In *International Conference on Learning Representations*, 2018. 249

77. Max Jaderberg, Wojciech M. Czarnecki, Iain Dunning, Luke Marris, Guy Lever, Antonio Garcia Castañeda, Charles Beattie, Neil C. Rabinowitz, Ari S. Morcos, Avraham Ruderman, Nicolas Sonnerat, Tim Green, Louise Deason, Joel Z. Leibo, David Silver, Demis Hassabis, Koray Kavukcuoglu, and Thore Graepel. Human-level performance in 3D multiplayer games with population-based reinforcement learning. *Science*, 364(6443):859–865, 2019. 242, 246, 247

78. Max Jaderberg, Valentin Dalibard, Simon Osindero, Wojciech M. Czarnecki, Jeff Donahue, Ali Razavi, Oriol Vinyals, Tim Green, Iain Dunning, Karen Simonyan, Chrisantha Fernando, and Koray Kavukcuoglu. Population based training of neural networks. *arXiv preprint arXiv:1711.09846*, 2017. 240, 241, 241, 246, 248, 252

79. Michael Johanson, Nolan Bard, Marc Lanctot, Richard G Gibson, and Michael Bowling. Efficient Nash equilibrium approximation through Monte Carlo counterfactual regret minimization. In *AAMAS*, pages 837–846, 2012. 231, 232, 251

80. Arthur Juliani, Ahmed Khalifa, Vincent-Pierre Berges, Jonathan Harper, Ervin Teng, Hunter Henry, Adam Crespi, Julian Togelius, and Danny Lange. Obstacle tower: A generalization challenge in vision, control, and planning. *arXiv preprint arXiv:1902.01378*, 2019. 252

81. Daniel Kahneman and Amos Tversky. Prospect theory: An analysis of decision under risk. In *Handbook of the Fundamentals of Financial Decision Making: Part I*, pages 99–127. World Scientific, 2013. 251

82. James Kennedy. Swarm intelligence. In *Handbook of Nature-Inspired and Innovative Computing*, pages 187–219. Springer, 2006. 251, 251

83. Shauharda Khadka, Somdeb Majumdar, Tarek Nassar, Zach Dwiel, Evren Tumer, Santiago Miret, Yinyin Liu, and Kagan Tumer. Collaborative evolutionary reinforcement learning. In *International Conference on Machine Learning*, pages 3341–3350. PMLR, 2019. 252

84. Shauharda Khadka and Kagan Tumer. Evolutionary reinforcement learning. *arXiv preprint arXiv:1805.07917*, 2018. 251

85. Daan Klijn and AE Eiben. A coevolutionairy approach to deep multi-agent reinforcement learning. *arXiv preprint arXiv:2104.05610*, 2021. 249, 252

86. Satwik Kottur, José MF Moura, Stefan Lee, and Dhruv Batra. Natural language does not emerge 'naturally' in multi-agent dialog. In *Proceedings of the 2017 Conference on Empirical Methods in Natural Language Processing, EMNLP 2017, Copenhagen*, pages 2962–2967, 2017. 236

87. Landon Kraemer and Bikramjit Banerjee. Multi-agent reinforcement learning as a rehearsal for decentralized planning. *Neurocomputing*, 190:82–94, 2016. 235, 235

88. Sarit Kraus and Daniel Lehmann. Diplomat, an agent in a multi agent environment: An overview. In *IEEE International Performance Computing and Communications Conference*, pages 434–438, 1988. 236

89. Steven Kuhn. *Prisoner's Dilemma.* The Stanford Encyclopedia of Philosophy, https://plato.stanford.edu/entries/prisoner-dilemma/, 1997. 227

90. Karol Kurach, Anton Raichuk, Piotr Stańczyk, Michał Zając, Olivier Bachem, Lasse Espeholt, Carlos Riquelme, Damien Vincent, Marcin Michalski, Olivier Bousquet, and Sylvain Gelly. Google research football: A novel reinforcement learning environment. In *Proceedings of the AAAI Conference on Artificial Intelligence*, volume 34, pages 4501–4510, 2020. 249

91. Marc Lanctot, Kevin Waugh, Martin Zinkevich, and Michael H Bowling. Monte Carlo sampling for regret minimization in extensive games. In *Advances in Neural Information Processing Systems*, pages 1078–1086, 2009. 224, 231, 234, 251

92. Marc Lanctot, Vinicius Zambaldi, Audrunas Gruslys, Angeliki Lazaridou, Karl Tuyls, Julien Pérolat, David Silver, and Thore Graepel. A unified game-theoretic approach to multiagent reinforcement learning. In *Advances in Neural Information Processing Systems*, pages 4190–4203, 2017. 222

93. Angeliki Lazaridou, Alexander Peysakhovich, and Marco Baroni. Multi-agent cooperation and the emergence of (natural) language. In *International Conference on Learning Representations*, 2017. 236

94. Joel Z Leibo, Edward Hughes, Marc Lanctot, and Thore Graepel. Autocurricula and the emergence of innovation from social interaction: A manifesto for multi-agent intelligence research. *arXiv preprint arXiv:1903.00742*, 2019. 240, 246

95. Joel Z Leibo, Vinicius Zambaldi, Marc Lanctot, Janusz Marecki, and Thore Graepel. Multi-agent reinforcement learning in sequential social dilemmas. In *Proceedings of the 16th Conference on Autonomous Agents and MultiAgent Systems, AAMAS 2017, São Paulo, Brazil*, pages 464–473, 2017. 234, 236, 237, 244

96. Sheng Li, Jayesh K Gupta, Peter Morales, Ross Allen, and Mykel J Kochenderfer. Deep implicit coordination graphs for multi-agent reinforcement learning. In *AAMAS '21: 20th International Conference on Autonomous Agents and Multiagent Systems*, 2021. 235

97. Michael L Littman. Markov games as a framework for multi-agent reinforcement learning. In *Machine Learning Proceedings 1994*, pages 157–163. Elsevier, 1994. 222, 251

98. Chunming Liu, Xin Xu, and Dewen Hu. Multiobjective reinforcement learning: A comprehensive overview. *IEEE Transactions on Systems, Man, and Cybernetics: Systems*, 45(3):385–398, 2014. 251

99. Siqi Liu, Guy Lever, Josh Merel, Saran Tunyasuvunakool, Nicolas Heess, and Thore Graepel. Emergent coordination through competition. In *International Conference on Learning Representations*, 2019. 240, 252, 252

100. Ryan Lowe, Yi Wu, Aviv Tamar, Jean Harb, Pieter Abbeel, and Igor Mordatch. Multi-agent Actor-Critic for mixed cooperative-competitive environments. In *Advances in Neural Information Processing Systems*, pages 6379–6390, 2017. 235, 242

101. Xiaoliang Ma, Xiaodong Li, Qingfu Zhang, Ke Tang, Zhengping Liang, Weixin Xie, and Zexuan Zhu. A survey on cooperative co-evolutionary algorithms. *IEEE Transactions on Evolutionary Computation*, 23(3):421–441, 2018. 238

102. Anuj Mahajan, Tabish Rashid, Mikayel Samvelyan, and Shimon Whiteson. Maven: Multi-agent variational exploration. In *Advances in Neural Information Processing Systems*, pages 7611–7622, 2019. 235

103. Somdeb Majumdar, Shauharda Khadka, Santiago Miret, Stephen McAleer, and Kagan Tumer. Evolutionary reinforcement learning for sample-efficient multiagent coordination. In *International Conference on Machine Learning*, 2020. 252

104. Julian N Marewski, Wolfgang Gaissmaier, and Gerd Gigerenzer. Good judgments do not require complex cognition. *Cognitive Processing*, 11(2):103–121, 2010. 237

105. Matej Moravčík, Martin Schmid, Neil Burch, Viliam Lisy̌, Dustin Morrill, Nolan Bard, Trevor Davis, Kevin Waugh, Michael Johanson, and Michael Bowling. DeepStack: Expert-level artificial intelligence in heads-up no-limit poker. *Science*, 356(6337):508–513, 2017. 242, 243, 252

106. Igor Mordatch and Pieter Abbeel. Emergence of grounded compositional language in multi-agent populations. In *Proceedings of the AAAI Conference on Artificial Intelligence*, volume 32, 2018. 240

107. Pol Moreno, Edward Hughes, Kevin R McKee, Bernardo Avila Pires, and Théophane Weber. Neural recursive belief states in multi-agent reinforcement learning. *arXiv preprint arXiv:2102.02274*, 2021. 236

108. David E Moriarty, Alan C Schultz, and John J Grefenstette. Evolutionary algorithms for reinforcement learning. *Journal of Artificial Intelligence Research*, 11:241–276, 1999. 251

109. Hossam Mossalam, Yannis M Assael, Diederik M Roijers, and Shimon Whiteson. Multi-objective deep reinforcement learning. *arXiv preprint arXiv:1610.02707*, 2016. 251

110. Sendhil Mullainathan and Richard H Thaler. Behavioral economics. Technical report, National Bureau of Economic Research, 2000. 251

111. Roger B Myerson. *Game Theory*. Harvard university press, 2013. 251

112. Sylvia Nasar. *A Beautiful Mind*. Simon and Schuster, 2011. 251

113. John Nash. Non-cooperative games. *Annals of mathematics*, pages 286–295, 1951. 251

114. John F Nash. Equilibrium points in *n*-person games. *Proceedings of the National Academy of Sciences*, 36(1):48–49, 1950.

115. John F Nash Jr. The bargaining problem. *Econometrica: Journal of the econometric society*, pages 155–162, 1950. 251

116. Frans A Oliehoek. Decentralized POMDPs. In *Reinforcement Learning*, pages 471–503. Springer, 2012. 251

117. Frans A Oliehoek and Christopher Amato. *A Concise Introduction to Decentralized POMDPs*. Springer, 2016. 225, 240, 251

118. Frans A Oliehoek, Matthijs TJ Spaan, Christopher Amato, and Shimon Whiteson. Incremental clustering and expansion for faster optimal planning in Dec-POMDPs. *Journal of Artificial Intelligence Research*, 46:449–509, 2013.

119. Shayegan Omidshafiei, Jason Pazis, Christopher Amato, Jonathan P How, and John Vian. Deep decentralized multi-task multi-agent reinforcement learning under partial observability. In *International Conference on Machine Learning*, pages 2681–2690. PMLR, 2017. 240

120. Santiago Ontanón, Gabriel Synnaeve, Alberto Uriarte, Florian Richoux, David Churchill, and Mike Preuss. A survey of real-time strategy game AI research and competition in StarCraft. *IEEE Transactions on Computational Intelligence and AI in Games*, 5(4):293–311, 2013. 247, 248, 252

121. Philip Paquette, Yuchen Lu, Steven Bocco, Max Smith, O-G Satya, Jonathan K Kummerfeld, Joelle Pineau, Satinder Singh, and Aaron C Courville. No-press diplomacy: Modeling multi-agent gameplay. In *Advances in Neural Information Processing Systems*, pages 4476–4487, 2019. 236, 252

122. Aske Plaat. *De vlinder en de mier / The butterfly and the ant—on modeling behavior in organizations*. Inaugural lecture. Tilburg University, 2010. 251

123. David Premack and Guy Woodruff. Does the chimpanzee have a theory of mind? *Behavioral and Brain Sciences*, 1(4):515–526, 1978. 236

124. Roxana Rădulescu, Patrick Mannion, Diederik M Roijers, and Ann Nowé. Multi-objective multi-agent decision making: a utility-based analysis and survey. *Autonomous Agents and Multi-Agent Systems*, 34(1):1–52, 2020. 226

125. Tabish Rashid, Mikayel Samvelyan, Christian Schroeder, Gregory Farquhar, Jakob Foerster, and Shimon Whiteson. QMIX: Monotonic value function factorisation for deep multi-agent reinforcement learning. In *International Conference on Machine Learning*, pages 4295–4304. PMLR, 2018. 235

126. Diederik M Roijers, Willem Röpke, Ann Nowé, and Roxana Rădulescu. On following pareto-optimal policies in multi-objective planning and reinforcement learning. In *Multi-Objective Decision Making Workshop*, 2021. 226

127. Diederik M Roijers, Peter Vamplew, Shimon Whiteson, and Richard Dazeley. A survey of multi-objective sequential decision-making. *Journal of Artificial Intelligence Research*, 48:67–113, 2013. 226

128. Willem Röpke, Roxana Radulescu, Diederik M Roijers, and Ann Ann Nowé. Communication strategies in multi-objective normal-form games. In *Adaptive and Learning Agents Workshop 2021*, 2021. 226

129. Jonathan Rubin and Ian Watson. Computer poker: A review. *Artificial intelligence*, 175(5-6):958–987, 2011. 251, 252

130. Jordi Sabater and Carles Sierra. Reputation and social network analysis in multi-agent systems. In *Proceedings of the First International Joint Conference on Autonomous Agents and Multiagent Systems: Part 1*, pages 475–482, 2002. 244

131. Tim Salimans, Jonathan Ho, Xi Chen, Szymon Sidor, and Ilya Sutskever. Evolution strategies as a scalable alternative to reinforcement learning. *arXiv:1703.03864*, 2017. 231, 240, 251

132. Mikayel Samvelyan, Tabish Rashid, Christian Schroeder De Witt, Gregory Farquhar, Nantas Nardelli, Tim GJ Rudner, Chia-Man Hung, Philip HS Torr, Jakob Foerster, and Shimon Whiteson. The StarCraft multi-agent challenge. In *Proceedings of the 18th International Conference on Autonomous Agents and MultiAgent Systems, AAMAS '19, Montreal*, 2019. 235, 248, 252, 252, 253

133. Tuomas Sandholm. The state of solving large incomplete-information games, and application to poker. *AI Magazine*, 31(4):13–32, 2010. 252

134. Tuomas Sandholm. Abstraction for solving large incomplete-information games. In *Proceedings of the AAAI Conference on Artificial Intelligence*, volume 29, 2015. 234

135. Thomas D Seeley. The honey bee colony as a superorganism. *American Scientist*, 77(6):546–553, 1989. 219

136. Lloyd S Shapley. Stochastic games. In *Proceedings of the National Academy of Sciences*, volume 39, pages 1095–1100, 1953. 222

137. Yoav Shoham and Kevin Leyton-Brown. *Multiagent Systems: Algorithmic, Game-Theoretic, and Logical Foundations*. Cambridge University Press, 2008. 221

138. Yoav Shoham, Rob Powers, and Trond Grenager. Multi-agent reinforcement learning: a critical survey. Technical report, Stanford University, 2003. 251

139. Robin C Sickles and Valentin Zelenyuk. *Measurement of productivity and efficiency*. Cambridge University Press, 2019. 225

140. David Silver, Satinder Singh, Doina Precup, and Richard S Sutton. Reward is enough. *Artificial Intelligence*, page 103535, 2021. 246

141. David Simões, Nuno Lau, and Luís Paulo Reis. Multi agent deep learning with cooperative communication. *Journal of Artificial Intelligence and Soft Computing Research*, 10, 2020. 236

142. Satinder Singh, Richard L Lewis, Andrew G Barto, and Jonathan Sorg. Intrinsically motivated reinforcement learning: An evolutionary perspective. *IEEE Transactions on Autonomous Mental Development*, 2(2):70–82, 2010. 252

143. Stephen J Smith, Dana Nau, and Tom Throop. Computer bridge: A big win for AI planning. *AI magazine*, 19(2):93–93, 1998. 236, 236

144. Kyunghwan Son, Daewoo Kim, Wan Ju Kang, David Earl Hostallero, and Yung Yi. QTRAN: Learning to factorize with transformation for cooperative multi-agent reinforcement learning. In *International Conference on Machine Learning*, pages 5887–5896. PMLR, 2019. 235

145. Felipe Petroski Such, Vashisht Madhavan, Edoardo Conti, Joel Lehman, Kenneth O Stanley, and Jeff Clune. Deep neuroevolution: Genetic algorithms are a competitive alternative for training deep neural networks for reinforcement learning. *arXiv preprint arXiv:1712.06567*, 2017. 251

146. Peter Sunehag, Guy Lever, Audrunas Gruslys, Wojciech Marian Czarnecki, Vinicius Zambaldi, Max Jaderberg, Marc Lanctot, Nicolas Sonnerat, Joel Z Leibo, Karl Tuyls, and Thore Graepel. Value-decomposition networks for cooperative multi-agent learning. In *Proceedings of the 17th International Conference on Autonomous Agents and MultiAgent Systems, AAMAS 2018, Stockholm, Sweden*, 2017. 235

147. Peter Sunehag, Guy Lever, Siqi Liu, Josh Merel, Nicolas Heess, Joel Z Leibo, Edward Hughes, Tom Eccles, and Thore Graepel. Reinforcement learning agents acquire flocking and symbiotic behaviour in simulated ecosystems. In *Artificial Life Conference Proceedings*, pages 103–110. MIT Press, 2019. 239

148. Oskari Tammelin. Solving large imperfect information games using CFR+. *arXiv preprint arXiv:1407.5042*, 2014. 231, 234

149. Ardi Tampuu, Tambet Matiisen, Dorian Kodelja, Ilya Kuzovkin, Kristjan Korjus, Juhan Aru, Jaan Aru, and Raul Vicente. Multiagent cooperation and competition with deep reinforcement learning. *PloS one*, 12(4):e0172395, 2017. 236, 251

150. Ming Tan. Multi-agent reinforcement learning: Independent vs. cooperative agents. In *International Conference on Machine Learning*, pages 330–337, 1993. 234

151. Shoshannah Tekofsky, Pieter Spronck, Martijn Goudbeek, Aske Plaat, and Jaap van den Herik. Past our prime: A study of age and play style development in Battlefield 3. *IEEE Transactions on Computational Intelligence and AI in Games*, 7(3):292–303, 2015. 252

152. Justin K Terry and Benjamin Black. Multiplayer support for the arcade learning environment. *arXiv preprint arXiv:2009.09341*, 2020. 249

153. Justin K Terry, Benjamin Black, Ananth Hari, Luis Santos, Clemens Dieffendahl, Niall L Williams, Yashas Lokesh, Caroline Horsch, and Praveen Ravi. Pettingzoo: Gym for multi-agent reinforcement learning. *arXiv preprint arXiv:2009.14471*, 2020. 249

154. Julian Togelius, Alex J Champandard, Pier Luca Lanzi, Michael Mateas, Ana Paiva, Mike Preuss, and Kenneth O Stanley. Procedural content generation: Goals, challenges and actionable steps. In *Artificial and Computational Intelligence in Games*. Schloss Dagstuhl-Leibniz-Zentrum für Informatik, 2013. 252

155. Armon Toubman, Jan Joris Roessingh, Pieter Spronck, Aske Plaat, and Jaap Van Den Herik. Dynamic scripting with team coordination in air combat simulation. In *International Conference on Industrial, Engineering and other Applications of Applied Intelligent Systems*, pages 440–449. Springer, 2014. 252

156. Thomas Trenner. Beating Kuhn poker with CFR using python. https://ai.plainenglish.io/building-a-poker-ai-part-6-beating-kuhn-poker-with-cfr-using-python-1b4172a6ab2d. 232, 233

157. Karl Tuyls, Julien Perolat, Marc Lanctot, Joel Z Leibo, and Thore Graepel. A generalised method for empirical game theoretic analysis. In *Proceedings of the 17th International Conference on Autonomous Agents and MultiAgent Systems, AAMAS 2018, Stockholm, Sweden*, 2018. 222

158. Karl Tuyls and Gerhard Weiss. Multiagent learning: Basics, challenges, and prospects. *AI Magazine*, 33(3):41–41, 2012. 251

159. Paul Tylkin, Goran Radanovic, and David C Parkes. Learning robust helpful behaviors in two-player cooperative Atari environments. In *Proceedings of the 20th International Conference on Autonomous Agents and MultiAgent Systems*, pages 1686–1688, 2021. 249

160. Wiebe Van der Hoek and Michael Wooldridge. Multi-agent systems. *Foundations of Artificial Intelligence*, 3:887–928, 2008. 251

161. Max J van Duijn. *The Lazy Mindreader: a Humanities Perspective on Mindreading and Multiple-Order Intentionality*. PhD thesis, Leiden University, 2016. 236

162. Max J Van Duijn, Ineke Sluiter, and Arie Verhagen. When narrative takes over: The representation of embedded mindstates in Shakespeare's Othello. *Language and Literature*, 24(2):148–166, 2015.

163. Max J Van Duijn and Arie Verhagen. Recursive embedding of viewpoints, irregularity, and the role for a flexible framework. *Pragmatics*, 29(2):198–225, 2019. 236

164. Kristof Van Moffaert and Ann Nowé. Multi-objective reinforcement learning using sets of pareto dominating policies. *Journal of Machine Learning Research*, 15(1):3483–3512, 2014. 226, 251

165. Oriol Vinyals, Igor Babuschkin, Wojciech M. Czarnecki, Michaël Mathieu, Andrew Dudzik, Junyoung Chung, David H. Choi, Richard Powell, Timo Ewalds, Petko Georgiev, Junhyuk Oh, Dan Horgan, Manuel Kroiss, Ivo Danihelka, Aja Huang, Laurent Sifre, Trevor Cai, John P. Agapiou, Max Jaderberg, Alexander Sasha Vezhnevets, Rémi Leblond, Tobias Pohlen, Valentin Dalibard, David Budden, Yury Sulsky, James Molloy, Tom Le Paine, Çaglar Gülçehre, Ziyu Wang, Tobias Pfaff, Yuhuai Wu, Roman Ring, Dani Yogatama, Dario Wünsch, Katrina McKinney, Oliver Smith, Tom Schaul, Timothy P. Lillicrap, Koray Kavukcuoglu, Demis Hassabis, Chris Apps, and David Silver. Grandmaster level in StarCraft II using multi-agent reinforcement learning. *Nature*, 575(7782):350–354, 2019. 242, 248, 248, 252

166. Oriol Vinyals, Timo Ewalds, Sergey Bartunov, Petko Georgiev, Alexander Sasha Vezhnevets, Michelle Yeo, Alireza Makhzani, Heinrich Küttler, John P. Agapiou, Julian Schrittwieser, John Quan, Stephen Gaffney, Stig Petersen, Karen Simonyan, Tom Schaul, Hado van Hasselt, David Silver, Timothy P. Lillicrap, Kevin Calderone, Paul Keet, Anthony Brunasso, David Lawrence, Anders Ekermo, Jacob Repp, and Rodney Tsing. StarCraft II: A new challenge for reinforcement learning. *arXiv:1708.04782*, 2017. 252

167. John Von Neumann and Oskar Morgenstern. *Theory of Games and Economic Behavior.* Princeton University Press, 1944. 222, 251

168. John Von Neumann, Oskar Morgenstern, and Harold William Kuhn. *Theory of Games and Economic Behavior (commemorative edition).* Princeton University Press, 2007. 222

169. Douglas Walker and Graham Walker. *The Official Rock Paper Scissors Strategy Guide.* Simon and Schuster, 2004. 251

170. Ying Wen, Yaodong Yang, Rui Luo, Jun Wang, and Wei Pan. Probabilistic recursive reasoning for multi-agent reinforcement learning. In *International Conference on Learning Representations*, 2019. 236

171. Shimon Whiteson. Evolutionary computation for reinforcement learning. In Marco A. Wiering and Martijn van Otterlo, editors, *Reinforcement Learning*, volume 12 of *Adaptation, Learning, and Optimization*, pages 325–355. Springer, 2012. 251

172. Marco A Wiering, Maikel Withagen, and Mădălina M Drugan. Model-based multi-objective reinforcement learning. In *2014 IEEE Symposium on Adaptive Dynamic Programming and Reinforcement Learning (ADPRL)*, pages 1–6. IEEE, 2014. 251

173. Daan Wierstra, Tom Schaul, Jan Peters, and Jürgen Schmidhuber. Natural evolution strategies. In *IEEE Congress on Evolutionary Computation*, pages 3381–3387, 2008. 251

174. Nick Wilkinson and Matthias Klaes. *An Introduction to Behavioral Economics.* Macmillan International Higher Education, 2017. 251

175. Annie Wong, Thomas Bäck, Anna V. Kononova, and Aske Plaat. Deep multiagent reinforcement learning: Challenges and directions. *Artificial Intelligence Review*, 2022. 234, 234, 234, 236, 251

176. Michael Wooldridge. *An Introduction to Multiagent Systems.* Wiley, 2009. 221, 251

177. Anita Williams Woolley, Christopher F Chabris, Alex Pentland, Nada Hashmi, and Thomas W Malone. Evidence for a collective intelligence factor in the performance of human groups. *science*, 330(6004):686–688, 2010. 219, 240, 244, 251

178. Yaodong Yang and Jun Wang. An overview of multi-agent reinforcement learning from game theoretical perspective. *arXiv preprint arXiv:2011.00583*, 2020. 251

179. Chao Yu, Akash Velu, Eugene Vinitsky, Yu Wang, Alexandre Bayen, and Yi Wu. The surprising effectiveness of PPO in cooperative, multi-agent games. *arXiv preprint arXiv:2103.01955*, 2021. 235

180. Kaiqing Zhang, Zhuoran Yang, and Tamer Başar. Multi-agent reinforcement learning: A selective overview of theories and algorithms. *arXiv preprint arXiv:1911.10635*, 2019. 222, 223, 223

181. Kaiqing Zhang, Zhuoran Yang, Han Liu, Tong Zhang, and Tamer Basar. Fully decentralized multi-agent reinforcement learning with networked agents. In *International Conference on Machine Learning*, pages 5872–5881. PMLR, 2018. 240

182. Yan Zheng, Zhaopeng Meng, Jianye Hao, Zongzhang Zhang, Tianpei Yang, and Changjie Fan. A deep Bayesian policy reuse approach against non-stationary agents. In *32nd Neural Information Processing Systems*, pages 962–972, 2018. 236

183. Martin Zinkevich, Michael Johanson, Michael Bowling, and Carmelo Piccione. Regret minimization in games with incomplete information. In *Advances in Neural Information Processing Systems*, pages 1729–1736, 2008.
231, 251

Chapter 8
Hierarchical Reinforcement Learning

The goal of artificial intelligence is to understand and create intelligent behavior; the goal of deep reinforcement learning is to find a behavior policy for ever larger sequential decision problems.

But how does real intelligence find these policies? One of the things that humans are good at is dividing a complex task into simpler subproblems, then solving those tasks, one by one, and combining them as the solution to the larger problem. These subtasks are of different scales, or granularity, than the original problem. For example, when planning a trip from your house to a hotel room in a far-away city, you typically only plan the start and the end in terms of footsteps taken in a certain direction. The in-between part may contain different modes of transportation that get you to your destination quicker, with macro steps, such as taking a train ride or a flight. During this macro, you do not try out footsteps in different directions. Our trip—our policy—is a combination of fine-grain primitive actions and coarse-grain macro-actions.

Hierarchical reinforcement learning studies this real-world-inspired approach to problem solving. It provides formalisms and algorithms to divide problems into larger subproblems and then plans with these subpolicies, as if they were subroutines.

In principle, the hierarchical approach can exploit structure in all sequential decision problems, although some problems are easier than others. Some environments can be subdivided into smaller problems in a natural way, such as navigational tasks on maps or path-finding tasks in mazes. Multi-agent problems also naturally divide into hierarchical teams and have large state spaces where hierarchical methods may help. For other problems, however, it can be hard to find efficient macros, or it can be computationally intensive to find good combinations of macro steps and primitive steps.

Another aspect of hierarchical methods is that since macro-actions take large steps, they may miss the global minimum. The best policies found by hierarchical

A. Plaat, *Deep Reinforcement Learning*,
https://doi.org/10.1007/978-981-19-0638-1_8

methods may be less optimal than those found by "flat" approaches (although they may get there much quicker).

In this chapter, we will start with an example to capture the flavor of hierarchical problem solving. Next, we will look at a theoretical framework that is used to model hierarchical algorithms and at a few examples of algorithms. Finally, we will look deeper at hierarchical environments.

The chapter ends with exercises, a summary, and pointers to further reading.

Core Concepts

- Solve large structured problems by divide and conquer
- Temporal abstraction of actions with options

Core Problem

- Find subgoals and subpolicies efficiently, to perform hierarchical abstraction

Core Algorithms

- Options framework (Sect. 8.2.1)
- Option Critic (Sect. 8.2.3.2)
- Hierarchical actor critic (Sect. 8.2.3.2)

Planning a Trip

Let us see how we plan a major trip to visit a friend that lives in another city, with a hierarchical method. The method would break up the trip in different parts. The first part would be to walk to your closet and get your things and then to get out and get your bike. You would go to the train station and park your bike. You would then take the train to the other city, possibly changing trains enroute if that would be necessary to get you there faster. Arriving in the city, your friend would meet you at the station and would drive you to their house.

A "flat" reinforcement learning method would have at its disposal actions consisting of footsteps in certain directions. This would make for a large space of possible policies, although the fine grain at which the policy would be planned—individual footsteps—would sure be able to find the optimal shortest route.

The hierarchical method has at its disposal a wider variety of actions—macro-actions: it can plan a bike ride, a train trip, and getting a ride by your friend. The route may not be the shortest possible (who knows if the train follows the shortest route between the two cities), but planning will be much faster than painstakingly optimizing footstep by footstep.

8.1 Granularity of the Structure of Problems

In hierarchical reinforcement learning, the granularity of abstractions is larger than the fine grain of the primitive actions of the environment. When we are preparing a meal, we reason in large action chunks: chop onion and cook spaghetti, instead of reasoning about the actuation of individual muscles in our hands and arms. Infants learn to use their muscles to perform certain tasks as they grow up until it becomes second nature.

We generate subgoals that act as temporal abstractions and subpolicies that are macros of multiple ordinary actions [54]. Temporal abstraction allows us to reason about actions of different time scales, sometimes with coarse-grain actions—taking a train—sometimes with fine grain actions—opening a door—mixing macro-actions with primitive actions.

Let us look at advantages and disadvantages of the hierarchical approach.

8.1.1 Advantages

We will start with the advantages of hierarchical methods [22]. First of all, hierarchical reinforcement learning simplifies problems through *abstraction*. Problems are abstracted into a higher level of aggregation. Agents create subgoals and solve fine-grain subtasks first. Actions are abstracted into larger macro-actions to solve these subgoals; agents use temporal abstraction.

Second, the temporal abstractions increase *sample efficiency*. The number of interactions with the environment is reduced because subpolicies are learned to solve subtasks, reducing the environment interactions. Since subtasks are learned, they can be transferred to other problems, supporting transfer learning.

Third, subtasks reduce brittleness due to overspecialization of policies. Policies become more *general* and are able to adapt to changes in the environment more easily.

Fourth, and most importantly, the higher level of abstraction allows agents to solve *larger* more complex problems. This is a reason why for complex multi-agent games such as StarCraft, where teams of agents must be managed, hierarchical approaches are used.

Multi-agent reinforcement learning often exhibits a hierarchical structure; problems can be organized such that each agent is assigned its own subproblem, or the agents themselves may be structured or organized in teams or groups. There can be cooperation within the teams or competition between the teams, or the behavior can be fully cooperative or fully competitive.

Flet-Berliac [22], in a recent overview, summarizes the promise of hierarchical reinforcement learning as follows: (1) achieve long-term credit assignment through faster learning and better generalization, (2) allow structured exploration by exploring with subpolicies rather than with primitive actions, and (3) perform transfer learning because different levels of hierarchy can encompass different knowledge.

8.1.2 Disadvantages

There are also disadvantages and challenges associated with hierarchical reinforcement learning. First of all, it works better when there is *domain knowledge* available about structure in the domain. Many hierarchical methods assume that domain knowledge is available to subdivide the environment so that hierarchical reinforcement learning can be applied.

Second, there is *algorithmic complexity* to be solved. Subgoals must be identified in the problem environment, subpolicies must be learned, and termination conditions are needed. These algorithms must be designed, which costs programmer effort.

Third, hierarchical approaches introduce a new type of actions, macro-actions. Macros are combinations of primitive actions, and their use can greatly improve the performance of the policy. On the other hand, the number of possible combinations of actions is exponentially large in their length [5]. For larger problems, enumeration of all possible macros is out of the question, and the overall-policy function has to be approximated. Furthermore, at each decision point in a hierarchical planning or learning algorithm, we now have the option to consider if any of the macro-actions improves the current policy. The *computational complexity* of the planning and learning choices increases by the introduction of the macro-actions [5], and approximation methods must be used. The efficiency gains of the hierarchical behavioral policy must outweigh the higher cost of finding this policy.

Fourth, the *quality* of a behavioral policy that includes macro-actions may be less than that of a policy consisting only of primitive actions. The macro-actions may skip over possible shorter routes, which the primitive actions would have found.

8.1.2.1 Conclusion

There are advantages and disadvantages to hierarchical reinforcement learning. Whether an efficient policy can be constructed and whether its accuracy is good enough depends on the problem at hand and also on the quality of the algorithms that are used to find this policy.

For a long time, finding good subgoals has been a major challenge. With recent algorithmic advances, especially in function approximation, important progress has been made. We will discuss these advances in the next section.

8.2 Divide and Conquer for Agents

To discuss hierarchical reinforcement learning, first we will discuss a model, the options framework, that formalizes the concepts of subgoals and subpolicies. Next, we will describe the main challenge of hierarchical reinforcement learning, which is sample efficiency. Then we will discuss the main part of this chapter: algorithms for finding subgoals and subpolicies, and finally we will provide an overview of algorithms that have been developed in the field.

8.2.1 The Options Framework

A hierarchical reinforcement learning algorithm tries to solve sequential decision problems more efficiently by identifying common substructures and reusing sub-policies to solve them. The hierarchical approach has three challenges [32, 48]: find subgoals, find a meta-policy over these subgoals, and find subpolicies for these subgoals.

Normally, in reinforcement learning, the agent follows in each state the action that is indicated by the policy. In 1999, Sutton, Precup, and Singh [60] introduced the options framework. This framework introduces formal constructs with which subgoals and subpolicies can be incorporated elegantly into the reinforcement learning setting. The idea of options is simple. Whenever a state is reached that is a subgoal, then, in addition to following a primitive action suggested by the main policy, the option can be taken. This means that not the main action policy is followed, but the option policy, a macro-action consisting of a different subpolicy specially aimed at satisfying the subgoal in one large step. In this way macros are incorporated into the reinforcement learning framework.

We have been using the terms *macro* and *option* somewhat loosely until now; there is, however, a difference between macros and options. A macro is any group of actions, possibly open-ended. An option is a group of actions with a termination condition. Options take in environment observations and output actions until a termination condition is met.

Formally, an option ω has three elements [2]. Each option $\omega = \langle I, \pi, \beta \rangle$ has the following triple:

I_ω	The initiation set $I \subseteq S$ is the states that the option can start from
$\pi_\omega(a\|s)$	The subpolicy $\pi : S \times A \to [0, 1]$ internal to this particular option
$\beta_\omega(s)$	The termination condition $\beta : S \to [0, 1]$ tells us if ω terminates in s

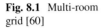

Fig. 8.1 Multi-room grid [60]

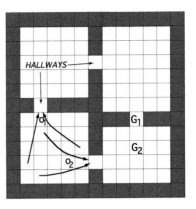

The set of all options is denoted as Ω. In the options framework, there are thus two types of policies: the (meta-)policy *over* options $\pi_\Omega(\omega|s)$ and the subpolicies $\pi_\omega(a|s)$. The subpolicies π_ω are short macros to get from I_ω to β_ω quickly, using the previously learned macro (subpolicy). Temporal abstractions mix actions of different granularity, short and long, primitive action and subpolicy. They allow traveling from I to β without additional learning, using a previously provided or learned subpolicy.

One of the problems for which the options framework works well is room navigation in a Grid world (Fig. 8.1). In a regular reinforcement learning problem, the agent would learn to move step by step. In hierarchical reinforcement learning, the doors between rooms are bottleneck states and are natural subgoals. Macro-actions (subpolicies) are to move to a door in one multi-step action (without considering alternative actions along the way). Then we can go to a different room, if we choose the appropriate option, using another macro, closer to where the main goal is located. The four-room problem from the figure is used in many research works in hierarchical reinforcement learning.

8.2.1.1 Universal Value Function

In the original options framework, the process of identifying the subgoals (the hallways, doors) is external. The subgoals have to be provided manually or by other methods [30, 33, 47, 57]. Subsequently, methods have been published to learn these subgoals.

Options are goal-conditioned subpolicies. More recently, a generalization to parameterized options has been presented in the universal value function, by Schaul et al. [53]. Universal value functions provide a unified theory for goal-conditioned parameterized value approximators $V(s, g, \theta)$.

8.2.2 Finding Subgoals

Whether the hierarchical method improves over a traditional flat method depends on a number of factors. First, there should be enough repeating structure in the domain to be exploited (are there many rooms?), second, the algorithm must find appropriate subgoals (can it find the doors?), third, the options that are found must repeat many times (is the puzzle played frequently enough for the option-finding cost to be offset?), and, finally, subpolicies must be found that give enough improvement (are the rooms large enough that options outweigh actions?).

The original options framework assumes that the structure of the domain is obvious and that the subgoals are given. When this is not the case, then the subgoals must be found by the algorithm. Let us look at an overview of approaches, both tabular and with deep function approximation.

8.2.3 Overview of Hierarchical Algorithms

The options framework provides a convenient formalism for temporal abstraction. In addition to the algorithms that can construct policies consisting of individual actions, we need algorithms that find the subgoals and learn the subpolicies. Finding efficient algorithms for the three tasks is important in order to be able to achieve an efficiency advantage over ordinary "flat" reinforcement learning.

Hierarchical reinforcement learning is based on subgoals. It implements a top-level policy over these subgoals and subpolicies to solve the subgoals. The landscape of subgoals determines to a great extent the efficiency of the algorithm [19]. In recent years new algorithms have been developed to find subpolicies for options, and the field has received a renewed interest [44]. Table 8.1 shows a list of approaches. The table starts with classic tabular approaches (above the line). It continues with more recent deep learning approaches. We will now look at some of the algorithms.

8.2.3.1 Tabular Methods

Divide and conquer is a natural method to exploit hierarchical problem structures. A famous early planning system is STRIPS, the Stanford Research Institute Problem Solver, designed by Richard Fikes and Nils Nilsson in the 1970s [21]. STRIPS created an extensive language for expressing planning problems and was quite influential. Concepts from STRIPS are at the basis of most modern planning systems, action languages, and knowledge representation systems [7, 25, 64]. The concept of macros as open-ended groups of actions was used in STRIPS to create higher-level primitives or subroutines.

Table 8.1 Hierarchical reinforcement learning approaches (tabular and deep)

Name	Agent	Environment	Find Subg	Find Subpol	Ref
STRIPS	Macro-actions	STRIPS planner	−	−	[21]
Abstraction Hier.	State abstraction	Scheduling/plan.	+	+	[31]
HAM	Abstract machines	MDP/maze	−	−	[42]
MAXQ	Value function decomposition	Taxi	−	−	[17]
HTN	Task networks	Block world	−	−	[13, 26]
Bottleneck	Randomized search	Four room	+	+	[57]
Feudal	Manager/worker, RNN	Atari	+	+	[15, 67]
Self p. goal emb.	Self-play subgoal	Mazebase, AntG	+	+	[58]
Deep Skill Netw.	Deep skill array, policy distillation	Minecraft	+	+	[63]
STRAW	End-to-end implicit plans	Atari	+	+	[66]
HIRO	Off-policy	Ant maze	+	+	[37]
Option Critic	Policy-gradient	Four room	+	+	[6]
HAC	Actor critic, hindsight exper. repl.	Four room ant	+	+	[3, 34]
Modul. pol. hier.	Bit-vector, intrinsic motivation	FetchPush	+	+	[43]
h-DQN	Intrinsic motivation	Montezuma's R.	−	+	[32]
Meta l. sh. hier.	Shared primitives, strength metric	Walk, crawl	+	+	[24]
CSRL	Model-based transition dynamics	Robot tasks	−	+	[35]
Learning Repr.	Unsup. subg. disc., intrinsic motiv.	Montezuma's R.	+	+	[48]
AMIGo	Adversarially intrinsic goals	MiniGrid PCG	+	+	[11]

Later planning-based approaches are Parr and Russell's hierarchical abstract machines [42] and Dieterich's MAXQ [17]. Typical applications of these systems are the blocks world, in which a robot arm has to manipulate blocks, stacking them on top of each other, and the taxi world, which we have seen in earlier chapters. An overview of these and other early approaches can be found in Barto et al. [8].

Many of these early approaches focused on macros (the subpolicies) and require that the experimenters identify the subgoals in a planning language. For problems where no such obvious subgoals are available, Knoblock [31] showed how abstraction hierarchies can be generated, although Backstrom et al. [5] found that doing so can be exponentially less efficient. For small room problems, however, Stolle and Precup [57] showed that subgoals can be found in a more efficient way, using a short randomized search to find bottleneck states that can be used as subgoals. This approach finds subgoals automatically, and efficiently, in a rooms-grid world.

Tabular hierarchical methods were mostly applied to small and low-dimensional problems and have difficulty finding subgoals, especially for large problems. The advent of deep function approximation methods attracted renewed interest in hierarchical methods.

8.2.3.2 Deep Learning

Function approximation can potentially reduce the problem of exponentially exploding search spaces that plague tabular methods, especially for subgoal discovery. Deep learning exploits similarities between states using commonalities between features and allows larger problems to be solved. Many new methods were developed. The deep learning approaches in hierarchical reinforcement learning typically are end-to-end: they generate both appropriate subgoals and their policies.

Feudal networks are an older idea from Dayan and Hinton in which an explicit control hierarchy is built of managers and workers that work on tasks and subtasks, organized as in a feudal fiefdom [15]. This idea was used 15 years later as a model for hierarchical deep reinforcement learning by Vezhnevets et al. [67], outperforming non-hierarchical A3C on Montezuma's Revenge, and performing well on other Atari games, achieving a similar score as Option Critic [6]. The approach uses a manager that sets abstract goals (in latent space) for workers. The feudal idea was also used as inspiration for a multi-agent cooperative reinforcement learning design [1], on proof of concept cooperative multi-agent problems on prespecified hierarchies.

Other deep learning approaches include deep skill networks [63], off-policy approaches [37], and self-play [58]. The latter uses an intrinsic motivation approach to learn both a low-level actor and the representation of the state space [45]. Subgoals are learned at the higher level, after which policies are trained at the lower level. Application environments for deep learning have become more challenging and now include Minecraft and robotic tasks such as ant navigation in multiple rooms and maze navigation. The approaches outperform basic non-hierarchical approaches such as DQN.

In STRAW, Vezhnevets et al. [66] learn a model of macro actions, and it is evaluated on text recognition tasks and on Atari games such as PacMan and Frostbite, showing promising results. Zhang et al. [71] use world models to learn latent landmarks (subgoals) for graph-based planning (see also Sect. 5.2.1.2).

Almost two decades after the options framework was introduced, Bacon et al. [6] introduced the Option Critic approach. Option Critic extends the options framework with methods to learn the option subgoal and subpolicy, so that it does not have to be provided externally anymore. The options are learned similar to actor critic using a gradient-based approach. The intra-option policies and termination functions, as

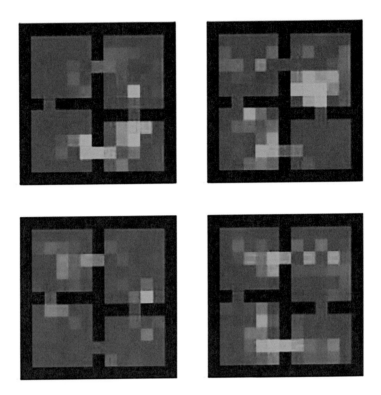

Fig. 8.2 Termination probabilities learned with 4 options by Option Critic [6]; options tend to favor squares close to doors

well as the policy over options, are learned simultaneously. The user of the Option Critic approach has to specify how many options have to be learned. The Option Critic paper reports good results for experiments in a four-room environment with 4 and with 8 options (Fig. 8.2). Option Critic learns options in an end-to-end fashion that scales to larger domains, outperforming DQN in four ALE games (Asterix, Seaquest, Ms. Pacman, Zaxxon) [6].

Levy et al. [34] presented an approach based on Option Critic, called hierarchical actor critic. This approach can learn the goal-conditioned policies at different levels concurrently, where previous approaches had to learn them in a bottom-up fashion. In addition, hierarchical actor critic uses a method for learning the multiple levels of policies for sparse rewards, using hindsight experience replay [3]. In typical robotics tasks, the reinforcement learning algorithm learns more from a successful outcome (bat hits the ball) than from an unsuccessful outcome (bat misses the ball, low). In this failure case, a human learner would draw the conclusion that we can now reach another goal, being bat misses the ball if we aim low. Hindsight experience replay allows learning to take place by incorporating such adjustments of the goal using the benefit of hindsight, so that the algorithm can now also learn from failures, by pretending that they were the goal that you wanted to reach, and learn from them as if they were.

Hierarchical actor critic has been evaluated on grid world tasks and more complex simulated robotics environments, using a 3-level hierarchy.

A final approach that we mention is AMIGo [11], which is related to intrinsic motivation. It uses a teacher to adversarially generate goals for a student. The student is trained with increasingly challenging goals to learn general skills. The system effectively builds up an automatic curriculum of goals. It is evaluated on MiniGrid, a parameterized world that is generated by procedural content generation [12, 49].

Conclusion

Looking back at the list of advantages and disadvantages at the start of this chapter, we see a range of interesting and creative ideas that achieve the advantages (Sect. 8.1.1) by providing methods to address the disadvantages (Sect. 8.1.2). In general, the tabular methods are restricted to smaller problems and often need to be provided with subgoals. Most of the newer deep learning methods find subgoals by themselves, for which then subpolicies are found. Many promising methods have been discussed, and most report to outperform one or more flat baseline algorithms.

The promising results stimulate further research in deep hierarchical methods, and more benchmark studies of large problems are needed. Let us have a closer look at the environments that have been used so far.

8.3   Hierarchical Environments

Many environments for hierarchical reinforcement learning exist, starting with mazes and the four-room environment from the options paper. Environments have evolved with the rest of the field of reinforcement learning; for hierarchical reinforcement learning, no clear favorite benchmark has emerged, although Atari en MuJoCo tasks are often used. In the following, we will review some of the environments that are used in algorithmic studies. Most hierarchical environments are smaller than typically used for model-free flat reinforcement learning, although some studies do use complex environments, such as StarCraft.

8.3.1   Four Rooms and Robot Tasks

Sutton et al. [60] presented the four-room problems to illustrate how the options model worked (Fig. 8.3, left panel). This environment has been used frequently in subsequent papers on reinforcement learning. The rooms are connected by

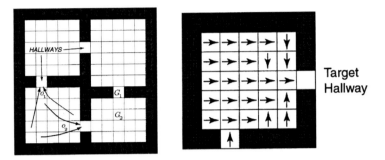

Fig. 8.3 Four rooms, and one room with subpolicy and subgoal [60]

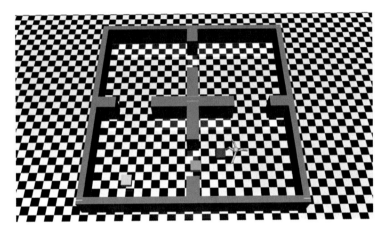

Fig. 8.4 Ant in four rooms [34]

hallways. Options point the way to these hallways, which lead to the goal G_2 of the environment. A hierarchical algorithm should identify the hallways as the subgoals and create subpolicies for each room to go to the hallway subgoal (Fig. 8.3, right panel).

The four-room environment is a toy environment with which algorithms can be explained. More complex versions can be created by increasing the dimensions of the grids and by increasing the number of rooms.

The hierarchical actor critic paper uses the four-room environment as a basis for a robot to crawl through. The agent has to learn both the locomotion task and solving the four-room problem (Fig. 8.4). Other environments that are used for hierarchical reinforcement learning are robot tasks, such as shown in Fig. 8.5 [51].

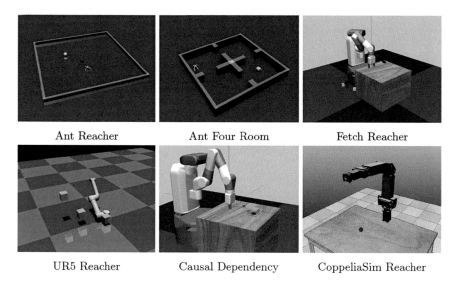

Ant Reacher   Ant Four Room   Fetch Reacher

UR5 Reacher   Causal Dependency   CoppeliaSim Reacher

Fig. 8.5 Six robot tasks [51]

8.3.2 Montezuma's Revenge

One of the most difficult situations for reinforcement learning is when there is little reward signal and when it is delayed. The game of Montezuma's Revenge consists of long stretches in which the agent has to walk without the reward changing. Without smart exploration methods, this game cannot be solved. Indeed, the game has long been a test-bed for research into goal-conditioned and exploration methods.

For the state in Fig. 8.6, the player has to go through several rooms while collecting items. However, to pass through doors (top right and top left corners), the player needs the key. To pick up the key, the player has to climb down the ladders and move toward the key. This is a long and complex sequence before receiving the reward increments for collecting the key. Next, the player has to go to the door to collect another increase in reward. Flat reinforcement learning algorithms struggle with this environment. For hierarchical reinforcement, the long stretches without a reward can be an opportunity to show the usefulness of the option, jumping through the space from states where the reward changes to another reward change. To do so, the algorithm has to be able to identify the key as a subgoal.

Rafati and Noelle [48] learn subgoals in Montezuma's Revenge and so do Kulkarni et al. [32]. Learning to choose promising subgoals is a challenging problem by itself. Once subgoals are found, the subpolicies can be learned by introducing a reward signal for achieving the subgoals. Such intrinsic rewards are related to *intrinsic motivation* and the psychological concept of curiosity [4, 39].

Figure 8.7 illustrates the idea behind intrinsic motivation. In ordinary reinforcement learning, a critic in the environment provides rewards to the agent. When

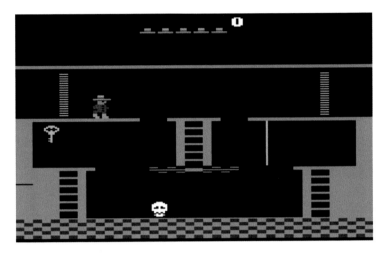

Fig. 8.6 Montezuma's revenge [9]

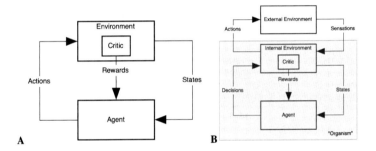

Fig. 8.7 Intrinsic motivation in reinforcement learning [56]

the agent has an internal environment where an internal critic provides rewards, these internal rewards provide an intrinsic motivation to the agent. This mechanism aims to more closely model exploration behavior in animals and humans [56]. For example, during curiosity-driven activities, children use knowledge to generate intrinsic goals while playing, building block structures, etc. While doing this, they construct subgoals such as putting a lighter entity on top of a heavier entity in order to build a tower [32, 56]. Intrinsic motivation is an active field of research. A recent survey is [4].

Montezuma's Revenge has also been used as benchmark for the Go-Explore algorithm, which has achieved good results in sparse reward problems, using a goal-conditioned policy with cell aggregation [20]. Go-Explore performs a planning-like form of backtracking, combining elements of planning and learning in a different way than AlphaZero.

8.3.3 Multi-Agent Environments

Many multi-agent problems are a natural match for hierarchical reinforcement learning, since agents often work together in teams or other hierarchical structure. For multi-agent hierarchical problems, a multitude of different environments are used.

Makar et al. [27, 36] study cooperative multi-agent learning and use small tasks such as a two-agent cooperative trash collection task, the dynamic rescheduling of automated guided vehicles in a factory, as well as an environment in which agents communicate among each other. Han et al. [29] use a multi-agent Taxi environment. Tang et al. [61] also use robotic trash collection.

Due to the computational complexity of multi-agent and hierarchical environments, many of the environments are of lower dimensionality than what we see in model-free and model-based single-agent reinforcement learning. There are a few exceptions, as we saw in the previous chapter (Capture the Flag and StarCraft). However, the algorithms that were used for these environments were based on population-based self-play algorithms that are well suited for parallelization; hierarchical reinforcement algorithms of the type that we have discussed in this chapter were of less importance [68].

8.3.4 Hands On: Hierarchical Actor Critic Example

The research reported in this chapter is of a more manageable scale than in some other chapters. Environments are smaller, and computational demands are more reasonable. Four-room experiments and experiments with movement of single robot arms invite experimentation and tweaking. Again, as in the other chapters, the code of most papers can be found online on GitHub.

Hierarchical reinforcement learning is well suited for experimentation because the environments are small, and the concepts of hierarchy, team, and subgoal are intuitively appealing. Debugging one's implementation should be just that bit easier when the desired behavior of the different pieces of code is clear.

To get you started with hierarchical reinforcement learning, we will go to HAC: hierarchical actor critic [34]. Algorithm 8.1 shows the pseudocode, where TBD is the subgoal in hindsight [34].

A blog[1] with animations has been written, a video[2] has been made of the results, and the code can be found in GitHub.[3]

---

[1] http://bigai.cs.brown.edu/2019/09/03/hac.html.

[2] https://www.youtube.com/watch?v=DYcVTveeNK0.

[3] https://github.com/andrew-j-levy/Hierarchical-Actor-Critc-HAC-.

Algorithm 8.1 Hierarchical actor critic [34]

Input:
    Key agent parameters: the number of levels in hierarchy k, maximum subgoal horizon H, and subgoal testing frequency λ
Output: k trained actor and critic functions $\pi_0, ..., \pi_{k-1}, Q_0, ..., Q_{k-1}$

for M episodes **do**   ▷ Train for M episodes
    $s \leftarrow S_{\text{init}}, g \leftarrow G_{k-1}$ ▷ Sample initial state and task goal
    train-level($k - 1, s, g$)   ▷ Begin training
    Update all actor and critic networks
end for

function TRAIN-LEVEL(i :: level, s :: state, g :: goal)
    $s_i \leftarrow s, g_i \leftarrow g$ ▷ Set current state and goal for level i
    for H attempts or until g_n, $i \leq n < k$ achieved **do**
        $a_i \leftarrow \pi_i(s_i, g_i)$ + noise (if not subgoal testing) ▷ Sample (noisy) action from policy
        if $i > 0$ **then**
            Determine whether to test subgoal a_i
            $s_i' \leftarrow$ train-level($i - 1, s_i, a_i$)   ▷ Train level $i - 1$ using subgoal a_i
        else
            Execute primitive action a_0 and observe next state s_0'
        end if
                                                                        ▷ Create replay transitions
        if $i > 0$ and a_i missed **then**
            if a_i was tested **then**   ▷ Penalize subgoal a_i
                Replay_Buffer$_i$ $\leftarrow$ [$s = s_i, a = a_i, r =$ Penalty, $s' = s_i', g = g_i, \gamma = 0$]
            end if
            $a_i \leftarrow s_i'$ ▷ Replace original action with action executed in hindsight
        end if
                                                                        ▷ Evaluate executed action on current goal and hindsight goals
        Replay_Buffer$_i$ $\leftarrow$ [$s = s_i, a = a_i, r \in \{-1, 0\}, s' = s_i', g = g_i, \gamma \in \{\gamma, 0\}$]
        HER_Storage$_i$ $\leftarrow$ [$s = s_i, a = a_i, r =$TBD, $s' = s_i', g =$TBD, $\gamma =$TBD]
        $s_i \leftarrow s_i'$
    end for
    Replay_Buffer$_i$ $\leftarrow$ Perform HER using HER_Storage$_i$ transitions
    return s_i' ▷ Output current state
end function

To run the hierarchical actor critic experiments, you need MuJoCo and the required Python wrappers. The code is TensorFlow 2 compatible. When you have cloned the repository, run the experiment with

```
python3 initialize_HAC.py --retrain
```

which will train an UR5 reacher agent with a 3-level hierarchy. Here is a video[4] that shows how it should look like after 450 training episodes. You can watch your trained agent with the command

```
python3 initialize_HAC.py --test --show
```

The README at the GitHub repository contains more suggestions on what to try. You can try different hyperparameters, and you can modify the designs, if you feel like it. Happy experimenting!

Summary and Further Reading

We will now summarize the chapter and provide pointers to further reading.

Summary

A typical reinforcement learning algorithm moves in small steps. For a state, it picks an action, gives it to the environment for a new state and a reward, and processes the reward to pick a new action. Reinforcement learning works step by small step. In contrast, consider the following problem: in the real world, when we plan a trip from A to B, we use abstraction to reduce the state space, to be able to reason at a higher level. We do not reason at the level of footsteps to take, but we first decide on the mode of transportation to get close to our goal, and then we fill in the different parts of the journey with small steps.

Hierarchical reinforcement learning tries to mimic this idea: conventional reinforcement learning works at the level of a single state; *hierarchical* reinforcement learning performs abstraction, solving subproblems in sequence. Temporal abstraction is described in a paper by Sutton et al. [60]. Hierarchical reinforcement learning uses the principles of divide and conquer to make solving large problems feasible. It finds subgoals in the space that it solves with subpolicies (macros or options).

Despite the appealing intuition, progress in hierarchical reinforcement learning was initially slow. Finding these new subgoals and subpolicies is a computationally intensive problems that is exponential in the number of actions, and in some situations it is quicker to use conventional "flat" reinforcement learning methods, unless domain knowledge can be exploited. The advent of deep learning provided

---

[4] https://www.youtube.com/watch?v=R86Vs9Vb6Bc.

a boost to hierarchical reinforcement learning, and much progress is being reported in important tasks such as learning subgoals automatically and finding subpolicies.

Although popular for single-agent reinforcement learning, hierarchical methods are also used in multi-agent problems. Multi-agent problems often feature agents that work in teams, cooperate within, and compete between the teams. Such an agent-hierarchy is a natural fit for hierarchical solution methods. Hierarchical reinforcement learning remains a promising technique.

Further Reading

Hierarchical reinforcement learning and subgoal finding have a rich and long history [8, 17, 23, 30, 33, 42, 44, 47, 60]; see also Table 8.1. Macro-actions are a basic approach [30, 50]. Others, using macros, are [18, 69, 70]. The options framework has provided a boost to the development of the field [60]. Other approaches are MAXQ [16] and Feudal networks [67].

Earlier tabular approaches are [17, 23, 30, 33, 42, 47, 60].

Recent methods are Option Critic [6] and hierarchical actor critic [34]. There are many deep learning methods for finding subgoals and subpolicies [14, 23, 24, 34, 37, 41, 46, 53, 59, 65]. Andrychowicz et al. [3] introduce hindsight experience replay, which can improve performance for hierarchical methods.

Intrinsic motivation is a concept from developmental neuroscience that has come to reinforcement learning with the purpose of providing learning signals in large spaces. It is related to curiosity. Botvinick et al. [10] have written an overview of hierarchical reinforcement learning and neuroscience. Aubret et al. [4] provide a survey of intrinsic motivation for reinforcement learning. Intrinsic motivation is used by [32, 48]. Intrinsic motivation is closely related to goal-driven reinforcement learning [38–40, 52, 53].

Exercises

It is time for the Exercises to test your knowledge.

Questions

Below are some quick questions to check your understanding of this chapter. For each question, a simple, single sentence answer should be sufficient.

1. Why can hierarchical reinforcement learning be faster?
2. Why can hierarchical reinforcement learning be slower?

3. Why may hierarchical reinforcement learning give an answer of lesser quality?
4. Is hierarchical reinforcement more general or less general?
5. What is an option?
6. What are the three elements that an option consists of?
7. What is a macro?
8. What is intrinsic motivation?
9. How do multi-agent and hierarchical reinforcement learning fit together?
10. What is so special about Montezuma's Revenge?

Exercises

Let us go to the programming exercises to become more familiar with the methods that we have covered in this chapter.

1. *Four Rooms* Implement a hierarchical solver for the four-room environment. You can code the hallway subgoals using domain knowledge. Use a simple tabular, planning, approach. How will you implement the subpolicies?
2. *Flat* Implement a flat planning or Q-learning-based solver for 4 rooms. Compare this program to the tabular hierarchical solver. Which is quicker? Which of the two does fewer environment actions?
3. *Sokoban* Implement a Sokoban solver using a hierarchical approach (challenging). The challenge in Sokoban is that there can be dead-ends in the game that you create rendering the game unsolvable (also see the literature [28, 55]). Recognizing these dead-end moves is important. What are the subgoals? Rooms or each box-task is one subgoal, or can you find a way to code dead-ends as subgoal? How far can you get? Find Sokoban levels.[5,6,7]
4. *Petting Zoo* Choose one of the easier multi-agent problems from the Petting Zoo [62], introduce teams, and write a hierarchical solver. First try a tabular planning approach, and then look at hierarchical actor critic (challenging).
5. *StarCraft* The same as the previous exercise, only now with StarCraft (very challenging).

References

1. Sanjeevan Ahilan and Peter Dayan. Feudal multi-agent hierarchies for cooperative reinforcement learning. *arXiv preprint arXiv:1901.08492*, 2019. 271
2. Safa Alver. The option-critic architecture. https://alversafa.github.io/blog/2018/11/28/optncrtc.html, 2018. 267

---

[5] http://sneezingtiger.com/sokoban/levels.html.

[6] http://www.sokobano.de/wiki/index.php?title=Level_format.

[7] https://www.sourcecode.se/sokoban/levels.

3. Marcin Andrychowicz, Filip Wolski, Alex Ray, Jonas Schneider, Rachel Fong, Peter Welinder, Bob McGrew, Josh Tobin, Pieter Abbeel, and Wojciech Zaremba. Hindsight experience replay. In *Advances in Neural Information Processing Systems*, pages 5048–5058, 2017. 270, 272, 280
4. Arthur Aubret, Laetitia Matignon, and Salima Hassas. A survey on intrinsic motivation in reinforcement learning. *arXiv preprint arXiv:1908.06976*, 2019. 275, 276, 280
5. Christer Backstrom and Peter Jonsson. Planning with abstraction hierarchies can be exponentially less efficient. In *Proceedings of the 14th International Joint Conference on Artificial Intelligence*, volume 2, pages 1599–1604, 1995. 266, 266, 270
6. Pierre-Luc Bacon, Jean Harb, and Doina Precup. The option-critic architecture. In *Proceedings of the AAAI Conference on Artificial Intelligence*, volume 31, 2017. 270, 271, 271, 272, 272, 280
7. Chitta Baral. *Knowledge Representation, Reasoning and Declarative Problem Solving*. Cambridge university press, 2003. 269
8. Andrew G Barto and Sridhar Mahadevan. Recent advances in hierarchical reinforcement learning. *Discrete Event Dynamic Systems*, 13(1-2):41–77, 2003. 270, 280
9. Marc G Bellemare, Yavar Naddaf, Joel Veness, and Michael Bowling. The Arcade Learning Environment: An evaluation platform for general agents. *Journal of Artificial Intelligence Research*, 47:253–279, 2013. 276
10. Matthew M Botvinick, Yael Niv, and Andew G Barto. Hierarchically organized behavior and its neural foundations: a reinforcement learning perspective. *Cognition*, 113(3):262–280, 2009. 280
11. Andres Campero, Roberta Raileanu, Heinrich Küttler, Joshua B Tenenbaum, Tim Rocktäschel, and Edward Grefenstette. Learning with AMIGo: Adversarially motivated intrinsic goals. In *International Conference on Learning Representations*, 2020. 270, 273
12. Maxime Chevalier-Boisvert, Lucas Willems, and Sumans Pal. Minimalistic gridworld environment for OpenAI Gym https://github.com/maximecb/gym-minigrid, 2018. 273
13. Ken Currie and Austin Tate. O-plan: the open planning architecture. *Artificial Intelligence*, 52(1):49–86, 1991. 270
14. Christian Daniel, Herke Van Hoof, Jan Peters, and Gerhard Neumann. Probabilistic inference for determining options in reinforcement learning. *Machine Learning*, 104(2):337–357, 2016. 280
15. Peter Dayan and Geoffrey E Hinton. Feudal reinforcement learning. In *Advances in Neural Information Processing Systems*, pages 271–278, 1993. 270, 271
16. Thomas G Dietterich. The MAXQ method for hierarchical reinforcement learning. In *International Conference on Machine Learning*, volume 98, pages 118–126, 1998. 280
17. Thomas G Dietterich. Hierarchical reinforcement learning with the MAXQ value function decomposition. *Journal of Artificial Intelligence Research*, 13:227–303, 2000. 270, 270, 280, 280
18. Ishan P Durugkar, Clemens Rosenbaum, Stefan Dernbach, and Sridhar Mahadevan. Deep reinforcement learning with macro-actions. *arXiv preprint arXiv:1606.04615*, 2016. 280
19. Zach Dwiel, Madhavun Candadai, Mariano Phielipp, and Arjun K Bansal. Hierarchical policy learning is sensitive to goal space design. *arXiv preprint arXiv:1905.01537*, 2019. 269
20. Adrien Ecoffet, Joost Huizinga, Joel Lehman, Kenneth O Stanley, and Jeff Clune. First return, then explore. *Nature*, 590(7847):580–586, 2021. 276
21. Richard E Fikes, Peter E Hart, and Nils J Nilsson. Learning and executing generalized robot plans. *Artificial Intelligence*, 3:251–288, 1972. 269, 270
22. Yannis Flet-Berliac. The promise of hierarchical reinforcement learning. https://thegradient.pub/the-promise-of-hierarchical-reinforcement-learning/, March 2019. 265, 266
23. Carlos Florensa, David Held, Xinyang Geng, and Pieter Abbeel. Automatic goal generation for reinforcement learning agents. In *International Conference on Machine Learning*, pages 1515–1528. PMLR, 2018. 280, 280, 280
24. Kevin Frans, Jonathan Ho, Xi Chen, Pieter Abbeel, and John Schulman. Meta learning shared hierarchies. In *International Conference on Learning Representations*, 2018. 270, 280

25. Michael Gelfond and Vladimir Lifschitz. Action languages. *Electronic Transactions on Artificial Intelligence*, 2(3–4):193–210, 1998. 269

26. Malik Ghallab, Dana Nau, and Paolo Traverso. *Automated Planning: theory and practice*. Elsevier, 2004. 270

27. Mohammad Ghavamzadeh, Sridhar Mahadevan, and Rajbala Makar. Hierarchical multi-agent reinforcement learning. *Autonomous Agents and Multi-Agent Systems*, 13(2):197–229, 2006. 277

28. Nathan Grinsztajn, Johan Ferret, Olivier Pietquin, Philippe Preux, and Matthieu Geist. There is no turning back: A self-supervised approach for reversibility-aware reinforcement learning. *arXiv preprint arXiv:2106.04480*, 2021. 281

29. Dongge Han, Wendelin Boehmer, Michael Wooldridge, and Alex Rogers. Multi-agent hierarchical reinforcement learning with dynamic termination. In *Pacific Rim International Conference on Artificial Intelligence*, pages 80–92. Springer, 2019. 277

30. Milos Hauskrecht, Nicolas Meuleau, Leslie Pack Kaelbling, Thomas L Dean, and Craig Boutilier. Hierarchical solution of Markov decision processes using macro-actions. In *UAI '98: Proceedings of the Fourteenth Conference on Uncertainty in Artificial Intelligence, University of Wisconsin Business School, Madison, Wisconsin*, 1998. 268, 280, 280, 280

31. Craig A Knoblock. Learning abstraction hierarchies for problem solving. In *AAAI*, pages 923–928, 1990. 270, 270

32. Tejas D Kulkarni, Karthik Narasimhan, Ardavan Saeedi, and Josh Tenenbaum. Hierarchical deep reinforcement learning: Integrating temporal abstraction and intrinsic motivation. In *Advances in Neural Information Processing Systems*, pages 3675–3683, 2016. 267, 270, 275, 276, 280

33. John E Laird, Paul S Rosenbloom, and Allen Newell. Chunking in Soar: the anatomy of a general learning mechanism. *Machine learning*, 1(1):11–46, 1986. 268, 280, 280

34. Andrew Levy, George Konidaris, Robert Platt, and Kate Saenko. Learning multi-level hierarchies with hindsight. In *International Conference on Learning Representations*, 2019. 270, 272, 274, 277, 277, 278, 280, 280

35. Zhuoru Li, Akshay Narayan, and Tze-Yun Leong. An efficient approach to model-based hierarchical reinforcement learning. In *Proceedings of the AAAI Conference on Artificial Intelligence*, volume 31, 2017. 270

36. Rajbala Makar, Sridhar Mahadevan, and Mohammad Ghavamzadeh. Hierarchical multi-agent reinforcement learning. In *Proceedings of the Fifth International Conference on Autonomous Agents*, pages 246–253. ACM, 2001. 277

37. Ofir Nachum, Shixiang Gu, Honglak Lee, and Sergey Levine. Data-efficient hierarchical reinforcement learning. In *Advances in Neural Information Processing Systems*, pages 3307–3317, 2018. 270, 271, 280

38. Pierre-Yves Oudeyer and Frederic Kaplan. How can we define intrinsic motivation? In *the 8th International Conference on Epigenetic Robotics: Modeling Cognitive Development in Robotic Systems*. Lund University Cognitive Studies, Lund: LUCS, Brighton, 2008. 280

39. Pierre-Yves Oudeyer and Frederic Kaplan. What is intrinsic motivation? A typology of computational approaches. *Frontiers in Neurorobotics*, 1:6, 2009. 275

40. Pierre-Yves Oudeyer, Frederic Kaplan, and Verena V Hafner. Intrinsic motivation systems for autonomous mental development. *IEEE Transactions on Evolutionary Computation*, 11(2):265–286, 2007. 280

41. Aleksandr I Panov and Aleksey Skrynnik. Automatic formation of the structure of abstract machines in hierarchical reinforcement learning with state clustering. *arXiv preprint arXiv:1806.05292*, 2018. 280

42. Ronald Parr and Stuart J Russell. Reinforcement learning with hierarchies of machines. In *Advances in Neural Information Processing Systems*, pages 1043–1049, 1998. 270, 270, 280, 280

43. Alexander Pashevich, Danijar Hafner, James Davidson, Rahul Sukthankar, and Cordelia Schmid. Modulated policy hierarchies. *arXiv preprint arXiv:1812.00025*, 2018. 270

44. Shubham Pateria, Budhitama Subagdja, Ah-hweewee Tan, and Chai Quek. Hierarchical reinforcement learning: A comprehensive survey. *ACM Computing Surveys (CSUR)*, 54(5):1–35, 2021. 269, 280

45. Alexandre Péré, Sébastien Forestier, Olivier Sigaud, and Pierre-Yves Oudeyer. Unsupervised learning of goal spaces for intrinsically motivated goal exploration. In *International Conference on Learning Representations*, 2018. 271

46. Karl Pertsch, Oleh Rybkin, Frederik Ebert, Shenghao Zhou, Dinesh Jayaraman, Chelsea Finn, and Sergey Levine. Long-horizon visual planning with goal-conditioned hierarchical predictors. In *Advances in Neural Information Processing Systems*, 2020. 280

47. Doina Precup, Richard S Sutton, and Satinder P Singh. Planning with closed-loop macro actions. In *Working notes of the 1997 AAAI Fall Symposium on Model-directed Autonomous Systems*, pages 70–76, 1997. 268, 280, 280

48. Jacob Rafati and David C Noelle. Learning representations in model-free hierarchical reinforcement learning. In *Proceedings of the AAAI Conference on Artificial Intelligence*, volume 33, pages 10009–10010, 2019. 267, 270, 275, 280

49. Roberta Raileanu and Tim Rocktäschel. RIDE: rewarding impact-driven exploration for procedurally-generated environments. In *International Conference on Learning Representations*, 2020. 273

50. Jette Randlov. Learning macro-actions in reinforcement learning. In *Advances in Neural Information Processing Systems*, pages 1045–1051, 1998. 280

51. Frank Röder, Manfred Eppe, Phuong DH Nguyen, and Stefan Wermter. Curious hierarchical actor-critic reinforcement learning. In *International Conference on Artificial Neural Networks*, pages 408–419. Springer, 2020. 274, 275

52. Richard M Ryan and Edward L Deci. Intrinsic and extrinsic motivations: Classic definitions and new directions. *Contemporary Educational Psychology*, 25(1):54–67, 2000. 280

53. Tom Schaul, Daniel Horgan, Karol Gregor, and David Silver. Universal value function approximators. In *International Conference on Machine Learning*, pages 1312–1320, 2015. 268, 280, 280

54. Jürgen Schmidhuber. Learning to generate sub-goals for action sequences. In *Artificial neural networks*, pages 967–972, 1991. 265

55. Yaron Shoham and Gal Elidan. Solving Sokoban with forward-backward reinforcement learning. In *Proceedings of the International Symposium on Combinatorial Search*, volume 12, pages 191–193, 2021. 281

56. Satinder Singh, Andrew G Barto, and Nuttapong Chentanez. Intrinsically motivated reinforcement learning. Technical report, University of Amherst, Mass, Department of Computer Science, 2005. 276, 276, 276

57. Martin Stolle and Doina Precup. Learning options in reinforcement learning. In *International Symposium on Abstraction, Reformulation, and Approximation*, pages 212–223. Springer, 2002. 268, 270, 270

58. Sainbayar Sukhbaatar, Emily Denton, Arthur Szlam, and Rob Fergus. Learning goal embeddings via self-play for hierarchical reinforcement learning. *arXiv preprint arXiv:1811.09083*, 2018. 270, 271

59. Peter Sunehag, Guy Lever, Audrunas Gruslys, Wojciech Marian Czarnecki, Vinicius Zambaldi, Max Jaderberg, Marc Lanctot, Nicolas Sonnerat, Joel Z Leibo, Karl Tuyls, and Thore Graepel. Value-decomposition networks for cooperative multi-agent learning. In *Proceedings of the 17th International Conference on Autonomous Agents and MultiAgent Systems, AAMAS 2018, Stockholm, Sweden*, 2017. 280

60. Richard S Sutton, Doina Precup, and Satinder Singh. Between MDPs and semi-MDPs: a framework for temporal abstraction in reinforcement learning. *Artificial Intelligence*, 112(1-2):181–211, 1999. 267, 268, 273, 274, 279, 280, 280, 280

61. Hongyao Tang, Jianye Hao, Tangjie Lv, Yingfeng Chen, Zongzhang Zhang, Hangtian Jia, Chunxu Ren, Yan Zheng, Zhaopeng Meng, Changjie Fan, and Li Wang. Hierarchical deep multiagent reinforcement learning with temporal abstraction. *arXiv preprint arXiv:1809.09332*, 2018. 277

62. Justin K Terry, Benjamin Black, Ananth Hari, Luis Santos, Clemens Dieffendahl, Niall L Williams, Yashas Lokesh, Caroline Horsch, and Praveen Ravi. PettingZoo: Gym for multi-agent reinforcement learning. *arXiv preprint arXiv:2009.14471*, 2020. 281

63. Chen Tessler, Shahar Givony, Tom Zahavy, Daniel Mankowitz, and Shie Mannor. A deep hierarchical approach to lifelong learning in Minecraft. In *Proceedings of the AAAI Conference on Artificial Intelligence*, volume 31, 2017. 270, 271

64. Frank Van Harmelen, Vladimir Lifschitz, and Bruce Porter. *Handbook of Knowledge Representation*. Elsevier, 2008. 269

65. Vivek Veeriah, Tom Zahavy, Matteo Hessel, Zhongwen Xu, Junhyuk Oh, Iurii Kemaev, Hado van Hasselt, David Silver, and Satinder Singh. Discovery of options via meta-learned subgoals. *arXiv preprint arXiv:2102.06741*, 2021. 280

66. Alexander Vezhnevets, Volodymyr Mnih, Simon Osindero, Alex Graves, Oriol Vinyals, John Agapiou, and Koray Kavukcuoglu. Strategic attentive writer for learning macro-actions. In *Advances in Neural Information Processing Systems*, pages 3486–3494, 2016. 270, 271

67. Alexander Vezhnevets, Simon Osindero, Tom Schaul, Nicolas Heess, Max Jaderberg, David Silver, and Koray Kavukcuoglu. Feudal networks for hierarchical reinforcement learning. In *Intl Conf on Machine Learning*, pages 3540–3549. PMLR, 2017. 270, 271, 280

68. Oriol Vinyals, Igor Babuschkin, Wojciech M. Czarnecki, Michaël Mathieu, Andrew Dudzik, Junyoung Chung, David H. Choi, Richard Powell, Timo Ewalds, Petko Georgiev, Junhyuk Oh, Dan Horgan, Manuel Kroiss, Ivo Danihelka, Aja Huang, Laurent Sifre, Trevor Cai, John P. Agapiou, Max Jaderberg, Alexander Sasha Vezhnevets, Rémi Leblond, Tobias Pohlen, Valentin Dalibard, David Budden, Yury Sulsky, James Molloy, Tom Le Paine, Çaglar Gülçehre, Ziyu Wang, Tobias Pfaff, Yuhuai Wu, Roman Ring, Dani Yogatama, Dario Wünsch, Katrina McKinney, Oliver Smith, Tom Schaul, Timothy P. Lillicrap, Koray Kavukcuoglu, Demis Hassabis, Chris Apps, and David Silver. Grandmaster level in StarCraft II using multi-agent reinforcement learning. *Nature*, 575(7782):350–354, 2019. 277

69. Yuchen Xiao, Joshua Hoffman, and Christopher Amato. Macro-action-based deep multi-agent reinforcement learning. In *Conference on Robot Learning*, pages 1146–1161. PMLR, 2020. 280

70. Sijia Xu, Hongyu Kuang, Zhuang Zhi, Renjie Hu, Yang Liu, and Huyang Sun. Macro action selection with deep reinforcement learning in StarCraft. In *AAAI Artificial Intelligence and Interactive Digital Entertainment*, volume 15, pages 94–99, 2019. 280

71. Lunjun Zhang, Ge Yang, and Bradly C Stadie. World model as a graph: Learning latent landmarks for planning. In *International Conference on Machine Learning*, pages 12611–12620. PMLR, 2021.

271

Chapter 9
Meta-Learning

Although current deep reinforcement learning methods have obtained great successes, training times for most interesting problems are high; they are often measured in weeks or months, consuming time and resources—as you may have noticed while doing some of the exercises at the end of the chapters.

Model-based methods aim to reduce the sample complexity in order to speed up learning—but still, for each new task a new network has to be trained from scratch. In this chapter we turn to another approach that aims to *reuse* information learned in earlier training tasks from a closely related problem. When humans learn a new task, they do not learn from a blank slate. Children learn to walk and then they learn to run; they follow a training curriculum, and they remember. Human learning builds on existing knowledge, using knowledge from previously learned tasks to facilitate the learning of new tasks. In machine learning such transfer of previously learned knowledge from one task to another is called *transfer learning*. We will study it in this chapter.

Humans learn continuously. When learning a new task, we do not start from scratch, zapping our minds first to emptiness. Previously learned task representations allow us to learn new representations for new tasks quickly; in effect, we have learned to learn. Understanding how we (learn to) learn has intrigued artificial intelligence researchers since the early days, and it is the topic of this chapter.

The fields of transfer learning and meta-learning are tightly related. For both, the goal is to speed up learning a new task, using previous knowledge. In transfer learning, we pretrain our parameter network with knowledge from a single task. In meta-learning, we use multiple related tasks.

In this chapter, we first discuss the concept of lifelong learning, something that is quite familiar to human beings. Then we discuss transfer learning, followed by meta-learning. Next, we discuss some of the benchmarks that are used to test transfer learning and meta-learning.

The chapter is concluded with exercises, a summary, and pointers to further reading.

A. Plaat, *Deep Reinforcement Learning*,
https://doi.org/10.1007/978-981-19-0638-1_9

Core Concepts

- Knowledge transfer
- Learning to learn

Core Problem

- Speed up learning with knowledge from related tasks.

Core Algorithms

- Pretraining (Listing 9.1)
- Model-Agnostic Meta-Learning (Algorithm 9.1)

Foundation Models

Humans are good at meta-learning. We learn new tasks more easily after we have learned other tasks. Teach us to walk, and we learn how to run. Teach us to play the violin, the viola, and the cello, and we more easily learn to play the double bass (Fig. 9.1).

Fig. 9.1 Violin, viola, cello, and double bass

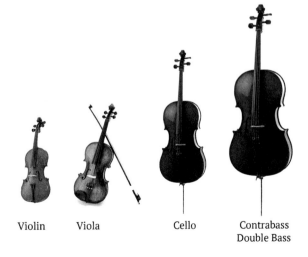

Violin   Viola   Cello   Contrabass
                                                    Double Bass

Current deep learning networks are large, with many layers of neurons and millions of parameters. For new problems, training large networks on large datasets or environments takes time, up to weeks, or months—both for supervised and for reinforcement learning. In order to shorten training times for subsequent networks, these are often pretrained, using foundation models [14]. With pretraining, some of the existing weights of another network are used as starting point for finetuning a network on a new dataset, instead of using a randomly initialized network.

Pretraining works especially well on deeply layered architectures. The reason is that the "knowledge" in the layers goes from generic to specific: lower layers contain generic filters such as lines and curves, and upper layers contain more specific filters such as ears, noses, and mouths (for a face recognition application) [66, 72]. These lower layers contain more generic information that is well suited for transfer to other tasks.

Foundation models are large models in a certain field, such as image recognition or natural language processing, which are trained extensively on large datasets. Foundation models contain general knowledge that can be specialized for a certain purpose. The world of applied deep learning has moved from training a net from scratch for a certain problem, to taking a part of an existing net that is trained for a related problem and then finetuning it on the new task. Nearly all state-of-the-art visual perception approaches rely on the same approach: (1) pretrain a convolutional network on a large, manually annotated image classification dataset and (2) finetune the network on a smaller, task-specific dataset [1, 27, 40, 47, 147], see Fig. 9.2 for some thumbnails of ImageNet [99]. For natural language recognition, pretraining is also the norm—for example, for large-scale pretrained language models such as Word2vec [73, 74], BERT [25], and GPT-3 [89].

In this chapter, we will study pretraining, and more.

Fig. 9.2 ImageNet thumbnails [99]

9.1 Learning to Learn Related Problems

Training times for modern deep networks are large. Training AlexNet for ImageNet took 5–6 days on 2 GPUs in 2012 [59], see Sect. B.3.1.1. In reinforcement learning, training AlphaGo took weeks [110, 111]; in natural language processing, training also takes a long time [25], even excessively long as in the case of GPT-3 [18]. Clearly, some solution is needed. Before we look closer at transfer learning, let us have a look at the bigger picture: lifelong learning.

When humans learn a new task, learning is based on previous experience. Initial learning by infants of elementary skills in vision, speech, and locomotion takes years. Subsequent learning of new skills builds on the previously acquired skills. Existing knowledge is adapted, and new skills are learned based on previous skills.

Lifelong learning remains a long-standing challenge for machine learning; in current methods the continuous acquisition of information often leads to interference of concepts or catastrophic forgetting [109]. This limitation represents a major drawback for deep networks that typically learn representations from stationary batches of training data. Although some advances have been made in narrow domains, significant advances are necessary to approach generally applicable lifelong learning.

Different approaches have been developed. Among the methods are meta-learning, domain adaptation, multi-task learning, and pretraining. Table 9.1 lists these approaches, together with regular single-task learning. The learning tasks are formulated using datasets, as in a regular supervised setting. The table shows how the lifelong learning methods differ in their training and test datasets and the different learning tasks.

The first line shows regular single-task learning. For single-task learning, the training and test datasets are both drawn from the same distribution (the datasets do not contain the same examples, but they are drawn from the same original dataset and the data distribution is expected to be the same), and the task to perform is the same for training and test.

In the next line, for transfer learning, networks trained on one dataset are used to speed up training for a different task, possibly using a much smaller dataset [82]. For example, having learned to recognize cars may be useful to speed up recognizing trucks. Since the datasets are not drawn from the same master dataset, their distribution will differ, and there typically is only an informal notion of

Table 9.1 Different kinds of supervised learning

Name	Dataset	Task
Single-task learning	$D_{train} \subseteq D, D_{test} \subseteq D$	$T = T_{train} = T_{test}$
Transfer learning	$D_1 \gg D_2$	$T_1 \neq T_2$
Multi-task learning	$D_{train} \subseteq D, D_{test} \subseteq D$	$T_1 \neq T_2$
Domain adaptation	$D_1 \neq D_2$	$T_1 = T_2$
Meta-learning	$\{D_1, \ldots, D_{N-1}\} \gg D_N$	$T_1, \ldots, T_{n-1} \neq T_N$

how "related" the datasets are. However, in practice transfer learning often provides significant speedups, and transfer learning, pretraining, and finetuning are currently used in many real-world training tasks, sometimes using large foundation models as a basis.

In multi-task learning, more than one task is learned from one dataset [20]. The tasks are often related, such as classification tasks of different, but related, classes of images, or learning spam filters for different email users. Regularization may be improved when a neural network is trained on related tasks at the same time [7, 22].

So far, our learning tasks were trying to speed up learning different tasks with related data. Domain adaptation switches this around: the task remains the same, but the data changes. In domain adaptation, a different dataset is used to perform the same task, such as recognizing pedestrians in different light conditions [127].

In meta-learning, both datasets and tasks are different, although not too different. In meta-learning, a sequence of datasets and learning tasks is generalized to learn a new (related) task quickly [17, 45, 48, 102]. The goal of meta-learning is to learn hyperparameters over a sequence of learning tasks. In deep meta-learning these hyperparameters include the initial network parameters; in this sense meta-learning can be considered to be multi-job transfer learning.

9.2 Transfer Learning and Meta-Learning Agents

We will now introduce transfer learning and meta-learning algorithms. Where in normal learning we would initialize our parameters randomly, in transfer learning we initialize them with (part of) the training results of another training task. This other task is related in some way to the new task, for example, a task to recognize images of dogs playing in a forest is initialized on a dataset of dogs playing in a park. The parameter transfer of the old task is called pretraining; the second phase, where the network learns the new task on the new dataset, is called finetuning. The pretraining will hopefully allow the new task to train faster.

Where transfer learning *transfers* knowledge from a *single* previous task, meta-learning aims to *generalize* knowledge from *multiple* previous learning tasks. Meta-learning tries to learn hyperparameters over these related learning tasks, which tell the algorithm how to learn the new task. Meta-learning thus aims to learn to learn. In deep meta-learning approaches, the initial network parameters are typically part of the hyperparameters. Note that in transfer learning we also use (part of) the parameters to speed up learning (finetuning) a new, related task. We can say that deep meta-learning generalizes transfer learning by learning the initial parameters over not one but a sequence of related tasks [45, 49]. (Definitions are still in flux, however, and different authors and different fields have different definitions.)

Transfer learning has become part of the standard approach in machine learning; meta-learning is still an area of active research. We will look into meta-learning shortly, after we have looked into transfer learning, multi-task learning, and domain adaptation.

9.2.1   Transfer Learning

Transfer learning aims to improve the process of learning new tasks using the experience gained by solving similar problems [83, 88, 123, 124]. Transfer learning aims to transfer past experience of source tasks and use it to boost learning in a related target task [82, 149].

In transfer learning, we first train a base network on a base dataset and task, and then we repurpose some of the learned features to a second target network to be trained on a target dataset and task. This process works better if the features are general, meaning suitable to both base and target tasks, instead of specific to the base task. This form of transfer learning is called inductive transfer. The scope of possible models (model bias) is narrowed in a beneficial way by using a model fit on a different but related task.

First we will look at task similarity, then at transfer learning, multi-task learning, and domain adaptation.

9.2.1.1   Task Similarity

Clearly, pretraining works better when the tasks are similar [20]. Learning to play the viola based on the violin is more similar than learning the tables of multiplication based on tennis. Different measures can be used to measure the similarity of examples and features in datasets, from linear one-dimensional measures to nonlinear multi-dimensional measures. Common measures are the cosine similarity for real-valued vectors and the radial basis function kernel [119, 131], but many more elaborate measures have been devised.

Similarity measures are also used to devise meta-learning algorithms, as we will see later.

9.2.1.2   Pretraining and Finetuning

When we want to transfer knowledge, we can transfer the weights of the network and then start re-training with the new dataset. Please refer back to Table 9.1. In pretraining the new dataset is smaller than the old dataset $D_1 \gg D_2$, and we train for a new task $T_1 \neq T_2$, which we want to train faster. This works when the new task is different, but similar, so that the old dataset D_1 contains useful information for the new task T_2.

To learn new image recognition problems, it is common to use a deep learning model pretrained for a large and challenging image classification task such as the ImageNet 1000-class photograph classification competition. Three examples of pretrained models include: the Oxford VGG Model, Google's Inception Model,

Microsoft's ResNet Model. For more examples, see the Caffe Model Zoo,[1] or other zoos[2] where more pretrained models are shared.

Transfer learning is effective because the images were trained on a corpus that requires the model to make predictions on a large number of classes, requiring the model to be general, and since it efficiently learns to extract features in order to perform well.

Convolutional neural network features are more generic in lower layers, such as color blobs or Gabor filters, and more specific to the original dataset in higher layers. Features must eventually transition from general to specific in the last layers of the network [145]. Pretraining copies some of the layers to the new task. Care should be taken how much of the old task network to copy. It is relatively safe to copy the more general lower layers. Copying the more specific higher layers may be detrimental to performance.

In natural language processing a similar situation occurs. In natural language processing, a word embedding is used that is a mapping of words to a high-dimensional continuous vector where different words with a similar meaning have a similar vector representation. Efficient algorithms exist to learn these word representations. Two examples of common pretrained word models trained on very large datasets of text documents include Google's Word2vec model [73] and Stanford's GloVe model [87].

9.2.1.3 Hands On: Pretraining Example

Transfer learning and pretraining have become a standard approach to learning new tasks, especially when only a small dataset is present, or when we wish to limit training time. Let us have a look at a hands-on example that is part of the Keras distribution (Sect. B.3.3.1). The Keras transfer learning example provides a basic ImageNet-based approach, getting data from TensorFlow DataSets (TFDS). The example follows a supervised learning approach, although the learning and finetuning phase can easily be substituted by a reinforcement learning setup. The Keras transfer learning example is at the Keras site[3] and can also be run in a Google Colab.

The most common incarnation of transfer learning in the context of deep learning is the following workflow:

1. Take layers from a previously trained model.
2. Freeze them, so as to avoid destroying any of the information they contain during future training rounds.

---

[1] https://caffe.berkeleyvision.org/model_zoo.html.

[2] https://modelzoo.co.

[3] https://keras.io/guides/transfer_learning/.

3. Add some new, trainable layers on top of the frozen layers. They will train on the new dataset using the old features as predictions.
4. Train the new layers on your new (small) dataset.
5. A last, optional, step is finetuning of the frozen layers, which consists of unfreezing the entire model you obtained above, and re-training it on the new data with a very low learning rate. This can potentially achieve meaningful improvements, by incrementally adapting the pretrained features to the new data.

Let us look at how this workflow works in practice in Keras. At the Keras site we find an accessible example code for pretraining (see Listing 9.1).

First, we instantiate a base model with pretrained weights. (We do not include the classifier on top.) Then, we freeze the base model (see Listing 9.2).

Next, we create a new model on top and train it.

The Keras example contains this example and others, including finetuning. Please go to the Keras site and improve your experience with pretraining in practice.

```
1  base_model = keras.applications.Xception(
2      weights='imagenet',  # Load weights pre-trained on ImageNet.
3      input_shape=(150, 150, 3),
4      include_top=False)  # Do not include the ImageNet classifier
           at the top.
5
6  base_model.trainable = False
```

Listing 9.1 Pretraining in Keras (1): instantiate model

```
1   inputs = keras.Input(shape=(150, 150, 3))
2   # The base_model is running in inference mode here,
3   # by passing 'training=False'. This is important for fine-tuning
4   x = base_model(inputs, training=False)
5   # Convert features of shape 'base_model.output_shape[1:]' to
         vectors
6   x = keras.layers.GlobalAveragePooling2D()(x)
7   # A Dense classifier with a single unit (binary classification)
8   outputs = keras.layers.Dense(1)(x)
9   model = keras.Model(inputs, outputs)
10
11
12  model.compile(optimizer=keras.optimizers.Adam(),
13                loss=keras.losses.BinaryCrossentropy(from_logits=
                      True),
14                metrics=[keras.metrics.BinaryAccuracy()])
15  model.fit(new_dataset, epochs=20, callbacks=..., validation_data
         =...)
```

Listing 9.2 Pretraining in Keras (2): create new model and train

9.2.1.4 Multi-Task Learning

Multi-task learning is related to transfer learning. In multi-task learning a single network is trained *at the same time* on multiple related tasks [20, 122].

In multi-task learning the learning process of one task benefits from the simultaneous learning of the related task. This approach is effective when the tasks have some commonality, such as learning to recognize breeds of dogs and breeds of cats. In multi-task learning, related learning tasks are learned at the same time, whereas in transfer learning they are learned in sequence by different networks. Multi-task learning improves regularization by requiring the algorithm to perform well on a related learning task instead of penalizing all overfitting uniformly [6, 32]. The two-headed AlphaGo Zero network optimizes for value and for policy at the same time in the same network [20, 82]. A multi-headed architecture is often used in multi-task learning, although in AlphaGo Zero the two heads are trained for two related aspects (policy and value) of the same task (playing Go games).

Multi-task learning has been applied with success to Atari games [55, 56].

9.2.1.5 Domain Adaptation

Domain adaptation is necessary when there is a change in the data distribution between the training dataset and the test dataset (domain shift). This problem is related to out-of-distribution learning [68]. Domain shifts are common in practical applications of artificial intelligence, such as when items must be recognized in different light conditions, or when the background changes. Conventional machine learning algorithms often have difficulty adapting to such changes.

The goal of domain adaptation is to compensate for the variation among two data distributions, to be able to reuse information from a source domain on a different target domain [127], see Fig. 9.3 for a backpack in different circumstances. As was indicated in Table 9.1, domain adaptation applies to the situation where the tasks are the same $T_1 = T_2$, but the datasets are different $D_1 \neq D_2$, although still somewhat similar. For example, the task may be to recognize a backpack, but in a different orientation, or to recognize a pedestrian, but in different lighting.

Domain adaptation can be seen as the opposite of pretraining. Pretraining uses the same dataset for a different task, while domain adaptation adapts to a new dataset for the same task [19].

In natural language processing, examples of domain shift are an algorithm that has been trained on news items that is then applied to a dataset of biomedical documents [24, 117], or a spam filter that is trained on a certain group of email users, which is deployed to a new target user [9]. Sudden changes in the environment (pandemics, severe weather) can also upset machine learning algorithms.

There are different techniques to overcome domain shift [23, 137, 142, 148]. In visual applications, adaptation can be achieved by reweighting the samples of the first dataset or clustering them for visually coherent sub-domains. Other approaches try to find transformations that map the source distribution to the target or learn

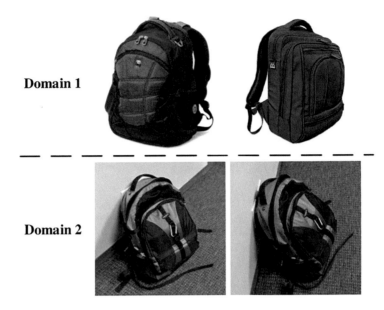

Fig. 9.3 Domain adaptation: recognizing items in different circumstances is difficult [42]

a classification model and a feature transformation jointly [127]. Adversarial techniques where feature representations are encouraged to be difficult to distinguish can be used to achieve adaptation [28, 129, 139], see also Sect. B.2.6.2.

9.2.2 Meta-Learning

Related to transfer learning is meta-learning. Where the focus in transfer learning is on transferring parameters from a single donor task to a receiver task for further finetuning, in meta-learning the focus is on using the knowledge of a number of tasks to learn how to learn a new task faster and better. Regular machine learning learns examples *for one* task, and meta-learning aims to learn *across* tasks. Machine learning learns parameters that approximate the function, and meta-learning learns *hyper*parameters *about* the learning function.[4] This is often expressed as *learning to learn*, a phrase that has defined the field ever since its introduction [103, 122, 124]. The term *meta-learning* has been used for many different contexts, not only for deep learning, but also for tasks ranging from hyperparameter optimization,

---

[4] Associating base learning with parameter learning and meta-learning with hyperparameter learning appears to give us a clear distinction; however, in practice the distinction is not so clear cut: in deep meta-learning the initialization of the regular parameters is considered to be an important "hyper"parameter.

Table 9.2 Meta-reinforcement learning approaches [49]

Name	Approach	Environment	Ref
Recurr. ML	Deploy recurrent networks on RL problems	–	[29, 135]
Meta Netw.	Fast reparam. of base learner by distinct meta-learner	O-glot, miniIm.	[78]
SNAIL	Attention mechanism coupled with temporal conv.	O-glot, miniIm.	[75]
LSTM ML	Embed base learner parameters in cell state of LSTM	miniImageNet	[95]
MAML	Learn initialization weights θ for fast adaptation	O-glot, miniIm.	[34]
iMAML	Approx. higher-order gradients, indep. of optim. path	O-glot, miniIm.	[92]
Meta SGD	Learn both the initialization and updates	O-glot, miniIm.	[70]
Reptile	Move init. toward task-specific updated weights	O-glot, miniIm.	[79]
BayesMAML	Learn multiple initializations Θ, jointly optim. SVGD	miniImagaNet	[144]

algorithm selection, to automated machine learning. (We will briefly look at these in Sect. 9.2.2.6.)

Deep meta reinforcement learning is an active area of research. Many algorithms are being developed, and much progress is being made. Table 9.2 lists nine algorithms that have been proposed for deep meta reinforcement learning (see the surveys [45, 49]).

9.2.2.1   Evaluating Few-Shot Learning Problems

One of the challenges of lifelong machine learning is to judge the performance of an algorithm. Regular train–test generalization does not capture the speed of adaptation of a meta-learning algorithm.

For this reason, meta-learning tasks are typically evaluated on their few-shot learning ability. In few-shot learning, we test if a learning algorithm can be made to recognize examples from classes from which it has seen only few examples in training. In few-shot learning prior knowledge is available in the network.

To translate few-shot learning to a human setting, we can think of a situation where a human plays the double bass after only a few minutes of training on the double bass, but after years on the violin, viola, or cello.

Meta-learning algorithms are often evaluated with few- shot learning tasks, in which the algorithm must recognize items of which it has only seen a few examples. This is formalized in the N-way-k-shot approach [21, 61, 136]. Figure 9.4 illustrates this process. Given a large dataset $\mathcal{D}$, a smaller training dataset D is sampled from this dataset. The N-way-k-shot classification problem constructs training dataset D such that it consists of N classes, of which, for each class, k examples are present in the dataset. Thus, the cardinality of $|D| = N \cdot k$.

Training task 1   **Training task 2** · · ·   **Test task 1** · · ·

Fig. 9.4 N-way-k-shot learning [15]

A full N-way-k-shot few-shot learning meta task $\mathcal{T}$ consists of many episodes in which base tasks $\mathcal{T}_i$ are performed. A base task consists of a training set and a test set, to test generalization. In few-shot terminology, the training set is called the support set, and the test set is called the query set. The support set has size $N \cdot k$, and the query set consists of a small number of examples. The meta-learning algorithm can learn from the episodes of N-way-k-shot query/support base tasks, until at meta-test time the generalization of the meta-learning algorithm is tested with another query, as is illustrated in the figure.

9.2.2.2   Deep Meta-Learning Algorithms

We will now turn our attention to deep meta-learning algorithms. The meta-learning field is still young and active; nevertheless, the field is converging on a set of definitions that we will present here. We start our explanation in a supervised setting.

Meta-learning is concerned with learning a task $\mathcal{T}$ from a set of base-learning tasks $\{\mathcal{T}_1, \mathcal{T}_2, \mathcal{T}_3, \dots\}$ so that a new (related) meta-test-task will reach a high accuracy quicker. Each base-learning task $\mathcal{T}_i$ consists of a dataset D_i and a learning objective, the loss function $\mathcal{L}_i$. Thus we get $\mathcal{T}_i = (D_i, \mathcal{L}_i)$. Each dataset consists of pairs of inputs and labels $D_i = \{(x_j, y_j)\}$ and is split into training and test sets $D_i = \{D_{\mathcal{T}_i, train}, D_{\mathcal{T}_i, test}\}$. On each training dataset a parameterized model $\hat{f}_{\theta_i}(D_{i, train})$ is approximated, with a loss function $\mathcal{L}_i(\theta_i, D_{i, train})$. The model $\hat{f}$ is approximated with a deep learning algorithm, which is governed by a set of hyperparameters ω. The particular hyperparameters vary from algorithm to algorithm, but frequently encountered hyperparameters are the learning rate α, the initial parameters θ_0, and algorithm constants.

This conventional machine learning algorithm is called the *base learner*. Each base learner task approximates a model $\hat{f}_i$ by finding the optimal parameters $\theta_i^{\star}$ to minimize the loss function on its dataset

$$\mathcal{T}_i = \hat{f}_{\theta_i^{\star}} = \arg\min_{\theta_i} \mathcal{L}_{i,\omega}(\theta_i, D_{i,train}),$$

while the learning algorithm is governed by hyperparameters ω.

9.2.2.3 Inner and Outer Loop Optimization

One of the most popular deep meta-learning approaches of the last few years is optimization-based meta-learning [49]. This approach optimizes the initial parameters θ of the network for fast learning of new tasks. Most optimization-based techniques do so by approaching meta-learning as a two-level optimization problem. At the *inner* level, a base learner makes task-specific updates to θ for the different observations in the training set. At the *outer* level, the meta-learner optimizes hyperparameters ω across a sequence of base tasks where the loss of each task is evaluated using the *test* data from the base tasks $D_{\mathcal{T}_i,test}$ [45, 67, 95].

The inner loop optimizes the parameters θ, and the outer loop optimizes the hyperparameters ω to find the best performance on the set of base tasks $i = 0, \ldots, M$ with the appropriate test data:

$$\omega^{\star} = \underbrace{\arg\min_{\omega} \mathcal{L}^{\text{meta}}}_{\text{outer loop}} (\underbrace{\arg\min_{\theta_i} \mathcal{L}_{\omega}^{\text{base}}(\theta_i, D_{i,train}), D_{i,test}}_{\text{inner loop}}).$$

The inner loop optimizes θ_i within the datasets D_i of the tasks $\mathcal{T}_i$, and the outer loop optimizes ω across the tasks and datasets.

The meta loss function optimizes for the meta objective, which can be accuracy, speed, or another goal over the set of base tasks (and datasets). The outcome of the meta optimization is a set of optimal hyperparameters $\omega^{\star}$.

In optimization-based meta-learning the most important hyperparameters ω are the optimal initial parameters $\theta_0^{\star}$. When the meta-learner only optimizes the initial parameters as hyperparameters ($\omega = \theta_0$), then the inner–outer formula simplifies as follows:

$$\theta_0^{\star} = \underbrace{\arg\min_{\theta_0} \mathcal{L}}_{\text{outer loop}} (\underbrace{\arg\min_{\theta_i} \mathcal{L}(\theta_i, D_{i,train}), D_{i,test}}_{\text{inner loop}}).$$

In this approach we meta-optimize the initial parameters θ_0, such that the loss function performs well on the *test* data of the base tasks. Section 9.2.2.5 describes MAML, a well-known example of this approach.

Deep meta-learning approaches are sometimes categorized as (1) similarity-metric-based, (2) model-based, and (3) optimization-based [49]. We will now have a closer look at two of the nine meta-reinforcement learning algorithms from Table 9.2. We will look at recurrent meta-learning and MAML; the former is a model-based approach, the latter optimization-based.

9.2.2.4 Recurrent Meta-Learning

For meta-reinforcement learning approaches to be able to learn, they must be able to remember what they have learned across subtasks. Let us see how recurrent meta-learning learns across tasks.

Recurrent meta-learning uses recurrent neural networks to remember this knowledge [29, 135]. The recurrent network serves as dynamic storage for the learned task embedding (weight vector). The recurrence can be implemented by an LSTM [135] or by gated recurrent units [29]. The choice of recurrent neural meta network (meta-RNN) determines how well it adapts to the subtasks, as it gradually accumulates knowledge about the base task structure.

Recurrent meta-learning tracks variables s, a, r, d that denote state, action, reward, and termination of the episode. For each task $\mathcal{T}_i$, recurrent meta-learning inputs the set of environment variables $\{s_{t+1}, a_t, r_t, d_t\}$ into a meta-RNN at each time step t. The meta-RNN outputs an action and a hidden state h_t. Conditioned on the hidden state h_t, the meta network outputs action a_t. The goal is to maximize the expected reward in each trial (Fig. 9.5). Since recurrent meta-learning embeds information from previously seen inputs in hidden state, it is regarded as a model-based meta-learner [49].

Recurrent meta-learners performed almost as well as model-free baselines on simple N-way-k-shot reinforcement learning tasks [29, 135]. However, the performance degrades in more complex problems, when dependencies span a longer horizon.

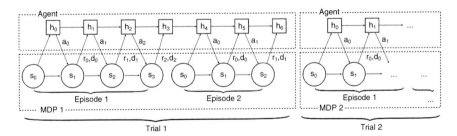

Fig. 9.5 Workflow of recurrent meta-learners in reinforcement learning contexts. State, action, reward, and termination flag at time step t are denoted by s_t, a_t, r_t, and d_t, and h_t refers to the hidden state [29]

Algorithm 9.1 MAML for reinforcement learning [34]

Require: $p(\mathcal{T})$: distribution over tasks
Require: α, β: step size hyperparameters
    randomly initialize θ
    while not done **do**
        Sample batch of tasks $\mathcal{T}_i \sim p(\mathcal{T})$
        for all $\mathcal{T}_i$ **do**
            Sample k trajectories $\mathcal{D} = \{(s_1, a_1, \ldots s_T)\}$ using f_θ in $\mathcal{T}_i$
            Evaluate $\nabla_\theta \mathcal{L}_{\mathcal{T}_i}(\pi_\theta)$ using $\mathcal{D}$ and $\mathcal{L}_{\mathcal{T}_i}$ in Eq. 9.1
            Compute adapted parameters with gradient descent: $\theta_i' = \theta - \alpha \nabla_\theta \mathcal{L}_{\mathcal{T}_i}(\pi_\theta)$
            Sample trajectories $\mathcal{D}_i' = \{(s_1, a_1, \ldots s_T)\}$ using $f_{\theta_i'}$ in $\mathcal{T}_i$
        end for
        Update $\theta \leftarrow \theta - \beta \nabla_\theta \sum_{\mathcal{T}_i \sim p(\mathcal{T})} \mathcal{L}_{\mathcal{T}_i}(\pi_{\theta_i'})$ using each $\mathcal{D}_i'$ and $\mathcal{L}_{\mathcal{T}_i}$ in Eq. 9.1
    end while

9.2.2.5 Model-Agnostic Meta-Learning

The model-agnostic meta-learning approach (MAML) [34] is an optimization approach that is model-agnostic: it can be used for different learning problems, such as classification, regression, and reinforcement learning.

As mentioned, the optimization view of meta-learning is especially focused on optimizing the initial parameters θ. The intuition behind the optimization view can be illustrated with a simple regression example (Fig. 9.6). Let us assume that we are faced with multiple linear regression problems $f_i(x)$. The model has two parameters: a and b, $\hat{f}(x) = a \cdot x + b$. When the meta-training set consists of four

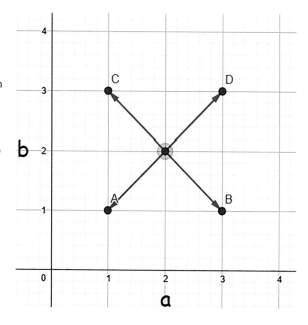

Fig. 9.6 The optimization approach aims to learn parameters from which other tasks can be learned quickly. The intuition behind optimization approaches such as MAML is that when our meta-training set consists of tasks A, B, C, and D, then if our meta-learning algorithm adjusts parameters a and b to (2, 2), then they are close to either of the four tasks and can be adjustly quickly to them, with few examples (after [34, 49])

tasks, A, B, C, and D, then we wish to optimize to a single set of parameters $\{a, b\}$ from which we can quickly learn the optimal parameters for each of the four tasks. In Fig. 9.6 the point in the middle represents this combination of parameters. The point is the closest to the four different tasks. This is how model-agnostic meta-learning works [34]: by exposing our model to various base tasks, we update the parameters $\theta = \{a, b\}$ to good initial parameters θ_0 that facilitate quick meta-adaptation.

Let us look at the process of training a deep learning model's parameters from a feature learning standpoint [34, 49], where the goal is that a few gradient steps can produce good results on a new task. We build a feature representation that is broadly suitable for many tasks, and by then finetuning the parameters slightly (primarily updating the top layer weights), we achieve good results—not unlike transfer learning. MAML finds parameters θ that are easy and fast to finetune, allowing the adaptation to happen in an embedding space that is well suited for fast learning. To put it another way, MAML's goal is to find the point in the middle of Fig. 9.6, from where the other tasks are easily reachable.

How does MAML work [34, 48, 90]? Please refer to the pseudocode for MAML in Algorithm 9.1. The learning task is an episodic Markov decision process with horizon T, where the learner is allowed to query a limited number of sample trajectories for few-shot learning. Each reinforcement learning task $\mathcal{T}_i$ contains an initial state distribution $p_i(s_1)$ and a transition distribution $p_i(s_{t+1}|s_t, a_t)$. The loss $\mathcal{L}_{\mathcal{T}_i}$ corresponds to the (negative) reward function R. The model being learned, π_θ, is a policy from states s_t to a distribution over actions a_t at each time step $t \in \{1, \ldots, T\}$. The loss for task $\mathcal{T}_i$ and policy π_θ takes the familiar form of the objective (Eq. 2.5):

$$\mathcal{L}_{\mathcal{T}_i}(\pi_\theta) = -\mathbb{E}_{s_t, a_t \sim \pi_\theta, p_{\mathcal{T}_i}} \left[\sum_{t=1}^{T} R_i(s_t, a_t, s_{t+1}) \right]. \qquad (9.1)$$

In k-shot reinforcement learning, k rollouts from π_θ and task $\mathcal{T}_i$, $(s_1, a_1, \ldots s_T)$, and the rewards $R(s_t, a_t)$, may be used for adaptation on a new task $\mathcal{T}_i$. MAML uses TRPO to estimate the gradient both for the policy gradient update(s) and the meta optimization [105].

The goal is to quickly learn new concepts, which is equivalent to achieving a minimal loss in few gradient update steps. The number of gradient steps has to be specified in advance. For a single gradient update step gradient descent produces updated parameters

$$\theta_i' = \theta - \alpha \nabla_\theta \mathcal{L}_{\mathcal{T}_i}(\pi_\theta)$$

specific to task i. The meta loss of one gradient step across tasks is

$$\theta \leftarrow \theta - \beta \nabla_\theta \sum_{\mathcal{T}_i \sim p(\mathcal{T})} \mathcal{L}_{\mathcal{T}_i}(\pi_{\theta_i'}) \qquad (9.2)$$

where $p(\mathcal{T})$ is a probability distribution over tasks. This expression contains an inner gradient $\nabla_\theta \mathcal{L}_{\mathcal{T}_i}(\pi_{\theta_i'})$. Optimizing this meta loss requires computing second-order gradients when backpropagating the meta gradient through the gradient operator in the meta objective (Eq. 9.2), which is computationally expensive [34]. Various algorithms have been inspired by MAML and aim to improve optimization-based meta-learning further [48, 79].

In meta-reinforcement learning the goal is to quickly find a policy for a new environment using only a small amount of experience. MAML has gained attention within the field of deep meta-learning, due to its simplicity (it requires two hyperparameters), its general applicability, and its strong performance.

9.2.2.6 Hyperparameter Optimization

Meta-learning has been around for a long time, long before deep learning became popular. It has been applied to classic machine learning tasks, such as regression, decision trees, support vector machines, clustering algorithms, Bayesian networks, evolutionary algorithms, and local search [13, 17, 132]. The hyperparameter view on meta-learning originated here.

Although this is a book about deep learning, it is interesting to briefly discuss this non-deep background, also because hyperparameter optimization is an important technology to find a good set of hyperparameters in reinforcement learning experiments.

Machine learning algorithms have hyperparameters that govern their behavior, and finding the optimal setting for these hyperparameters has long been called meta-learning. A naive approach is to enumerate all combinations and run the machine learning problem for them. For all but the smallest hyperparameter spaces such a grid search will be prohibitively slow. Among the smarter meta-optimization approaches are random search, Bayesian optimization, gradient-based optimization, and evolutionary optimization.

This meta-algorithm approach has given rise to algorithm configuration research, such as SMAC [50],[5] ParamILS [51],[6] irace [71],[7] and algorithm selection research [57, 96], such as SATzilla [141].[8] Well-known hyperparameter optimization packages include scikit-learn [86],[9] scikit-optimize,[10] nevergrad [94],[11] and Optuna [3].[12] Hyperparameter optimization and algorithm configuration have

---

[5] https://github.com/automl/SMAC3.

[6] http://www.cs.ubc.ca/labs/beta/Projects/ParamILS/.

[7] http://iridia.ulb.ac.be/irace/.

[8] http://www.cs.ubc.ca/labs/beta/Projects/SATzilla/.

[9] https://scikit-learn.org/stable/.

[10] https://scikit-optimize.github.io/stable/index.html.

[11] https://code.fb.com/ai-research/nevergrad/.

[12] https://optuna.org.

grown into the field of automated machine learning or AutoML.[13] The AutoML field is a large and active field of meta-learning, and an overview book is [52].

All machine learning algorithms have a bias, and different algorithms perform better on different types of problems. Hyperparameters constrain this algorithm bias. This so-called inductive bias reflects the set of assumptions about the data on which the algorithms are based. Learning algorithms perform better when this bias matches the learning problem (for example: CNNs work well on image problems, RNNs on language problems). Meta-learning changes this inductive bias, either by choosing a different learning algorithm, by changing the network initialization, or by other means, allowing the algorithm to be adjusted to work well on different problems.

9.2.2.7   Meta-Learning and Curriculum Learning

There is an interesting connection between meta-learning and curriculum learning. Both approaches aim to improve the speed and accuracy of learning, by learning from a set of subtasks.

In meta-learning, knowledge is gained from subtasks, so that the learning of a new, related, task can be quick. In curriculum learning (Sect. 6.2.3.1) we aim to learn quicker by dividing a large and difficult learning task into a set of subtasks, ordered from easy to hard.

Thus we can conclude that curriculum learning is a form of meta-learning where the subtasks are ordered from easy to hard, or, equivalently, that meta-learning is unordered curriculum learning.

9.2.2.8   From Few-Shot to Zero-Shot Learning

Meta-learning uses information from previous learning tasks to learn new tasks quicker [124]. Meta-learning algorithms are often evaluated in a few-shot setting, to see how well they do in image classification problems when they are shown only few training examples. This *few-shot learning* problem aims to correctly classify queries with little previous support for the new class. In the previous sections we have discussed how meta-learning algorithms aim to achieve few-shot learning. A discussion of meta-learning would not be complete without mentioning zero-shot learning.

Zero-shot learning (ZSL) goes a step further than few-shot learning. In zero-shot learning an example has to be recognized as belonging to a class without ever having been trained on an example of this class [65, 81]. In zero-shot learning the classes covered by training instances and the classes we aim to classify are disjoint. This may sound like an impossibility—how can you recognize something you have never seen before?—yet it is something that we, humans, do all the time: having learned

---

[13] https://www.automl.org.

to pour coffee in a cup, we can also pour tea, even if we never have seen tea before. (Or, having learned to play the violin, the viola, and the cello, we can play the double bass, to some degree, even if we have never played the double bass before.)

Whenever we recognize something that we have not seen before, we are actually using extra information (or features). If we recognize a red beak in a picture of a bird species that we have never seen before, then the concepts "red" and "beak" are known to us, because we have learned them in other contexts.

Zero-shot learning recognizes new categories of instances without training examples. Attribute-based zero-shot learning uses separate high-level attribute descriptions of the new categories, based on categories previously learned in the dataset. Attributes are an intermediate representation that enables parameter sharing between classes [2]. The extra information can be in the form of textual description of the class—red, or beak—in addition to the visual information [64]. The learner must be able to match text with image information.

Zero-shot learning approaches are designed to learn this intermediate semantic layer, the attributes, and apply them at inference time to predict new classes, when they are provided with descriptions in terms of these attributes. Attributes correspond to high-level properties of the objects that are shared across multiple classes (which can be detected by machines and which can be understood by humans). Attribute-based image classification is a label-embedding problem where each class is embedded in the space of attribute vectors. As an example, if the classes correspond to animals, possible attributes include *has paws*, *has stripes*, or *is black*.

9.3 Meta-Learning Environments

Now that we have seen how transfer learning and meta-learning can be implemented, it is time to look at some of the environments that are used to evaluate the algorithms. We will list important datasets, environments, and foundation models for images, behavior, and text, expanding our scope beyond pure reinforcement learning. We will look at how well the approaches succeed in generalizing quickly to new machine learning tasks.

Many benchmarks have been introduced to test transfer and meta-learning algorithms. Benchmarks for conventional machine learning algorithms aim to offer a variety of *challenging* learning tasks. Benchmarks for meta-learning, in contrast, aim to offer *related* learning tasks. Some benchmarks are parameterized, where the difference between tasks can be controlled.

Meta-learning aims to learn new and related tasks quicker, trading off speed versus accuracy. This raises the question how fast and how accurate meta-learning algorithms are under different circumstances. In answering these questions we must keep in mind that the closer the learning tasks are, the easier the task is, and the quicker and the more accurate results will be. Hence, we should carefully look at which dataset a benchmark uses when we compare results.

Table 9.3 Datasets, environments, and models for meta-learning in images, behavior, and text

Name	Type	Domain	Ref
ALE	Single	Games	[8]
MuJoCo	Single	Robot	[126]
DeepMind Control	Single	Robot	[120]
BERT	Transfer	Text	[25]
GPT-3	Transfer	Text	[89]
ImageNet	Transfer	Image	[33]
Omniglot	Meta	Image	[61]
Mini-ImageNet	Meta	Image	[133]
Meta-Dataset	Meta	Image	[128]
Meta-World	Meta	Robot	[146]
Alchemy	Meta	Unity	[134]

Table 9.3 lists some of the environments that are often used for meta-learning experiments. Some are regular deep learning environments designed for single-task learning ("single"), some are transfer learning and pretraining datasets ("transfer"), and some datasets and environments are specifically designed for meta-learning experiments ("meta").

We will now describe them in more detail. ALE (Sect. 3.1.1), MuJoCo (Sect. 4.1.3), and the DeepMind control suite (Sect. 4.3.2) are originally single- task deep learning environments. They are also being used in meta-learning experiments and few-shot learning, often with moderate results, since the tasks are typically not very similar (Pong is not like Pac-Man).

9.3.1 Image Processing

Traditionally, two datasets have emerged as de facto benchmarks for few-shot image learning: Omniglot [62], and mini-ImageNet [99, 133].

Omniglot is a dataset for one shot learning. This dataset contains 1623 different handwritten characters from 50 different alphabets and contains 20 examples per class (character) [62, 63]. Most recent methods obtain very high accuracy on Omniglot, rendering comparisons between them mostly uninformative.

Mini-ImageNet uses the same setup as Omniglot for testing, consisting of 60,000 color images of size 84 × 84 with 100 classes (64/16/20 for train/validation/test) and contains 600 examples per class [133]. Albeit harder than Omniglot, most recent methods achieve similar accuracy when controlling for model capacity. Meta-learning algorithms such as Bayesian Program Learning and MAML achieved accuracies comparable to human performance on Omniglot and ImageNet, with accuracies in the high nineties and error rates as low as a few percent [63].

Models trained on the largest datasets, such as ImageNet, are used as foundation models [14]. Pretrained models can be downloaded from the Model Zoo.[14]

These benchmarks may be too homogeneous for testing meta-learning. In contrast, real-life learning experiences are heterogeneous: they vary in terms of the number of classes and examples per class and are unbalanced. Furthermore, the Omniglot and Mini-ImageNet benchmarks measure within dataset generalization. For meta-learning, we are eventually after models that can generalize to entirely new distributions. For this reason, new datasets are being developed specifically for meta-learning.

9.3.2 Natural Language Processing

In natural language processing BERT is a well-known pretrained model. BERT stands for bidirectional encoder representations from transformers [25]. It is designed to pretrain deep bidirectional representations from unlabeled text by jointly conditioning on two contexts. BERT has shown that transfer learning can work well in natural language tasks. BERT can be used for classification tasks such as sentiment analysis, question answering tasks, and named entity recognition. BERT is a large model, with 345 million parameters [98].

An even larger pretrained transformer model is the generative pretrained transformer 3, or GPT-3 [89], with 175 billion parameters. The quality of the text generated by GPT-3 is exceptionally good, and it is difficult to distinguish from that written by humans. In this case, it appears that size matters. OpenAI provides a public interface where you can see for yourself how well it performs.[15]

BERT and GPT-3 are large models that are used more and more as a foundation model as a basis for other experiments to pretrain on.

9.3.3 Meta-Dataset

A recent dataset specifically designed for meta-learning is meta-dataset [128].[16] Meta-dataset is a set of datasets, consisting of: ImageNet, Omniglot, Aircraft, Birds, Textures, Quick Draw, Fungi, Flower, Traffic Signs, and MSCOCO. Thus, the datasets provide a more heterogeneous challenge than earlier single-dataset experiments.

Triantafillou et al. report results with matching networks [133], prototypical networks [112], first-order MAML [34], and relation networks [118]. As can be

---

[14] https://modelzoo.co.

[15] https://openai.com/blog/openai-api/.

[16] https://github.com/google-research/meta-dataset.

expected, the accuracy on the larger (meta) dataset is much lower than that on previous homogeneous datasets. They find that a variant of MAML performs best, although for classifiers that are trained on other datasets most accuracies are between 40% and 60%, except for Birds and Flowers, which scores in the 70s and 80s, closer to the single-dataset results for Omniglot and ImageNet. Meta-learning for heterogeneous datasets remains a challenging task (see also [21, 125]).

9.3.4   Meta-World

For deep reinforcement learning two traditionally popular environments are ALE and MuJoCo. The games in the ALE benchmarks typically differ considerably, which makes the ALE test set challenging for meta-learning, and little success has been reported. (There are a few exceptions that apply transfer learning (pretraining) to DQN [76, 84, 113] succeeding in multitask learning in a set of Atari games that all move a ball.)

Robotics tasks, on the other hand, are more easily parameterizable. Test tasks can be generated with the desired level of similarity, making robotic tasks amenable to meta-learning testing. Typical tasks such as reaching and learning different walking gaits are more related than two Atari games such as, for example, Breakout and Space Invaders.

To provide a better benchmark that is more challenging for meta-reinforcement learning, Yu et al. introduced Meta-World [146],[17] a benchmark for multi-task and meta reinforcement learning (see Fig. 9.7 for their pictorial explanation of the difference between multi-task and meta-learning). Meta-World consists of 50 distinct manipulation tasks with a robotic arm (Fig. 9.8). The tasks are designed to be different and contain structure, which can be leveraged for transfer to new tasks.

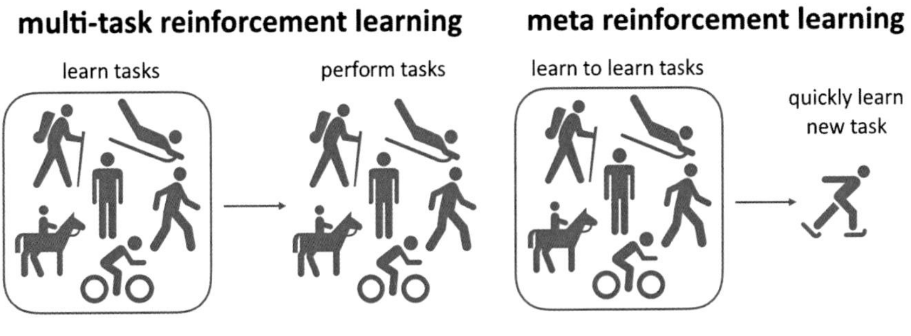

Fig. 9.7   Multi-task and meta-reinforcement learning [146]

---

[17] https://meta-world.github.io.

Fig. 9.8 Meta-world tasks [146]

When the authors of Meta-World evaluated six state-of-the-art meta-reinforcement and multi-task learning algorithms on these tasks, they found that generalization of existing algorithms to heterogeneous tasks is limited. They tried PPO, TRPO, SAC, RL^2[29], MAML, and PEARL [93]. Small variations of tasks such as different object positions can be learned with reasonable success, but the algorithms struggled to learn multiple tasks at the same time, even with as few as ten distinct training tasks. In contrast to more limited meta-learning benchmarks, Meta-World emphasizes generalization to new tasks and interaction scenarios, not just a parametric variation in goals.

9.3.5 Alchemy

A final meta-reinforcement learning benchmark that we will discuss is Alchemy [134]. The Alchemy benchmark is a procedurally generated 3D video game [107], implemented in Unity [54]. Task generation is parameterized, and varying degrees of similarity and hidden structure can be chosen. The process by which Alchemy levels are created is accessible to researchers, and a perfect Bayesian ideal observer can be implemented.

Experiments with two agents are reported, VMPO and IMPALA. The VMPO [85, 115] agent is based on a gated transformer network. The IMPALA [31, 53] agent is based on population-based training with an LSTM core network. Both agents are strong deep learning methods, although not necessarily for meta-learning. Again, in both agents meta-learning became more difficult as learning tasks became more diverse. The reported performance on meta-learning was weak.

The Alchemy platform can be found on GitHub.[18] A human-playable interface is part of the environment.

9.3.6   Hands On: Meta-World Example

Let us get experience with running Meta-World benchmark environments. We will run a few popular agent algorithms on them, such as PPO and TRPO. Meta-World is as easy to use as Gym. The code can be found on GitHub,[19] and the accompanying implementations of a suite of agent algorithms are called Garage [37].[20] Garage runs on PyTorch and TensorFlow.

Standard Meta-World benchmarks include multi-task and meta-learning setups, named MT1, MT10, MT50, and ML1, ML10, ML50. The Meta-World benchmark can be installed with pip

```
pip install git+https://github.com/rlworkgroup/
metaworld.git@master\#egg=metaworld
```

Note that Meta-World is a robotics benchmark and needs MuJoCo, so you have to install that too.[21]

The GitHub site contains brief example instructions on the usage of the benchmark, please refer to Listing 9.3. The benchmark can be used to test the meta-learning performance of your favorite algorithm, or you can use one of the baselines provided in Garage.

Conclusion

In this chapter we have seen different approaches to learning new and different tasks with few or even zero examples. Impressive results have been achieved, although major challenges remain to learn general adaptation when tasks are more diverse. As is often the case in new fields, many different approaches have been tried. The field of meta-learning is an active field of research, aiming to reduce one of the main problems of machine learning, and many new methods will continue to be developed.

---

[18] https://github.com/deepmind/dm_alchemy.

[19] https://github.com/rlworkgroup/metaworld.

[20] https://github.com/rlworkgroup/garage.

[21] https://github.com/openai/mujoco-py#install-mujoco.

```
1   import metaworld
2   import random
3
4   print(metaworld.ML1.ENV_NAMES)   # Try available environments
5
6   ml1 = metaworld.ML1('pick-place-v1')  # Construct the benchmark
7
8   env = ml1.train_classes['pick-place-v1']()
9   task = random.choice(ml1.train_tasks)
10  env.set_task(task)   # Set task
11
12  obs = env.reset()   # Reset environment
13  a = env.action_space.sample()   # Sample an action
14  obs, reward, done, info = env.step(a)   # Step the environoment
```

Listing 9.3 Using meta-world

Summary and Further Reading

We will now summarize the chapter and provide pointers to further reading.

Summary

This chapter is concerned with learning new tasks faster and with smaller datasets or lower sample complexity. Transfer learning is concerned with transferring knowledge that has been learned to solve a task, to another task, to allow quicker learning. A popular transfer learning approach is pretraining, where some network layers are copied to initialize a network for a new task, followed by finetuning, to improve performance on the new task, but with a smaller dataset.

Another approach is meta-learning, or learning to learn. Here knowledge of how a sequence of previous tasks is learned is used to learn a new task quicker. Meta-learning learns hyperparameters of the different tasks. In deep meta-learning, the set of initial network parameters is usually considered to be such a hyperparameter. Meta-learning aims to learn hyperparameters that can learn a new task with only a few training examples, often using N-way-k-shot learning. For deep few-shot learning the Model-Agnostic Meta-Learning (MAML) approach is well known and has inspired follow-up work.

Meta-learning is of great importance in machine learning. For tasks that are related, good results are reported. For more challenging benchmarks, where tasks are less related (such as pictures of animals from very different species), results are reported that are weaker.

Further Reading

Meta-learning is a highly active field of research. Good entry points are [16, 45, 48, 49, 150]. Meta-learning has attracted much attention in artificial intelligence, both in supervised learning and in reinforcement learning. Many books and surveys have been written about the field of meta-learning, see, for example [17, 102, 106, 130, 138].

There has been active research interest in meta-learning algorithms for some time, see, for example, [10, 103, 104, 132]. Research into transfer learning and meta-learning has a long history, starting with Pratt and Thrun [88, 124]. Early surveys into the field are [82, 121, 137]; more recent surveys are [148, 149]. Huh et al. focus on ImageNet [47, 91]. Yang et al. study the relation between transfer learning and curriculum learning with Sokoban [143].

Early principles of meta-learning are described by Schmidhuber [103, 104]. Meta-learning surveys are [17, 43, 45, 48, 49, 102, 106, 130, 132, 138]. Papers on similarity-metric meta-learning are [38, 58, 108, 112, 118, 133]. Papers on model-based meta-learning are [29, 30, 39, 75, 78, 101, 135]. Papers on optimization-based meta-learning are many [5, 11, 34–36, 41, 69, 70, 79, 92, 95, 100, 144].

Domain adaptation is studied in [4, 23, 27, 77, 148]. Zero-shot learning is an active and promising field. Interesting papers are [2, 12, 26, 46, 64, 65, 81, 91, 97, 116, 140]. Like zero-shot learning, few-shot learning is also a popular area of meta-learning research [44, 60, 80, 114, 116]. Benchmark papers are [21, 125, 128, 134, 146].

Exercises

It is time to test our understanding of transfer learning and meta-learning with some exercises and questions.

Questions

Below are some quick questions to check your understanding of this chapter. For each question, a simple, single sentence answer is sufficient.

1. What is the reason for the interest in meta-learning and transfer learning?
2. What is transfer learning?
3. What is meta-learning?
4. How is meta-learning different from multi-task learning?
5. Zero-shot learning aims to identify classes that it has not seen before. How is that possible?
6. Is pretraining a form of transfer learning?
7. Can you explain learning to learn?

8. Are the initial network parameters also hyperparameters? Explain.
9. What is an approach for zero-shot learning?
10. As the diversity of tasks increases, does meta-learning achieve good results?

Exercises

Let us go to the programming exercises to become more familiar with the methods that we have covered in this chapter. Meta-learning and transfer learning experiments are often very computationally expensive. You may need to scale down dataset sizes or skip some exercises as a last resort.

1. *Pretraining* Implement pretraining and finetuning in the Keras pretraining example from Sect. 9.2.1.3.[22] Do the exercises as suggested, including finetuning on the cats and dogs training set. Note the uses of preprocessing, data augmentation, and regularization (dropout and batch normalization). See the effects of increasing the number of layers that you transfer on training performance and speed.

2. *MAML* Reptile [79] is a meta-learning approach inspired by MAML, but first order, and faster, specifically designed for few-shot learning. The Keras website contains a segment on Reptile.[23] At the start a number of hyperparameters are defined: learning rate, step size, batch size, the number of meta-learning iterations, the number of evaluation iterations, how many shots, classes, etcetera. Study the effect of tuning different hyperparameters, especially the ones related to few-shot learning: the classes, the shots, and the number of iterations.
   To delve deeper into few-shot learning, also have a look at the MAML code, which has a section on reinforcement learning.[24] Try different environments.

3. *Meta-World* As we have seen in Sect. 9.3.4, Meta-World [146] is an elaborate benchmark suite for meta- reinforcement learning. Re-read the section, go to GitHub, and install the benchmark.[25] Also go to Garage to install the agent algorithms so that you are able to test their performance.[26] See that they work with your PyTorch or TensorFlow setup. First try running the ML1 meta benchmark for PPO. Then try MAML and RL2. Next, try the more elaborate meta-learning benchmarks. Read the Meta-World paper, and see if you can reproduce their results.

---

[22] https://keras.io/guides/transfer_learning/.

[23] https://keras.io/examples/vision/reptile/.

[24] https://github.com/cbfinn/maml_rl/tree/master/rllab.

[25] https://github.com/rlworkgroup/metaworld.

[26] https://github.com/rlworkgroup/garage.

4. *ZSL* We go from few-shot learning to zero-shot learning. One of the ways in which zero-shot learning works is by learning attributes that are shared by classes. Read the papers *Label-Embedding for Image Classification* [2], and *An embarrassingly simple approach to zero-shot learning* [97], and go to the code [12].[27] Implement it, and try to understand how attribute learning works. Print the attributes for the classes, and use the different datasets. Does MAML work for few-shot learning? (challenging)

References

1. Pulkit Agrawal, Ross Girshick, and Jitendra Malik. Analyzing the performance of multilayer neural networks for object recognition. In *European Conference on Computer Vision*, pages 329–344. Springer, 2014. 289
2. Zeynep Akata, Florent Perronnin, Zaid Harchaoui, and Cordelia Schmid. Label-embedding for attribute-based classification. In *Proceedings of the IEEE Conference on Computer Vision and Pattern Recognition*, pages 819–826, 2013. 305, 312, 314
3. Takuya Akiba, Shotaro Sano, Toshihiko Yanase, Takeru Ohta, and Masanori Koyama. Optuna: A next-generation hyperparameter optimization framework. In *Proceedings of the 25th ACM SIGKDD International Conference on Knowledge Discovery & Data Mining*, pages 2623–2631, 2019. 303
4. Dario Amodei, Chris Olah, Jacob Steinhardt, Paul Christiano, John Schulman, and Dan Mané. Concrete problems in AI safety. *arXiv preprint arXiv:1606.06565*, 2016. 312
5. Antreas Antoniou, Harrison Edwards, and Amos Storkey. How to train your MAML. *arXiv preprint arXiv:1810.09502*, 2018. 312
6. Andreas Argyriou, Theodoros Evgeniou, and Massimiliano Pontil. Multi-task feature learning. In *Advances in Neural Information Processing Systems*, pages 41–48, 2007. 295
7. Jonathan Baxter. A model of inductive bias learning. *Journal of Artificial Intelligence Research*, 12:149–198, 2000. 291
8. Marc G Bellemare, Yavar Naddaf, Joel Veness, and Michael Bowling. The Arcade Learning Environment: An evaluation platform for general agents. *Journal of Artificial Intelligence Research*, 47:253–279, 2013. 306
9. Shai Ben-David, John Blitzer, Koby Crammer, and Fernando Pereira. Analysis of representations for domain adaptation. *Advances in Neural Information Processing Systems*, 19:137, 2007. 295
10. Yoshua Bengio, Samy Bengio, and Jocelyn Cloutier. Learning a synaptic learning rule. Technical report, Montreal, 1990. 312
11. Luca Bertinetto, Joao F Henriques, Philip HS Torr, and Andrea Vedaldi. Meta-learning with differentiable closed-form solvers. In *International Conference on Learning Representations*, 2018. 312
12. Shrisha Bharadwaj. Embarrassingly simple zero shot learning. https://github.com/chichilicious/embarrsingly-simple-zero-shot-learning, 2018. 312, 314
13. Christopher M Bishop. *Pattern Recognition and Machine Learning*. Information science and statistics. Springer Verlag, Heidelberg, 2006. 303
14. Rishi Bommasani, Drew A. Hudson, Ehsan Adeli, Russ Altman, Simran Arora, Sydney von Arx, Michael S. Bernstein, Jeannette Bohg, Antoine Bosselut, Emma Brunskill, Erik Brynjolfsson, Shyamal Buch, Dallas Card, Rodrigo Castellon, Niladri Chatterji, Annie S.

---

[27] https://github.com/sbharadwajj/embarrassingly-simple-zero-shot-learning.

Chen, Kathleen Creel, Jared Quincy Davis, Dorottya Demszky, Chris Donahue, Moussa Doumbouya, Esin Durmus, Stefano Ermon, John Etchemendy, Kawin Ethayarajh, Li Fei-Fei, Chelsea Finn, Trevor Gale, Lauren Gillespie, Karan Goel, Noah D. Goodman, Shelby Grossman, Neel Guha, Tatsunori Hashimoto, Peter Henderson, John Hewitt, Daniel E. Ho, Jenny Hong, Kyle Hsu, Jing Huang, Thomas Icard, Saahil Jain, Dan Jurafsky, Pratyusha Kalluri, Siddharth Karamcheti, Geoff Keeling, Fereshte Khani, Omar Khattab, Pang Wei Koh, Mark S. Krass, Ranjay Krishna, and Rohith Kuditipudi. On the opportunities and risks of foundation models. *arXiv preprint arXiv:2108.07258*, 2021. 289, 307

15. Borealis. Few shot learning tutorial https://www.borealisai.com/en/blog/tutorial-2-few-shot-learning-and-meta-learning-i/. 298

16. Matthew Botvinick, Sam Ritter, Jane X Wang, Zeb Kurth-Nelson, Charles Blundell, and Demis Hassabis. Reinforcement learning, fast and slow. *Trends in Cognitive Sciences*, 23(5):408–422, 2019. 312

17. Pavel Brazdil, Christophe Giraud Carrier, Carlos Soares, and Ricardo Vilalta. *Metalearning: Applications to data mining*. Springer Science & Business Media, 2008. 291, 303, 312, 312

18. Tom B. Brown, Benjamin Mann, Nick Ryder, Melanie Subbiah, Jared Kaplan, Prafulla Dhariwal, Arvind Neelakantan, Pranav Shyam, Girish Sastry, Amanda Askell, Sandhini Agarwal, Ariel Herbert-Voss, Gretchen Krueger, Tom Henighan, Rewon Child, Aditya Ramesh, Daniel M. Ziegler, Jeffrey Wu, Clemens Winter, Christopher Hesse, Mark Chen, Eric Sigler, Mateusz Litwin, Scott Gray, Benjamin Chess, Jack Clark, Christopher Berner, Sam McCandlish, Alec Radford, Ilya Sutskever, and Dario Amodei. Language models are few-shot learners. In *Advances in Neural Information Processing Systems*, 2020. 290

19. Thomas Carr, Maria Chli, and George Vogiatzis. Domain adaptation on the Atari. In *Proceedings of the 18th International Conference on Autonomous Agents and MultiAgent Systems, AAMAS '19, Montreal*, pages 1859–1861, 2018. 295

20. Rich Caruana. Multitask learning. *Machine Learning*, 28(1):41–75, 1997. 291, 292, 295, 295

21. Wei-Yu Chen, Yen-Cheng Liu, Zsolt Kira, Yu-Chiang Frank Wang, and Jia-Bin Huang. A closer look at few-shot classification. In *International Conference on Learning Representations*, 2019. 297, 308, 312

22. Carlo Ciliberto, Youssef Mroueh, Tomaso Poggio, and Lorenzo Rosasco. Convex learning of multiple tasks and their structure. In *International Conference on Machine Learning*, pages 1548–1557. PMLR, 2015. 291

23. Gabriela Csurka. Domain adaptation for visual applications: A comprehensive survey. In *Domain Adaptation in Computer Vision Applications*, Advances in Computer Vision and Pattern Recognition, pages 1–35. Springer, 2017. 295, 312

24. Hal Daumé III. Frustratingly easy domain adaptation. In *ACL 2007, Proceedings of the 45th Annual Meeting of the Association for Computational Linguistics, June 23–30, 2007, Prague*, 2007. 295

25. Jacob Devlin, Ming-Wei Chang, Kenton Lee, and Kristina Toutanova. BERT: pre-training of deep bidirectional transformers for language understanding. In *Proceedings of the 2019 Conference of the North American Chapter of the Association for Computational Linguistics: Human Language Technologies, NAACL-HLT 2019, Minneapolis*, 2018. 289, 290, 306, 307

26. Chuong B Do and Andrew Y Ng. Transfer learning for text classification. *Advances in Neural Information Processing Systems*, 18:299–306, 2005. 312

27. Jeff Donahue, Yangqing Jia, Oriol Vinyals, Judy Hoffman, Ning Zhang, Eric Tzeng, and Trevor Darrell. DeCAF: A deep convolutional activation feature for generic visual recognition. In *International Conference on Machine Learning*, pages 647–655. PMLR, 2014. 289, 312

28. Yuntao Du, Zhiwen Tan, Qian Chen, Xiaowen Zhang, Yirong Yao, and Chongjun Wang. Dual adversarial domain adaptation. *arXiv preprint arXiv:2001.00153*, 2020. 296

29. Yan Duan, John Schulman, Xi Chen, Peter L Bartlett, Ilya Sutskever, and Pieter Abbeel. RL^2: Fast reinforcement learning via slow reinforcement learning. *arXiv preprint arXiv:1611.02779*, 2016. 297, 300, 300, 300, 300, 309, 312

30. Harrison Edwards and Amos Storkey. Towards a neural statistician. In *International Conference on Learning Representations*, 2017. 312
31. Lasse Espeholt, Hubert Soyer, Remi Munos, Karen Simonyan, Vlad Mnih, Tom Ward, Yotam Doron, Vlad Firoiu, Tim Harley, Iain Dunning, Shane Legg, and Koray Kavukcuoglu. IMPALA: Scalable distributed deep-RL with importance weighted actor-learner architectures. In *International Conference on Machine Learning*, pages 1407–1416. PMLR, 2018. 309
32. Theodoros Evgeniou and Massimiliano Pontil. Regularized multi-task learning. In *Proceedings of the tenth ACM SIGKDD International Conference on Knowledge Discovery and Data Mining*, pages 109–117. ACM, 2004. 295
33. Li Fei-Fei, Jia Deng, and Kai Li. ImageNet: Constructing a large-scale image database. *Journal of Vision*, 9(8):1037–1037, 2009. 306
34. Chelsea Finn, Pieter Abbeel, and Sergey Levine. Model-Agnostic Meta-Learning for fast adaptation of deep networks. In *International Conference on Machine Learning*, pages 1126–1135. PMLR, 2017. 297, 301, 301, 301, 302, 302, 302, 303, 307, 312
35. Chelsea Finn, Aravind Rajeswaran, Sham Kakade, and Sergey Levine. Online meta-learning. In *International Conference on Machine Learning*, pages 1920–1930. PMLR, 2019.
36. Chelsea Finn, Kelvin Xu, and Sergey Levine. Probabilistic Model-Agnostic Meta-Learning. In *Advances in Neural Information Processing Systems*, pages 9516–9527, 2018. 312
37. The garage contributors. Garage: A toolkit for reproducible reinforcement learning research. https://github.com/rlworkgroup/garage, 2019. 310
38. Victor Garcia and Joan Bruna. Few-shot learning with graph neural networks. In *International Conference on Learning Representations*, 2017. 312
39. Marta Garnelo, Dan Rosenbaum, Christopher Maddison, Tiago Ramalho, David Saxton, Murray Shanahan, Yee Whye Teh, Danilo Rezende, and SM Ali Eslami. Conditional neural processes. In *International Conference on Machine Learning*, pages 1704–1713. PMLR, 2018. 312
40. Ross Girshick, Jeff Donahue, Trevor Darrell, and Jitendra Malik. Rich feature hierarchies for accurate object detection and semantic segmentation. In *Proceedings of the IEEE Conference on Computer Vision and Pattern Recognition*, pages 580–587, 2014. 289
41. Erin Grant, Chelsea Finn, Sergey Levine, Trevor Darrell, and Thomas Griffiths. Recasting gradient-based meta-learning as hierarchical bayes. In *International Conference on Learning Representations*, 2018. 312
42. Xifeng Guo, Wei Chen, and Jianping Yin. A simple approach for unsupervised domain adaptation. In *2016 23rd International Conference on Pattern Recognition (ICPR)*, pages 1566–1570. IEEE, 2016. 296
43. Abhishek Gupta, Russell Mendonca, YuXuan Liu, Pieter Abbeel, and Sergey Levine. Meta-reinforcement learning of structured exploration strategies. In *Advances in Neural Information Processing Systems*, pages 5307–5316, 2018. 312
44. Irina Higgins, Arka Pal, Andrei Rusu, Loic Matthey, Christopher Burgess, Alexander Pritzel, Matthew Botvinick, Charles Blundell, and Alexander Lerchner. DARLA: Improving zero-shot transfer in reinforcement learning. In *International Conference on Machine Learning*, pages 1480–1490. PMLR, 2017. 312
45. Timothy Hospedales, Antreas Antoniou, Paul Micaelli, and Amos Storkey. Meta-learning in neural networks: A survey. *arXiv preprint arXiv:2004.05439*, 2020. 291, 291, 297, 299, 312, 312
46. R Lily Hu, Caiming Xiong, and Richard Socher. Zero-shot image classification guided by natural language descriptions of classes: A meta-learning approach. In *Advances in Neural Information Processing Systems*, 2018. 312
47. Minyoung Huh, Pulkit Agrawal, and Alexei A Efros. What makes ImageNet good for transfer learning? *arXiv preprint arXiv:1608.08614*, 2016. 289, 312
48. Mike Huisman, Jan van Rijn, and Aske Plaat. Metalearning for deep neural networks. In Pavel Brazdil et al., editors, *Metalearning: Applications to data mining*. Springer, 2022. 291, 302, 303, 312, 312

49. Mike Huisman, Jan N. van Rijn, and Aske Plaat. A survey of deep meta-learning. *Artificial Intelligence Review*, 2021. 291, 297, 297, 299, 300, 300, 301, 302, 312, 312

50. Frank Hutter, Holger H Hoos, and Kevin Leyton-Brown. Sequential model-based optimization for general algorithm configuration. In *International Conference on Learning and Intelligent Optimization*, pages 507–523. Springer, 2011. 303

51. Frank Hutter, Holger H Hoos, Kevin Leyton-Brown, and Thomas Stützle. ParamILS: an automatic algorithm configuration framework. *Journal of Artificial Intelligence Research*, 36:267–306, 2009. 303

52. Frank Hutter, Lars Kotthoff, and Joaquin Vanschoren. *Automated Machine Learning: Methods, Systems, Challenges*. Springer Nature, 2019. 304

53. Max Jaderberg, Valentin Dalibard, Simon Osindero, Wojciech M. Czarnecki, Jeff Donahue, Ali Razavi, Oriol Vinyals, Tim Green, Iain Dunning, Karen Simonyan, Chrisantha Fernando, and Koray Kavukcuoglu. Population based training of neural networks. *arXiv preprint arXiv:1711.09846*, 2017. 309

54. Arthur Juliani, Vincent-Pierre Berges, Ervin Teng, Andrew Cohen, Jonathan Harper, Chris Elion, Chris Goy, Yuan Gao, Hunter Henry, Marwan Mattar, and Danny Lange. Unity: A general platform for intelligent agents. *arXiv preprint arXiv:1809.02627*, 2018. 309

55. Stephen Kelly and Malcolm I Heywood. Multi-task learning in Atari video games with emergent tangled program graphs. In *Proceedings of the Genetic and Evolutionary Computation Conference*, pages 195–202. ACM, 2017. 295

56. Stephen Kelly and Malcolm I Heywood. Emergent tangled program graphs in multi-task learning. In *IJCAI*, pages 5294–5298, 2018. 295

57. Pascal Kerschke, Holger H Hoos, Frank Neumann, and Heike Trautmann. Automated algorithm selection: Survey and perspectives. *Evolutionary Computation*, 27(1):3–45, 2019. 303

58. Gregory Koch, Richard Zemel, and Ruslan Salakhutdinov. Siamese neural networks for one-shot image recognition. In *ICML Deep Learning workshop*, volume 2. Lille, 2015. 312

59. Alex Krizhevsky, Ilya Sutskever, and Geoffrey E Hinton. ImageNet classification with deep convolutional neural networks. In *Advances in Neural Information Processing Systems*, pages 1097–1105, 2012. 290

60. Yen-Ling Kuo, Boris Katz, and Andrei Barbu. Encoding formulas as deep networks: Reinforcement learning for zero-shot execution of LTL formulas. *arXiv preprint arXiv:2006.01110*, 2020. 312

61. Brenden Lake, Ruslan Salakhutdinov, Jason Gross, and Joshua Tenenbaum. One shot learning of simple visual concepts. In *Proceedings of the Annual Meeting of the Cognitive Science Society*, volume 33, 2011. 297, 306

62. Brenden M Lake, Ruslan Salakhutdinov, and Joshua B Tenenbaum. Human-level concept learning through probabilistic program induction. *Science*, 350(6266):1332–1338, 2015. 306, 306

63. Brenden M Lake, Ruslan Salakhutdinov, and Joshua B Tenenbaum. The Omniglot challenge: a 3-year progress report. *Current Opinion in Behavioral Sciences*, 29:97–104, 2019. 306, 306

64. Christoph H Lampert, Hannes Nickisch, and Stefan Harmeling. Learning to detect unseen object classes by between-class attribute transfer. In *2009 IEEE Conference on Computer Vision and Pattern Recognition*, pages 951–958. IEEE, 2009. 305, 312

65. Hugo Larochelle, Dumitru Erhan, and Yoshua Bengio. Zero-data learning of new tasks. In *AAAI*, volume 1, page 3, 2008. 304, 312

66. Yann LeCun, Yoshua Bengio, and Geoffrey Hinton. Deep learning. *Nature*, 521(7553):436, 2015. 289

67. Yoonho Lee and Seungjin Choi. Gradient-based meta-learning with learned layerwise metric and subspace. In *International Conference on Machine Learning*, pages 2927–2936. PMLR, 2018. 299

68. Sergey Levine, Aviral Kumar, George Tucker, and Justin Fu. Offline reinforcement learning: Tutorial, review, and perspectives on open problems. *arXiv preprint arXiv:2005.01643*, 2020. 295

69. Ke Li and Jitendra Malik. Learning to optimize neural nets. *arXiv preprint arXiv:1703.00441*, 2017. 312

70. Zhenguo Li, Fengwei Zhou, Fei Chen, and Hang Li. Meta-SGD: learning to learn quickly for few-shot learning. *arXiv preprint arXiv:1707.09835*, 2017. 297, 312

71. Manuel López-Ibáñez, Jérémie Dubois-Lacoste, Leslie Pérez Cáceres, Mauro Birattari, and Thomas Stützle. The irace package: Iterated racing for automatic algorithm configuration. *Operations Research Perspectives*, 3:43–58, 2016. 303

72. Dhruv Mahajan, Ross Girshick, Vignesh Ramanathan, Kaiming He, Manohar Paluri, Yixuan Li, Ashwin Bharambe, and Laurens van der Maaten. Exploring the limits of weakly supervised pretraining. In *Proceedings of the European Conference on Computer Vision (ECCV)*, pages 181–196, 2018. 289

73. Tomas Mikolov, Kai Chen, Greg Corrado, and Jeffrey Dean. Efficient estimation of word representations in vector space. In *International Conference on Learning Representations*, 2013. 289, 293

74. Tomas Mikolov, Ilya Sutskever, Kai Chen, Greg Corrado, and Jeffrey Dean. Distributed representations of words and phrases and their compositionality. In *Advances in Neural Information Processing Systems*, 2013. 289

75. Nikhil Mishra, Mostafa Rohaninejad, Xi Chen, and Pieter Abbeel. A simple neural attentive meta-learner. In *International Conference on Learning Representations*, 2018. 297, 312

76. Akshita Mittel and Purna Sowmya Munukutla. Visual transfer between Atari games using competitive reinforcement learning. In *Proceedings of the IEEE/CVF Conference on Computer Vision and Pattern Recognition Workshops*, pages 0–0, 2019. 308

77. Matthias Müller-Brockhausen, Mike Preuss, and Aske Plaat. Procedural content generation: Better benchmarks for transfer reinforcement learning. In *Conference on Games*, 2021. 312

78. Tsendsuren Munkhdalai and Hong Yu. Meta networks. In *International Conference on Machine Learning*, pages 2554–2563. PMLR, 2017. 297, 312

79. Alex Nichol, Joshua Achiam, and John Schulman. On first-order meta-learning algorithms. *arXiv preprint arXiv:1803.02999*, 2018. 297, 303, 312, 313

80. Junhyuk Oh, Satinder Singh, Honglak Lee, and Pushmeet Kohli. Zero-shot task generalization with multi-task deep reinforcement learning. In *International Conference on Machine Learning*, pages 2661–2670. PMLR, 2017. 312

81. Mark M Palatucci, Dean A Pomerleau, Geoffrey E Hinton, and Tom Mitchell. Zero-shot learning with semantic output codes. In *Advances in Neural Information Processing Systems 22*, 2009. 304, 312

82. Sinno Jialin Pan and Qiang Yang. A survey on transfer learning. *IEEE Transactions on Knowledge and Data Engineering*, 22(10):1345–1359, 2010. 290, 292, 295, 312

83. German I Parisi, Ronald Kemker, Jose L Part, Christopher Kanan, and Stefan Wermter. Continual lifelong learning with neural networks: A review. *Neural Networks*, 113:54–71, 2019. 292

84. Emilio Parisotto, Jimmy Lei Ba, and Ruslan Salakhutdinov. Actor-mimic: Deep multitask and transfer reinforcement learning. *arXiv preprint arXiv:1511.06342*, 2015. 308

85. Emilio Parisotto, Francis Song, Jack Rae, Razvan Pascanu, Caglar Gulcehre, Siddhant Jayakumar, Max Jaderberg, Raphael Lopez Kaufman, Aidan Clark, Seb Noury, et al. Stabilizing transformers for reinforcement learning. In *International Conference on Machine Learning*, pages 7487–7498. PMLR, 2020. 309

86. Fabian Pedregosa, Gaël Varoquaux, Alexandre Gramfort, Vincent Michel, Bertrand Thirion, Olivier Grisel, Mathieu Blondel, Peter Prettenhofer, Ron Weiss, Vincent Dubourg, Jake VanderPlas, Alexandre Passos, David Cournapeau, Matthieu Brucher, Matthieu Perrot, and Edouard Duchesnay. Scikit-learn: Machine learning in Python. *Journal of Machine Learning Research*, 12(Oct):2825–2830, 2011. 303

87. Jeffrey Pennington, Richard Socher, and Christopher D Manning. GloVe: Global vectors for word representation. In *Proceedings of the 2014 Conference on Empirical Methods in Natural Language Processing (EMNLP)*, pages 1532–1543, 2014. 293

88. Lorien Y Pratt. Discriminability-based transfer between neural networks. In *Advances in Neural Information Processing Systems*, pages 204–211, 1993. 292, 312

89. Alec Radford, Karthik Narasimhan, Tim Salimans, and Ilya Sutskever. Improving language understanding by generative pre-training. https://openai.com/blog/language-unsupervised/, 2018. 289, 306, 307

90. Aniruddh Raghu, Maithra Raghu, Samy Bengio, and Oriol Vinyals. Rapid learning or feature reuse? Towards understanding the effectiveness of MAML. In *International Conference on Learning Representations*, 2020. 302

91. Rajat Raina, Andrew Y Ng, and Daphne Koller. Constructing informative priors using transfer learning. In *Proceedings of the 23rd international conference on Machine learning*, pages 713–720, 2006. 312, 312

92. Aravind Rajeswaran, Chelsea Finn, Sham Kakade, and Sergey Levine. Meta-learning with implicit gradients. In *Advances in Neural Information Processing Systems*, 2019. 297, 312

93. Kate Rakelly, Aurick Zhou, Chelsea Finn, Sergey Levine, and Deirdre Quillen. Efficient off-policy meta-reinforcement learning via probabilistic context variables. In *International Conference on Machine Learning*, pages 5331–5340. PMLR, 2019. 309

94. J. Rapin and O. Teytaud. Nevergrad - A gradient-free optimization platform. https://GitHub.com/FacebookResearch/Nevergrad, 2018. 303

95. Sachin Ravi and Hugo Larochelle. Optimization as a model for few-shot learning. In *International Conference on Learning Representations*, 2017. 297, 299, 312

96. John R. Rice. The algorithm selection problem. *Advances in Computers*, 15(65–118):5, 1976. 303

97. Bernardino Romera-Paredes and Philip Torr. An embarrassingly simple approach to zero-shot learning. In *International Conference on Machine Learning*, pages 2152–2161, 2015. 312, 314

98. Denis Rothman. *Transformers for Natural Language Processing*. Packt Publishing, 2021. 307

99. Olga Russakovsky, Jia Deng, Hao Su, Jonathan Krause, Sanjeev Satheesh, Sean Ma, Zhiheng Huang, Andrej Karpathy, Aditya Khosla, Michael S. Bernstein, Alexander C. Berg, and Li Fei-Fei. ImageNet large scale visual recognition challenge. *International Journal of Computer Vision*, 115(3):211–252, 2015. 289, 289, 306

100. Andrei A Rusu, Dushyant Rao, Jakub Sygnowski, Oriol Vinyals, Razvan Pascanu, Simon Osindero, and Raia Hadsell. Meta-learning with latent embedding optimization. In *International Conference on Learning Representations*, 2019. 312

101. Adam Santoro, Sergey Bartunov, Matthew Botvinick, Daan Wierstra, and Timothy Lillicrap. Meta-learning with memory-augmented neural networks. In *International Conference on Machine Learning*, pages 1842–1850, 2016. 312

102. Tom Schaul and Jürgen Schmidhuber. Metalearning. *Scholarpedia*, 5(6):4650, 2010. 291, 312, 312

103. Jürgen Schmidhuber. *Evolutionary Principles in Self-Referential Learning, or on Learning how to Learn: the Meta-Meta-...Hook*. PhD thesis, Technische Universität München, 1987. 296, 312, 312

104. Jürgen Schmidhuber, Jieyu Zhao, and MA Wiering. Simple principles of metalearning. Technical report, IDSIA, 1996. 312, 312

105. John Schulman, Sergey Levine, Pieter Abbeel, Michael Jordan, and Philipp Moritz. Trust region policy optimization. In *International Conference on Machine Learning*, pages 1889–1897, 2015. 302

106. Nicolas Schweighofer and Kenji Doya. Meta-learning in reinforcement learning. *Neural Networks*, 16(1):5–9, 2003. 312, 312

107. Noor Shaker, Julian Togelius, and Mark J Nelson. *Procedural Content Generation in Games*. Springer, 2016. 309

108. Pranav Shyam, Shubham Gupta, and Ambedkar Dukkipati. Attentive recurrent comparators. In *International Conference on Machine Learning*, pages 3173–3181. PMLR, 2017. 312

109. Daniel L Silver, Qiang Yang, and Lianghao Li. Lifelong machine learning systems: Beyond learning algorithms. In *2013 AAAI Spring Symposium Series*, 2013. 290

110. David Silver, Aja Huang, Chris J. Maddison, Arthur Guez, Laurent Sifre, George van den Driessche, Julian Schrittwieser, Ioannis Antonoglou, Veda Panneershelvam, Marc Lanctot, Sander Dieleman, Dominik Grewe, John Nham, Nal Kalchbrenner, Ilya Sutskever, Timothy Lillicrap, Madeleine Leach, Koray Kavukcuoglu, Thore Graepel, and Demis Hassabis. Mastering the game of Go with deep neural networks and tree search. *Nature*, 529(7587):484, 2016. 290

111. David Silver, Julian Schrittwieser, Karen Simonyan, Ioannis Antonoglou, Aja Huang, Arthur Guez, Thomas Hubert, Lucas Baker, Matthew Lai, Adrian Bolton, Yutian Chen, Timothy Lillicrap, Fan Hui, Laurent Sifre, George van den Driessche, Thore Graepel, and Demis Hassabis. Mastering the game of Go without human knowledge. *Nature*, 550(7676):354, 2017. 290

112. Jake Snell, Kevin Swersky, and Richard Zemel. Prototypical networks for few-shot learning. In *Advances in Neural Information Processing Systems*, pages 4077–4087, 2017. 307, 312

113. Doron Sobol, Lior Wolf, and Yaniv Taigman. Visual analogies between Atari games for studying transfer learning in RL. *arXiv preprint arXiv:1807.11074*, 2018. 308

114. Sungryull Sohn, Junhyuk Oh, and Honglak Lee. Hierarchical reinforcement learning for zero-shot generalization with subtask dependencies. In *Advances in Neural Information Processing Systems*, pages 7156–7166, 2018. 312

115. H. Francis Song, Abbas Abdolmaleki, Jost Tobias Springenberg, Aidan Clark, Hubert Soyer, Jack W. Rae, Seb Noury, Arun Ahuja, Siqi Liu, Dhruva Tirumala, Nicolas Heess, Dan Belov, Martin A. Riedmiller, and Matthew M. Botvinick. V-MPO: on-policy maximum a posteriori policy optimization for discrete and continuous control. In *International Conference on Learning Representations*, 2019. 309

116. Lise Stork, Andreas Weber, Jaap van den Herik, Aske Plaat, Fons Verbeek, and Katherine Wolstencroft. Large-scale zero-shot learning in the wild: Classifying zoological illustrations. *Ecological Informatics*, 62:101222, 2021. 312, 312

117. Baochen Sun, Jiashi Feng, and Kate Saenko. Return of frustratingly easy domain adaptation. In *Proceedings of the AAAI Conference on Artificial Intelligence*, volume 30, 2016. 295

118. Flood Sung, Yongxin Yang, Li Zhang, Tao Xiang, Philip HS Torr, and Timothy M Hospedales. Learning to compare: Relation network for few-shot learning. In *Proceedings of the IEEE conference on computer vision and pattern recognition*, pages 1199–1208, 2018. 307, 312

119. Pang-Ning Tan, Michael Steinbach, and Vipin Kumar. *Introduction to Data Mining*. Pearson Education India, 2016. 292

120. Yuval Tassa, Yotam Doron, Alistair Muldal, Tom Erez, Yazhe Li, Diego de Las Casas, David Budden, Abbas Abdolmaleki, Josh Merel, Andrew Lefrancq, Timothy Lillicrap, and Martin Riedmiller. DeepMind control suite. *arXiv preprint arXiv:1801.00690*, 2018. 306

121. Matthew E Taylor and Peter Stone. Transfer learning for reinforcement learning domains: A survey. *Journal of Machine Learning Research*, 10(Jul):1633–1685, 2009. 312

122. Sebastian Thrun. Is learning the *n*-th thing any easier than learning the first? In *Advances in Neural Information Processing Systems*, pages 640–646. Morgan Kaufmann, 1996. 295, 296

123. Sebastian Thrun. *Explanation-based neural network learning: A lifelong learning approach*, volume 357. Springer, 2012. 292

124. Sebastian Thrun and Lorien Pratt. *Learning to Learn*. Springer, 2012. 292, 296, 304, 312

125. Yonglong Tian, Yue Wang, Dilip Krishnan, Joshua B Tenenbaum, and Phillip Isola. Rethinking few-shot image classification: a good embedding is all you need? In *European Conference on Computer Vision*, 2020. 308, 312

126. Emanuel Todorov, Tom Erez, and Yuval Tassa. MuJoCo: A physics engine for model-based control. In *IEEE/RSJ International Conference on Intelligent Robots and Systems (IROS)*, pages 5026–5033, 2012. 306

127. Tatiana Tommasi, Martina Lanzi, Paolo Russo, and Barbara Caputo. Learning the roots of visual domain shift. In *European Conference on Computer Vision*, pages 475–482. Springer, 2016. 291, 295, 296

128. Eleni Triantafillou, Tyler Zhu, Vincent Dumoulin, Pascal Lamblin, Utku Evci, Kelvin Xu, Ross Goroshin, Carles Gelada, Kevin Swersky, Pierre-Antoine Manzagol, and Hugo Larochelle. Meta-dataset: A dataset of datasets for learning to learn from few examples. In *International Conference on Learning Representations*, 2020. 306, 307, 312

129. Eric Tzeng, Judy Hoffman, Kate Saenko, and Trevor Darrell. Adversarial discriminative domain adaptation. In *Proceedings of the IEEE Conference on Computer Vision and Pattern Recognition*, pages 7167–7176, 2017. 296

130. Joaquin Vanschoren. Meta-learning: A survey. *arXiv preprint arXiv:1810.03548*, 2018. 312, 312

131. Jean-Philippe Vert, Koji Tsuda, and Bernhard Schölkopf. A primer on kernel methods. In *Kernel Methods in Computational Biology*, volume 47, pages 35–70. MIT Press Cambridge, MA, 2004. 292

132. Ricardo Vilalta and Youssef Drissi. A perspective view and survey of meta-learning. *Artificial Intelligence Review*, 18(2):77–95, 2002. 303, 312, 312

133. Oriol Vinyals, Charles Blundell, Timothy Lillicrap, Daan Wierstra, et al. Matching networks for one shot learning. In *Advances in Neural Information Processing Systems*, pages 3630–3638, 2016. 306, 306, 306, 307, 312

134. Jane X. Wang, Michael King, Nicolas Porcel, Zeb Kurth-Nelson, Tina Zhu, Charlie Deck, Peter Choy, Mary Cassin, Malcolm Reynolds, H. Francis Song, Gavin Buttimore, David P. Reichert, Neil C. Rabinowitz, Loic Matthey, Demis Hassabis, Alexander Lerchner, and Matthew Botvinick. Alchemy: A structured task distribution for meta-reinforcement learning. *arXiv:2102.02926*, 2021. 306, 309, 312

135. Jane X. Wang, Zeb Kurth-Nelson, Dhruva Tirumala, Hubert Soyer, Joel Z. Leibo, Rémi Munos, Charles Blundell, Dharshan Kumaran, and Matthew Botvinick. Learning to reinforcement learn. *arXiv preprint arXiv:1611.05763*, 2016. 297, 300, 300, 300, 312

136. Yaqing Wang, Quanming Yao, James T Kwok, and Lionel M Ni. Generalizing from a few examples: A survey on few-shot learning. *ACM Computing Surveys*, 53(3):1–34, 2020. 297

137. Karl Weiss, Taghi M Khoshgoftaar, and DingDing Wang. A survey of transfer learning. *Journal of Big data*, 3(1):1–40, 2016. 295, 312

138. Lilian Weng. Meta-learning: Learning to learn fast. Lil'Log https://lilianweng.github.io/lil-log/2018/11/30/meta-learning.html, November 2018. 312, 312

139. Markus Wulfmeier, Alex Bewley, and Ingmar Posner. Addressing appearance change in outdoor robotics with adversarial domain adaptation. In *2017 IEEE/RSJ International Conference on Intelligent Robots and Systems (IROS)*, pages 1551–1558. IEEE, 2017. 296

140. Yongqin Xian, Christoph H Lampert, Bernt Schiele, and Zeynep Akata. Zero-shot learning—a comprehensive evaluation of the good, the bad and the ugly. *IEEE Transactions on Pattern Analysis and Machine Intelligence*, 41(9):2251–2265, 2018. 312

141. Lin Xu, Frank Hutter, Holger H Hoos, and Kevin Leyton-Brown. SATzilla: portfolio-based algorithm selection for SAT. *Journal of Artificial Intelligence Research*, 32:565–606, 2008. 303

142. Wen Xu, Jing He, and Yanfeng Shu. Transfer learning and deep domain adaptation. In *Advances in Deep Learning*. IntechOpen, 2020. 295

143. Zhao Yang, Mike Preuss, and Aske Plaat. Transfer learning and curriculum learning in Sokoban. *arXiv preprint arXiv:2105.11702*, 2021. 312

144. Jaesik Yoon, Taesup Kim, Ousmane Dia, Sungwoong Kim, Yoshua Bengio, and Sungjin Ahn. Bayesian model-agnostic meta-learning. In *Proceedings of the 32nd International Conference on Neural Information Processing Systems*, pages 7343–7353, 2018. 297, 312

145. Jason Yosinski, Jeff Clune, Yoshua Bengio, and Hod Lipson. How transferable are features in deep neural networks? In *Neural Information Processing Systems*, pages 3320–3328, 2014. 293

146. Tianhe Yu, Deirdre Quillen, Zhanpeng He, Ryan Julian, Karol Hausman, Chelsea Finn, and Sergey Levine. Meta-world: A benchmark and evaluation for multi-task and meta reinforcement learning. In *Conference on Robot Learning*, pages 1094–1100. PMLR, 2020. 306, 308, 308, 309, 312, 313

147. Matthew D Zeiler and Rob Fergus. Visualizing and understanding convolutional networks. In *European Conference on Computer Vision*, pages 818–833. Springer, 2014. 289

148. Lei Zhang. Transfer adaptation learning: A decade survey. *arXiv:1903.04687*, 2019. 295, 312, 312

149. Fuzhen Zhuang, Zhiyuan Qi, Keyu Duan, Dongbo Xi, Yongchun Zhu, Hengshu Zhu, Hui Xiong, and Qing He. A comprehensive survey on transfer learning. *Proceedings of the IEEE*, 109(1):43–76, 2020. 292, 312

150. Luisa Zintgraf, Kyriacos Shiarli, Vitaly Kurin, Katja Hofmann, and Shimon Whiteson. Fast context adaptation via meta-learning. In *International Conference on Machine Learning*, pages 7693–7702. PMLR, 2019. 312

Chapter 10
Further Developments

We have come to the end of this book. We will reflect on what we have learned. In this chapter we will review the main themes and essential lessons, and we will look to the future.

Why do we study deep reinforcement learning? Our inspiration is the dream of artificial intelligence; to understand human intelligence and to create intelligent behavior that can supplement our own, so that together we can grow. For reinforcement learning our goal is to learn from the world, to learn increasingly complex behaviors for increasingly complex sequential decision problems. The preceding chapters have shown us that many successful algorithms were inspired by how humans learn.

Currently, many environments consist of games and simulated robots; in the future this may include human–computer interactions and collaborations in teams with real humans.

10.1 Development of Deep Reinforcement Learning

Reinforcement learning has made a remarkable transition, from a method that was used to learn small tabular toy problems, to learning simulated robots how to walk, to playing the largest multi-agent real-time strategy games, and beating the best humans in Go and poker. The reinforcement learning paradigm is a framework in which many learning algorithms have been developed. The framework is able to incorporate powerful ideas from other fields, such as deep learning, and autoencoders.

To appreciate the versatility of reinforcement learning, and now that we have studied the field in great detail, let us have a closer look at how the developments in the field have proceeded over time.

10.1.1 Tabular Methods

Reinforcement learning starts with a simple agent/environment loop, where an environment performs the agent's actions and returns a reward (the model-free approach). We use a Markov decision process to formalize reinforcement learning. The value and policy functions are initially implemented in a tabular fashion, limiting the method to small environments, since the agent has to fit the function representations in memory. Typical environments are Grid world, Cartpole, and Mountain car; a typical algorithm to find the optimal policy function is tabular Q-learning. Basic principles in the design of these algorithms are exploration, exploitation, and on-policy/off-policy learning. Furthermore, imagination, as a form of model-based reinforcement learning, was developed.

This part of the field forms a well-established and stable basis, that is, however, only suitable for learning small single-agent problems. The advent of deep learning caused the field to shift into a higher gear.

10.1.2 Model-Free Deep Learning

Inspired by breakthroughs in supervised image recognition, deep learning was also applied to Q-learning, causing the Atari breakthrough for deep reinforcement learning. The basis of the success of deep learning in reinforcement learning are methods to break correlations and improve convergence (replay buffer and a separate target network). The DQN algorithm [51] has become quite well known. Policy-based and actor critic approaches work well with deep learning and are also applicable to continuous action spaces. Many model-free actor critic variants have been developed [32, 48, 50, 70]; they are often tested on simulated robot applications. Algorithms often reach good quality optima, but model-free algorithms have a high sample complexity. Tables 3.1 and 4.1 list these algorithms.

This part of the field—deep model-free value and policy-based algorithms—can now be considered as well-established, with mature algorithms whose behavior is well understood, and with good results for high-dimensional single-agent environments. Typical high-dimensional environments are the Arcade Learning Environment and MuJoCo simulated physics locomotion tasks.

10.1.3 Multi-Agent Methods

Next, more advanced methods are covered. In Chap. 5 model-based algorithms combine planning and learning to improve sample efficiency. For high-dimensional visual environments they use uncertainty modeling and latent models or world models, to reduce the dimensionality for planning. The algorithms are listed in Table 5.2.

Furthermore, the step from single agent to multi-agent is made, enlarging the type of problems that can be modeled, getting closer to real-world problems. The strongest human Go players are beaten with a model-based self-play combination of MCTS and a deep actor critic algorithm. The self-play setup performs curriculum learning, learning from previous learning tasks ordered from easy to hard, and is in that sense a form of meta-learning. Variants are shown in Table 6.2.

For multi-agent and imperfect information problems, deep reinforcement learning is used to study competition, emergent collaboration, and hierarchical team learning. In these areas reinforcement learning comes close to multi-agent systems and population-based methods, such as swarm computing. Parallel population-based methods may be able to learn the policy quicker than gradient-based methods, and they may be a good fit for multi-agent problems.

Also, imperfect information multi-agent problems are studied, such as poker; for competitive games, counterfactual regret minimization was developed. Research into cooperation is continuing. Early strong results are reported in StarCraft using team collaboration and team competition. In addition, research is being performed into emergent social behavior, connecting the fields of reinforcement learning to swarm computing and multi-agent systems. Algorithms and experiments are listed in Table 7.2. A list of hierarchical approaches is shown in Table 8.1.

In human learning, new concepts are learned based on old concepts. Transfer learning from foundation models and meta-learning aim to reuse the existing knowledge, or even learn to learn. Meta-learning, curriculum learning, and hierarchical learning are emerging as techniques to conquer ever larger state spaces. Table 9.2 shows meta-learning approaches.

All these areas should be considered as advanced reinforcement learning, where active research is still very much occurring. New algorithms are being developed, and experiments typically require large amounts of compute power. Furthermore, results are less robust and require much hyperparameter tuning. More advances are needed and expected.

10.1.4   Evolution of Reinforcement Learning

In contrast to supervised learning, which learns from a fixed dataset, reinforcement learning is a mechanism for learning by doing, just as children learn. The agent/environment framework has turned out to be a versatile approach that can be augmented and enhanced when we try new problem domains, such as high dimensions, multi-agent, or imperfect information. Reinforcement learning has encompassed methods from supervised learning (deep learning) and unsupervised learning (autoencoders), as well as from population-based optimization.

In this sense reinforcement learning has evolved from being a single-agent Markov decision process to a framework for learning, based on agent and environment. Other approaches can be hooked into this framework, to learn new fields, and to improve performance. These additions can interpret high-dimensional states (as in DQN) or shrink a state space (as with latent models). When accommodating self-play, the framework provided us with a curriculum learning sequence, yielding world class levels of play in two-agent games.

10.2 Main Challenges

Deep reinforcement learning is being used to understand more real-world sequential decision making situations. Among the applications that motivate these developments are self-driving cars and other autonomous operations, image and speech recognition, decision making, and, in general, acting naturally.

What will the future bring for deep reinforcement learning? The main challenge for deep reinforcement learning is to manage the combinatorial explosion that occurs when a sequence of decisions is chained together. Finding the right kind of inductive bias can exploit structure in this state space.

We list three major challenges for current and future research in deep reinforcement learning:

1. Solving larger problems faster
2. Solving problems with more agents
3. Interacting with people

The following techniques address these challenges:

1. Solving larger problems faster

   - Reducing sample complexity with latent models
   - Curriculum learning in self-play methods
   - Hierarchical reinforcement learning
   - Learn from previous tasks with transfer learning and meta-learning
   - Better exploration through intrinsic motivation

2. Solving problems with more agents

   - Hierarchical reinforcement learning
   - Population-based self-play league methods

3. Interacting with people

   - Explainable AI
   - Generalization

Let us have a closer look at these techniques, to see what future developments can be expected for them.

10.2.1 Latent Models

Chapter 5 discussed model-based deep reinforcement learning methods. In model-based methods a transition model is learned that is then used with planning to augment the policy function, reducing sample complexity. A problem for model-based methods in high-dimensional problems is that high-capacity networks need many observations in order to prevent overfitting, negating the potential reduction in sample complexity.

One of the most promising model-based methods is the use of autoencoders to create latent models that compress or abstract from irrelevant observations, yielding a lower-dimensional latent state model that can be used for planning in a reduced state space. The reduction in sample complexity of model-based approaches can thus be maintained. Latent models create compact world representations that are also used in hierarchical and multi-agent problems, and further work is ongoing.

A second development in model-based deep reinforcement learning is the use of end-to-end planning and learning of the transition model. Especially for elaborate self-play designs such as AlphaZero, where an MCTS planner is integrated in self-learning, the use of end-to-end learning is advantageous, as the work on MuZero has shown. Research is ongoing in this field where planning and learning are combined [17, 22, 34, 36, 52, 68, 69].

10.2.2 Self-Play

Chapter 6 discussed learning by self-play in two-agent games. In many two-agent games the transition function is given. When the environment of an agent is played by the opponent with the exact same transition function, a self-learning self-play system can be constructed in which both agent and environment improve each other. We have discussed examples of cycles of continuous improvement, from tabula rasa to world champion level.

After earlier results in backgammon [79], AlphaZero achieved landmark results in Go, chess, and shogi [73, 74]. The AlphaZero design includes an MCTS planner in a self-play loop that improves a dual-headed deep residual network [73]. The design has spawned much further research, as well as inspired general interest in artificial intelligence and reinforcement learning [21, 26, 40, 43, 45, 55, 68, 72].

10.2.3 Hierarchical Reinforcement Learning

Team play is important in multi-agent problems, and hierarchical approaches can structure the environment in a hierarchy of agents. Hierarchical reinforcement learning methods are also applied to single-agent problems, using principles of divide and conquer. Many large single-agent problems are hierarchically structured. Hierarchical methods aim to make use of this structure by dividing large problems into smaller subproblems; they group primitive actions into macro-actions. When a policy has been found with a solution for a certain subproblem, then this can be reused when the subproblem surfaces again. Note that for some problems it is difficult to find a hierarchical structure that can be exploited efficiently.

Hierarchical reinforcement learning has been studied for some time [10, 27, 77]. Recent work has been reported on successful methods for deep hierarchical learning and population-based training [46, 47, 53, 62], and more is to be expected.

10.2.4   Transfer Learning and Meta-Learning

Among the major challenges of deep reinforcement learning is the long training time. Transfer learning and meta-learning aim to reduce the long training times, by transferring learned knowledge from existing to new (but related) tasks, and by learning to learn from the training of previous tasks, to speed up learning new (but related) tasks.

In the fields of image recognition and natural language processing it has become common practice to use networks that are pretrained on ImageNet [18, 25] or BERT [19] or other large pretrained networks [5, 54]. Optimization-based methods such as MAML learn better initial network parameters for new tasks and have spawned much further research.

Zero-shot learning is a meta-learning approach where outside information is learned, such as attributes or a textual description for image content, which is then used to recognize individuals from a new class [1, 63, 76, 85]. Meta-learning is a highly active field where more results can be expected.

Foundation models are large models, such as ImageNet for image recognition, and GPT-3 in natural language processing, which are trained extensively on large datasets. They contain general knowledge that can be specialized for a certain more specialized task. They can also be used for multi-modal tasks, where text and image information are combined. The DALL-E project is able to create images that go with textual descriptions. See Fig. 10.1 for amusing or beautiful examples ("an armchair in the shape of an avocado") [60]. GPT-3 has also been used to study zero-shot learning, with success, in the CLIP project [61].

10.2.5   Population-Based Methods

Most reinforcement learning has focused on one- or two-agent problems. However, the world around us is full of many-agent problems. The major problem of multi-agent reinforcement learning is to model the large, nonstationary, problem space.

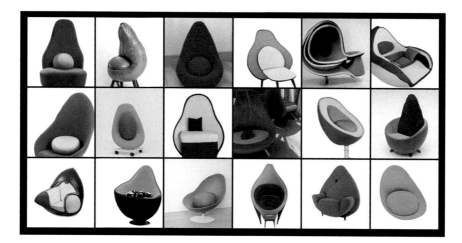

Fig. 10.1 DALL-E, an algorithm that draws pictures based on textual commands [60]

Recent work applies self-play methods in a multi-agent setting, where entire populations of agents are trained against each other. This approach combines aspects of evolutionary algorithms (combining and mutating policies, as well as culling of underperforming agent policies) and hierarchical methods (modeling of team collaboration).

Population-based training of leagues of agents has achieved success in highly complex multi-agent games: StarCraft [82] and Capture the Flag [37]. Population-based methods, combining evolutionary principles with self-play, curriculum learning, and hierarchical methods are active areas of research [38, 41, 49, 65, 83].

10.2.6 Exploration and Intrinsic Motivation

The prime motivator for learning in reinforcement learning is reward. However, reward is sparse in many sequential decision problems. Reward shaping tries to augment the reward function with heuristic knowledge. Developmental psychology argues that learning is (also) based on curiosity, or intrinsically motivated. Intrinsic motivation is a basic curiosity drive to explore for exploration's sake, deriving satisfaction from the exploration process itself.

The field of intrinsic motivation is relatively new in reinforcement learning. Links with hierarchical reinforcement learning and the options framework are being explored, as well as with models of curiosity [14, 66, 67].

Intrinsic motivation in reinforcement learning can be used for exploring open-ended environments [3, 75] (see also Sect. 8.3.2). Intrinsically motivated goal-conditioned algorithms can train agents to learn to represent, generate, and pursue their own goals [15]. The success of the Go-Explore algorithm in domains with sparse rewards also stresses the importance of exploration in reinforcement learning [23].

10.2.7   Explainable AI

Explainable AI (XAI) is closely related to the topics of planning and learning that we discuss in this book and to natural language processing.

When a human expert suggests an answer, this expert can be questioned to explain the reasoning behind the answer. This is a desirable property, and enhances how much we trust the answer. Most clients receiving advice, be it financial or medical, put greater trust in a well-reasoned explanation than in a *yes* or *no* answer without any explanation.

Decision support systems that are based on classic symbolic AI can often be made to provide such reasoning easily. For example, interpretable models [58], decision trees [59], graphical models [39, 44], and search trees [13] can be traversed, and the choices at decision points can be recorded and used to translate in a human-understandable argument.

Connectionist approaches such as deep learning, in contrast, are less interpretable. Their accuracy, however, is typically much higher than the classical approaches. Explainable AI aims to combine the ease of interpreting symbolic AI with the accuracy of connectionist approaches [7, 20, 30].

The work on soft decision trees [28, 35] and adaptive neural trees [78] has shown how hybrid approaches of planning and learning can try to build an explanatory decision tree based on a neural network. These works build in part on model compression [8, 9] and belief networks [2, 4, 6, 16, 33, 56, 71, 80]. Unsupervised methods can be used to find interpretable models [58, 64, 81]. Model-based reinforcement learning methods aim to perform deep sequential planning in learned world models [31].

10.2.8   Generalization

Benchmarks drive algorithmic progress in artificial intelligence. Chess and poker have given us deep heuristic search; ImageNet has driven deep supervised learning; ALE has given us deep Q-learning; MuJoCo and the DeepMind control suite have driven actor critic methods; Omniglot, MiniImagenet, and Meta-World drive meta-learning; and StarCraft and other multi-agent games drive hierarchical and population-based methods.

As the work on explainable AI indicates, there is a trend in reinforcement learning to study problems that are closer to the real world, through model-based methods, multi-agent methods, meta-learning, and hierarchical methods.

As deep reinforcement learning will be more widely applied to real-world problems, generalization becomes important, as is argued by Zhang et al. [87]. Where supervised learning experiments have a clear training set/test set separation to measure generalization, in reinforcement learning agents often fail to generalize beyond the environment they were trained in [12, 57]. Reinforcement learning agents often overfit on their training environment [24, 29, 84, 86, 88]. This becomes especially difficult in sim-to-real transfer [89]. One benchmark specifically designed to increase generalization is Procgen. It aims to increase environment diversity through procedural content generation, providing 16 parameterizable environments [11].

Benchmarks will continue to drive progress in artificial intelligence, especially for generalization [42].

10.3 The Future of Artificial Intelligence

This book has covered the stable basis of deep reinforcement learning, as well as active areas of research. Deep reinforcement learning is a highly active field, and many more developments will follow.

We have seen complex methods for solving sequential decision problems, some of which are easily solved on a daily basis by humans in the world around us. In certain problems, such as backgammon, chess, checkers, poker, and Go, computational methods have now surpassed human ability. In most other endeavors, such as pouring water from a bottle in a cup, writing poetry, or falling in love, humans still reign supreme.

Reinforcement learning is inspired by biological learning, yet computational and biological methods for learning are still far apart. Human intelligence is general and broad—we know much about many different topics, and we use our general knowledge of previous tasks when learning new things. Artificial intelligence is specialized and deep—computers can be good at certain tasks, but their intelligence is narrow, and learning from other tasks is a challenge.

Two conclusions are clear. First, for humans, hybrid intelligence, where human general intelligence is augmented by specialized artificial intelligence, can be highly beneficial. Second, for AI, the field of deep reinforcement learning is taking cues from human learning in hierarchical methods, curriculum learning, learning to learn, and multi-agent cooperation.

The future of artificial intelligence is human.

References

1. Zeynep Akata, Florent Perronnin, Zaid Harchaoui, and Cordelia Schmid. Label-embedding for attribute-based classification. In *Proceedings of the IEEE Conference on Computer Vision and Pattern Recognition*, pages 819–826, 2013. 328
2. John Asmuth, Lihong Li, Michael L Littman, Ali Nouri, and David Wingate. A Bayesian sampling approach to exploration in reinforcement learning. In *Proceedings of the Twenty-Fifth Conference on Uncertainty in Artificial Intelligence*, pages 19–26. AUAI Press, 2009. 330
3. Arthur Aubret, Laetitia Matignon, and Salima Hassas. A survey on intrinsic motivation in reinforcement learning. *arXiv preprint arXiv:1908.06976*, 2019. 329
4. Marc Bellemare, Joel Veness, and Michael Bowling. Bayesian learning of recursively factored environments. In *International Conference on Machine Learning*, pages 1211–1219, 2013. 330
5. Steven Bird, Ewan Klein, and Edward Loper. *Natural language processing with Python: analyzing text with the natural language toolkit*. O'Reilly Media, Inc., 2009. 328
6. Eric Brochu, Vlad M Cora, and Nando De Freitas. A tutorial on Bayesian optimization of expensive cost functions, with application to active user modeling and hierarchical reinforcement learning. *arXiv preprint arXiv:1012.2599*, 2010. 330
7. Cameron Browne, Dennis JNJ Soemers, and Eric Piette. Strategic features for general games. In *KEG@ AAAI*, pages 70–75, 2019. 330
8. Cristian Buciluă, Rich Caruana, and Alexandru Niculescu-Mizil. Model compression. In *Proceedings of the 12th ACM SIGKDD International Conference on Knowledge Discovery and Data Mining*, pages 535–541, 2006. 330
9. Yu Cheng, Duo Wang, Pan Zhou, and Tao Zhang. A survey of model compression and acceleration for deep neural networks. *arXiv preprint arXiv:1710.09282*, 2017. 330
10. Junyoung Chung, Sungjin Ahn, and Yoshua Bengio. Hierarchical multiscale recurrent neural networks. In *International Conference on Learning Representations*, 2016. 328
11. Karl Cobbe, Chris Hesse, Jacob Hilton, and John Schulman. Leveraging procedural generation to benchmark reinforcement learning. In *International Conference on Machine Learning*, pages 2048–2056. PMLR, 2020. 331
12. Karl Cobbe, Oleg Klimov, Chris Hesse, Taehoon Kim, and John Schulman. Quantifying generalization in reinforcement learning. In *International Conference on Machine Learning*, pages 1282–1289, 2018. 331
13. Helder Coelho and Luis Moniz Pereira. Automated reasoning in geometry theorem proving with prolog. *Journal of Automated Reasoning*, 2(4):329–390, 1986. 330
14. Cédric Colas, Pierre Fournier, Mohamed Chetouani, Olivier Sigaud, and Pierre-Yves Oudeyer. Curious: intrinsically motivated modular multi-goal reinforcement learning. In *International Conference on Machine Learning*, pages 1331–1340. PMLR, 2019. 329
15. Cédric Colas, Tristan Karch, Olivier Sigaud, and Pierre-Yves Oudeyer. Intrinsically motivated goal-conditioned reinforcement learning: a short survey. *arXiv preprint arXiv:2012.09830*, 2020. 329
16. Luis M De Campos, Juan M Fernandez-Luna, José A Gámez, and José M Puerta. Ant colony optimization for learning Bayesian networks. *International Journal of Approximate Reasoning*, 31(3):291–311, 2002. 330
17. Joery A. de Vries, Ken S. Voskuil, Thomas M. Moerland, and Aske Plaat. Visualizing MuZero models. *arXiv preprint arXiv:2102.12924*, 2021. 327
18. Jia Deng, Wei Dong, Richard Socher, Li-Jia Li, Kai Li, and Li Fei-Fei. ImageNet: A large-scale hierarchical image database. In *2009 IEEE Conference on Computer Vision and Pattern Recognition*, pages 248–255. IEEE, 2009. 328
19. Jacob Devlin, Ming-Wei Chang, Kenton Lee, and Kristina Toutanova. BERT: pre-training of deep bidirectional transformers for language understanding. In *Proceedings of the 2019 Conference of the North American Chapter of the Association for Computational Linguistics: Human Language Technologies, NAACL-HLT 2019, Minneapolis*, 2018. 328

20. Derek Doran, Sarah Schulz, and Tarek R Besold. What does explainable AI really mean? A new conceptualization of perspectives. *arXiv preprint arXiv:1710.00794*, 2017. 330

21. Yan Duan, John Schulman, Xi Chen, Peter L Bartlett, Ilya Sutskever, and Pieter Abbeel. RL^2: Fast reinforcement learning via slow reinforcement learning. *arXiv preprint arXiv:1611.02779*, 2016. 327

22. Werner Duvaud and Aurèle Hainaut. MuZero general: Open reimplementation of MuZero. https://github.com/werner-duvaud/muzero-general, 2019. 327

23. Adrien Ecoffet, Joost Huizinga, Joel Lehman, Kenneth O Stanley, and Jeff Clune. First return, then explore. *Nature*, 590(7847):580–586, 2021. 330

24. Jesse Farebrother, Marlos C Machado, and Michael Bowling. Generalization and regularization in DQN. *arXiv preprint arXiv:1810.00123*, 2018. 331

25. Li Fei-Fei, Jia Deng, and Kai Li. ImageNet: Constructing a large-scale image database. *Journal of Vision*, 9(8):1037–1037, 2009. 328

26. Dieqiao Feng, Carla P Gomes, and Bart Selman. Solving hard AI planning instances using curriculum-driven deep reinforcement learning. *arXiv preprint arXiv:2006.02689*, 2020. 327

27. Yannis Flet-Berliac. The promise of hierarchical reinforcement learning. https://thegradient.pub/the-promise-of-hierarchical-reinforcement-learning/, March 2019. 328

28. Nicholas Frosst and Geoffrey Hinton. Distilling a neural network into a soft decision tree. In *Proceedings of the First International Workshop on Comprehensibility and Explanation in AI and ML*, 2017. 330

29. Dibya Ghosh, Jad Rahme, Aviral Kumar, Amy Zhang, Ryan P Adams, and Sergey Levine. Why generalization in RL is difficult: Epistemic POMDPs and implicit partial observability. *Advances in Neural Information Processing Systems*, 34, 2021. 331

30. David Gunning. Explainable artificial intelligence (XAI). *Defense Advanced Research Projects Agency (DARPA)*, 2, 2017. 330

31. David Ha and Jürgen Schmidhuber. World models. *arXiv preprint arXiv:1803.10122*, 2018. 330

32. Tuomas Haarnoja, Aurick Zhou, Pieter Abbeel, and Sergey Levine. Soft actor-critic: Off-policy maximum entropy deep reinforcement learning with a stochastic actor. In *International Conference on Machine Learning*, pages 1861–1870. PMLR, 2018. 324

33. David Heckerman, Dan Geiger, and David M Chickering. Learning Bayesian networks: The combination of knowledge and statistical data. *Machine Learning*, 20(3):197–243, 1995. 330

34. Matteo Hessel, Ivo Danihelka, Fabio Viola, Arthur Guez, Simon Schmitt, Laurent Sifre, Theophane Weber, David Silver, and Hado van Hasselt. Muesli: Combining improvements in policy optimization. In *International Conference on Machine Learning*, pages 4214–4226, 2021. 327

35. Geoffrey Hinton, Oriol Vinyals, and Jeff Dean. Distilling the knowledge in a neural network. *arXiv preprint arXiv:1503.02531*, 2015. 330

36. Thomas Hubert, Julian Schrittwieser, Ioannis Antonoglou, Mohammadamin Barekatain, Simon Schmitt, and David Silver. Learning and planning in complex action spaces. In *International Conference on Machine Learning*, pages 4476–4486, 2021. 327

37. Max Jaderberg, Wojciech M. Czarnecki, Iain Dunning, Luke Marris, Guy Lever, Antonio Garcia Castañeda, Charles Beattie, Neil C. Rabinowitz, Ari S. Morcos, Avraham Ruderman, Nicolas Sonnerat, Tim Green, Louise Deason, Joel Z. Leibo, David Silver, Demis Hassabis, Koray Kavukcuoglu, and Thore Graepel. Human-level performance in 3D multiplayer games with population-based reinforcement learning. *Science*, 364(6443):859–865, 2019. 329

38. Max Jaderberg, Valentin Dalibard, Simon Osindero, Wojciech M. Czarnecki, Jeff Donahue, Ali Razavi, Oriol Vinyals, Tim Green, Iain Dunning, Karen Simonyan, Chrisantha Fernando, and Koray Kavukcuoglu. Population based training of neural networks. *arXiv preprint arXiv:1711.09846*, 2017. 329

39. Michael Irwin Jordan. *Learning in Graphical Models*, volume 89. Springer Science & Business Media, 1998. 330

40. John Jumper, Richard Evans, Alexander Pritzel, Tim Green, Michael Figurnov, Olaf Ronneberger, Kathryn Tunyasuvunakool, Russ Bates, Augustin Žídek, Anna Potapenko, Alex

Bridgland, Clemens Meyer, Simon A. A. Kohl, Andrew J. Ballard, Andrew Cowie, Bernardino Romera-Paredes, Stanislav Nikolov, Rishub Jain, Jonas Adler, Trevor Back, Stig Petersen, David Reiman, Ellen Clancy, Michal Zielinski, Martin Steinegger, Michalina Pacholska, Tamas Berghammer, Sebastian Bodenstein, David Silver, Oriol Vinyals, Andrew W. Senior, Koray Kavukcuoglu, Pushmeet Kohli, and Demis Hassabis. Highly accurate protein structure prediction with AlphaFold. *Nature*, 596(7873):583–589, 2021. 327

41. Shauharda Khadka, Somdeb Majumdar, Tarek Nassar, Zach Dwiel, Evren Tumer, Santiago Miret, Yinyin Liu, and Kagan Tumer. Collaborative evolutionary reinforcement learning. In *International Conference on Machine Learning*, pages 3341–3350. PMLR, 2019. 329

42. Robert Kirk, Amy Zhang, Edward Grefenstette, and Tim Rocktäschel. A survey of generalisation in deep reinforcement learning. *arXiv preprint arXiv:2111.09794*, 2021. 331

43. Alexandre Laterre, Yunguan Fu, Mohamed Khalil Jabri, Alain-Sam Cohen, David Kas, Karl Hajjar, Torbjorn S Dahl, Amine Kerkeni, and Karim Beguir. Ranked reward: Enabling self-play reinforcement learning for combinatorial optimization. *arXiv preprint arXiv:1807.01672*, 2018. 327

44. Steffen L Lauritzen. *Graphical Models*, volume 17. Clarendon Press, 1996. 330

45. Joel Z Leibo, Edward Hughes, Marc Lanctot, and Thore Graepel. Autocurricula and the emergence of innovation from social interaction: A manifesto for multi-agent intelligence research. *arXiv preprint arXiv:1903.00742*, 2019. 327

46. Andrew Levy, George Konidaris, Robert Platt, and Kate Saenko. Learning multi-level hierarchies with hindsight. In *International Conference on Learning Representations*, 2019. 328

47. Siyuan Li, Rui Wang, Minxue Tang, and Chongjie Zhang. Hierarchical reinforcement learning with advantage-based auxiliary rewards. In *Advances in Neural Information Processing Systems*, pages 1407–1417, 2019. 328

48. Timothy P Lillicrap, Jonathan J Hunt, Alexander Pritzel, Nicolas Heess, Tom Erez, Yuval Tassa, David Silver, and Daan Wierstra. Continuous control with deep reinforcement learning. In *International Conference on Learning Representations*, 2016. 324

49. Risto Miikkulainen, Jason Liang, Elliot Meyerson, Aditya Rawal, Daniel Fink, Olivier Francon, Bala Raju, Hormoz Shahrzad, Arshak Navruzyan, Nigel Duffy, and Babak Hodjat. Evolving deep neural networks. In *Artificial Intelligence in the Age of Neural Networks and Brain Computing*, pages 293–312. Elsevier, 2019. 329

50. Volodymyr Mnih, Adria Puigdomenech Badia, Mehdi Mirza, Alex Graves, Timothy Lillicrap, Tim Harley, David Silver, and Koray Kavukcuoglu. Asynchronous methods for deep reinforcement learning. In *International Conference on Machine Learning*, pages 1928–1937, 2016. 324

51. Volodymyr Mnih, Koray Kavukcuoglu, David Silver, Andrei A. Rusu, Joel Veness, Marc G. Bellemare, Alex Graves, Martin A. Riedmiller, Andreas Fidjeland, Georg Ostrovski, Stig Petersen, Charles Beattie, Amir Sadik, Ioannis Antonoglou, Helen King, Dharshan Kumaran, Daan Wierstra, Shane Legg, and Demis Hassabis. Human-level control through deep reinforcement learning. *Nature*, 518(7540):529–533, 2015. 324

52. Thomas M Moerland. *The Intersection of Planning and Learning*. PhD thesis, Delft University of Technology, 2021. 327

53. Ofir Nachum, Shixiang Gu, Honglak Lee, and Sergey Levine. Data-efficient hierarchical reinforcement learning. In *Advances in Neural Information Processing Systems*, pages 3307–3317, 2018. 328

54. Prakash M Nadkarni, Lucila Ohno-Machado, and Wendy W Chapman. Natural language processing: an introduction. *Journal of the American Medical Informatics Association*, 18(5):544–551, 2011. 328

55. Sanmit Narvekar, Bei Peng, Matteo Leonetti, Jivko Sinapov, Matthew E Taylor, and Peter Stone. Curriculum learning for reinforcement learning domains: A framework and survey. *Journal Machine Learning Research*, 2020. 327

56. Richard E Neapolitan. *Learning Bayesian networks*. Pearson Prentice Hall, Upper Saddle River, NJ, 2004. 330

57. Charles Packer, Katelyn Gao, Jernej Kos, Philipp Krähenbühl, Vladlen Koltun, and Dawn Song. Assessing generalization in deep reinforcement learning. *arXiv preprint arXiv:1810.12282*, 2018. 331

58. Hugo M Proença and Matthijs van Leeuwen. Interpretable multiclass classification by MDL-based rule lists. *Information Sciences*, 512:1372–1393, 2020. 330, 330

59. J Ross Quinlan. Induction of decision trees. *Machine Learning*, 1(1):81–106, 1986. 330

60. Alec Radford, Jong Wook Kim, Chris Hallacy, Aditya Ramesh, Gabriel Goh, Sandhini Agarwal, Girish Sastry, Amanda Askell, Pamela Mishkin, Jack Clark, Gretchen Krueger, and Ilya Sutskever. Learning transferable visual models from natural language supervision. In *International Conference on Machine Learning*, 2021. 328, 329

61. Aditya Ramesh, Mikhail Pavlov, Gabriel Goh, Scott Gray, Chelsea Voss, Alec Radford, Mark Chen, and Ilya Sutskever. Zero-shot text-to-image generation. In *International Conference on Machine Learning*, 2021. 328

62. Frank Röder, Manfred Eppe, Phuong DH Nguyen, and Stefan Wermter. Curious hierarchical actor-critic reinforcement learning. In *International Conference on Artificial Neural Networks*, pages 408–419. Springer, 2020. 328

63. Bernardino Romera-Paredes and Philip Torr. An embarrassingly simple approach to zero-shot learning. In *International Conference on Machine Learning*, pages 2152–2161, 2015. 328

64. Cynthia Rudin. Stop explaining black box machine learning models for high stakes decisions and use interpretable models instead. *Nature Machine Intelligence*, 1(5):206–215, 2019. 330

65. Tim Salimans, Jonathan Ho, Xi Chen, Szymon Sidor, and Ilya Sutskever. Evolution strategies as a scalable alternative to reinforcement learning. *arXiv:1703.03864*, 2017. 329

66. Vieri Giuliano Santucci, Pierre-Yves Oudeyer, Andrew Barto, and Gianluca Baldassarre. Intrinsically motivated open-ended learning in autonomous robots. *Frontiers in Neurorobotics*, 13:115, 2020. 329

67. Jürgen Schmidhuber. A possibility for implementing curiosity and boredom in model-building neural controllers. In *Proc. of the international conference on simulation of adaptive behavior: From animals to animats*, pages 222–227, 1991. 329

68. Julian Schrittwieser, Ioannis Antonoglou, Thomas Hubert, Karen Simonyan, Laurent Sifre, Simon Schmitt, Arthur Guez, Edward Lockhart, Demis Hassabis, Thore Graepel, Timothy Lillicrap, and David Silver. Mastering Atari, go, chess and shogi by planning with a learned model. *Nature*, 588(7839):604–609, 2020. 327, 327

69. Julian Schrittwieser, Thomas Hubert, Amol Mandhane, Mohammadamin Barekatain, Ioannis Antonoglou, and David Silver. Online and offline reinforcement learning by planning with a learned model. *arXiv preprint arXiv:2104.06294*, 2021. 327

70. John Schulman, Filip Wolski, Prafulla Dhariwal, Alec Radford, and Oleg Klimov. Proximal policy optimization algorithms. *arXiv preprint arXiv:1707.06347*, 2017. 324

71. Marco Scutari. Learning Bayesian networks with the bnlearn R package. *Journal of Statistical Software*, 35(i03), 2010. 330

72. Marwin HS Segler, Mike Preuss, and Mark P Waller. Planning chemical syntheses with deep neural networks and symbolic AI. *Nature*, 555(7698):604, 2018. 327

73. David Silver, Thomas Hubert, Julian Schrittwieser, Ioannis Antonoglou, Matthew Lai, Arthur Guez, Marc Lanctot, Laurent Sifre, Dharshan Kumaran, Thore Graepel, Timothy Lillicrap, Karen Simonyan, and Demis Hassabis. A general reinforcement learning algorithm that masters chess, shogi, and Go through self-play. *Science*, 362(6419):1140–1144, 2018. 327, 327

74. David Silver, Julian Schrittwieser, Karen Simonyan, Ioannis Antonoglou, Aja Huang, Arthur Guez, Thomas Hubert, Lucas Baker, Matthew Lai, Adrian Bolton, Yutian Chen, Timothy Lillicrap, Fan Hui, Laurent Sifre, George van den Driessche, Thore Graepel, and Demis Hassabis. Mastering the game of Go without human knowledge. *Nature*, 550(7676):354, 2017. 327

75. Satinder Singh, Andrew G Barto, and Nuttapong Chentanez. Intrinsically motivated reinforcement learning. Technical report, University of Amherst, Mass, Department of Computer Science, 2005. 329

76. Sungryull Sohn, Junhyuk Oh, and Honglak Lee. Hierarchical reinforcement learning for zero-shot generalization with subtask dependencies. In *Advances in Neural Information Processing Systems*, pages 7156–7166, 2018. 328

77. Richard S Sutton, Doina Precup, and Satinder Singh. Between MDPs and semi-MDPs: a framework for temporal abstraction in reinforcement learning. *Artificial Intelligence*, 112(1–2):181–211, 1999. 328

78. Ryutaro Tanno, Kai Arulkumaran, Daniel C Alexander, Antonio Criminisi, and Aditya Nori. Adaptive neural trees. In *International Conference on Machine Learning*, pages 6166–6175, 2019. 330

79. Gerald Tesauro. TD-Gammon: A self-teaching backgammon program. In *Applications of Neural Networks*, pages 267–285. Springer, 1995. 327

80. Marc Teyssier and Daphne Koller. Ordering-based search: A simple and effective algorithm for learning Bayesian networks. *arXiv preprint arXiv:1207.1429*, 2012. 330

81. Alfredo Vellido, José David Martín-Guerrero, and Paulo JG Lisboa. Making machine learning models interpretable. In *ESANN*, volume 12, pages 163–172, 2012. 330

82. Oriol Vinyals, Igor Babuschkin, Wojciech M. Czarnecki, Michaël Mathieu, Andrew Dudzik, Junyoung Chung, David H. Choi, Richard Powell, Timo Ewalds, Petko Georgiev, Junhyuk Oh, Dan Horgan, Manuel Kroiss, Ivo Danihelka, Aja Huang, Laurent Sifre, Trevor Cai, John P. Agapiou, Max Jaderberg, Alexander Sasha Vezhnevets, Rémi Leblond, Tobias Pohlen, Valentin Dalibard, David Budden, Yury Sulsky, James Molloy, Tom Le Paine, Çaglar Gülçehre, Ziyu Wang, Tobias Pfaff, Yuhuai Wu, Roman Ring, Dani Yogatama, Dario Wünsch, Katrina McKinney, Oliver Smith, Tom Schaul, Timothy P. Lillicrap, Koray Kavukcuoglu, Demis Hassabis, Chris Apps, and David Silver. Grandmaster level in Starcraft II using multi-agent reinforcement learning. *Nature*, 575(7782):350–354, 2019. 329

83. Vanessa Volz, Jacob Schrum, Jialin Liu, Simon M Lucas, Adam Smith, and Sebastian Risi. Evolving Mario levels in the latent space of a deep convolutional generative adversarial network. In *Proceedings of the Genetic and Evolutionary Computation Conference*, pages 221–228, 2018. 329

84. Shimon Whiteson, Brian Tanner, Matthew E Taylor, and Peter Stone. Protecting against evaluation overfitting in empirical reinforcement learning. In *2011 IEEE Symposium on Adaptive Dynamic Programming and Reinforcement Learning (ADPRL)*, pages 120–127. IEEE, 2011. 331

85. Yongqin Xian, Christoph H Lampert, Bernt Schiele, and Zeynep Akata. Zero-shot learning—a comprehensive evaluation of the good, the bad and the ugly. *IEEE Transactions on Pattern Analysis and Machine Intelligence*, 41(9):2251–2265, 2018. 328

86. Amy Zhang, Nicolas Ballas, and Joelle Pineau. A dissection of overfitting and generalization in continuous reinforcement learning. *arXiv preprint arXiv:1806.07937*, 2018. 331

87. Chiyuan Zhang, Samy Bengio, Moritz Hardt, Benjamin Recht, and Oriol Vinyals. Understanding deep learning (still) requires rethinking generalization. *Communications of the ACM*, 64(3):107–115, 2021. 331

88. Chiyuan Zhang, Oriol Vinyals, Remi Munos, and Samy Bengio. A study on overfitting in deep reinforcement learning. *arXiv preprint arXiv:1804.06893*, 2018. 331

89. Wenshuai Zhao, Jorge Peña Queralta, and Tomi Westerlund. Sim-to-real transfer in deep reinforcement learning for robotics: a survey. In *2020 IEEE Symposium Series on Computational Intelligence (SSCI)*, pages 737–744. IEEE, 2020. 331

Appendix A
Mathematical Background

This appendix provides essential mathematical background and establishes the notation that we use in this book. We start with notation of sets and functions, and then we discuss probability distributions, expectations, and information theory. You will most likely have seen some of these in previous courses. We will also discuss how to differentiate through an expectation, which frequently appears in machine learning.

This appendix is based on Moerland [3].

A.1 Sets and Functions

We start at the beginning, with sets and functions.

A.1.1 Sets

A.1.1.1 Discrete Set

A discrete set is a set of countable elements.

Examples:

- $X = \{1, 2, .., n\}$ (integers)
- $X = \{\text{up, down, left, right}\}$ (arbitrary elements)
- $X = \{0, 1\}^d$ (d-dimensional binary space)

© The Author(s), under exclusive license to Springer Nature Singapore Pte Ltd. 2022
A. Plaat, *Deep Reinforcement Learning*,
https://doi.org/10.1007/978-981-19-0638-1

A.1.1.2 Continuous Set

A continuous set is a set of connected elements.

Examples:

- $X = [2, 11]$ (bounded interval)
- $X = \mathbb{R}$ (real line)
- $X = [0, 1]^d$ (d-dimensional hypercube)

A.1.1.3 Conditioning a Set

We can also condition within a set, by using : or |. For example, the *discrete probability k-simplex*, which is what we actually use to define a discrete probability distribution over k categories, is given by

$$X = \{x \in [0, 1]^k : \sum_k x_k = 1\}.$$

This means that x is a vector of length k, consisting of entries between 0 and 1, with the restriction that the vector sums to 1.

A.1.1.4 Cardinality and Dimensionality

It is important to distinguish the cardinality and dimensionality of a set:

- The *cardinality* (size) counts the number of elements in a vector space, for which we write $|X|$.
- The *dimensionality* counts the number of dimensions in the vector space X, for which we write $\text{Dim}(X)$.

Examples:

- The discrete space $X = \{0, 1, 2\}$ has cardinality $|X| = 3$ and dimensionality $\text{Dim}(X) = 1$.
- The discrete vector space $X = \{0, 1\}^4$ has cardinality $|X| = 2^4 = 16$ and dimensionality $\text{Dim}(X) = 4$.

A.1.1.5   Cartesian Product

We can combine two spaces by taking the Cartesian product, denoted by $\times$, which consists of all the possible combinations of elements in the first and second sets:

$$X \times Z = \{(x, z) : x \in X, z \in Z\}$$

We can also combine discrete and continuous spaces through Cartesian products.

Example: Assume $X = \{20, 30\}$ and $Z = \{0, 1\}$. Then

$$X \times Z = \{(20, 0), (20, 1), (30, 0), (30, 1)\}$$

Assume $X = \mathbb{R}$ and $Z = \mathbb{R}$. Then $X \times Z = \mathbb{R}^2$.

A.1.2   Functions

- A function f maps a value in the function's *domain* X to a (unique) value in the function's *co-domain/range* Y, where X and Y can be discrete or continuous sets.
- We write the statement that f is a function from X to Y as

$$f : X \to Y.$$

Examples:

- $y = x^2$ maps every value in domain $X \in \mathbb{R}$ to range $Y \in \mathbb{R}^+$ (see Fig. A.1).

A.2   Probability Distributions

A probability distribution is a mathematical function that gives the probability of the occurrence of a set of possible outcomes. The set of possible outcomes is called the *sample space*, which can be discrete or continuous, and is denoted by X. For example, for flipping a coin $X = \{\text{heads, tails}\}$. When we actually sample the variable, we get a particular value $x \in X$. For example, for the first coin flip $x_1 = \text{heads}$. Before we actually sample the outcome, the particular outcome value

Fig. A.1 $y = x^2$

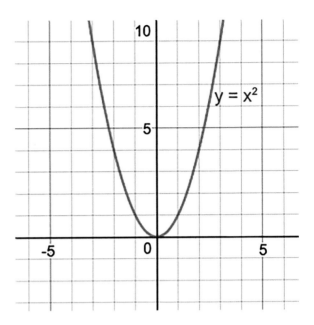

is still unknown. We say that it is a *random variable*, denoted by X, which always has an associated probability distribution $p(X)$.

Sample space (a set)	X
Random variable	X
Particular value	x

Depending on whether the sample space is a discrete or continuous set, the distribution $p(X)$ and the way to represent it differ. We detail both below, see also Fig. A.2.

A.2.1 Discrete Probability Distributions

- A *discrete variable* X can take values in a discrete set $X = \{1, 2, .., n\}$. A particular value that X takes is denoted by x.
- Discrete variable X has an associated *probability mass function*: $p(X)$, where $p : X \to [0, 1]$. Each possible value x that the variable can take is associated with a probability $p(X = x) \in [0, 1]$. (For example, $p(X = 1) = 0.2$, the probability that X is equal to 1 is 20%.)
- Probability distributions always sum to 1: $\sum_{x \in X} p(x) = 1$.

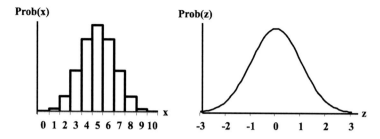

Fig. A.2 Examples of discrete (left) versus continuous (right) probability distribution [2]

A.2.1.1   Parameters

We represent a probability distribution with *parameters*. For a discrete distribution of size n, we need $n-1$ parameters, $\{p_{x=1}, .., p_{x=n-1}\}$, where $p_{x=1} = p(x=1)$. The probability of the last category follows from the sum to one constraint, $p_{x=n} = 1 - \sum_{i=1}^{n-1} p_{x=i}$.

> **Example**: A discrete variable X that can take three values ($X = \{1, 2, 3\}$), with associated probability distribution $p(X = x)$:
>
> $$\frac{p(X = 1)\ \ p(X = 2)\ \ p(X = 3)}{0.2 \qquad\ \ 0.4 \qquad\ \ 0.4}.$$

A.2.1.2   Representing Discrete Random Variables

It is important to realize that we always represent a discrete variable *as a vector of probabilities*. Therefore, the above variable X does not really take values $X = \{1, 2, 3\}$, because 1, 2, and 3 are arbitrary categories (category two is not twice as much as the first category). We could just as well have written $X = \{a, b, c\}$. Always think of the possible values of a discrete variable as separate entries. Therefore, we should represent the value of a discrete variable as a vector of probabilities. In the data, when we observe the ground truth, this becomes a *one-hot encoding*, where we put all mass on the observed class.

> **Example**: In the above example, we had ($X = \{1, 2, 3\}$). Imagine we sample X three times and observe 1, 2, and 3, respectively. We would actually

(continued)

represent these observations as

Observed category	Representation
1	$(1, 0, 0)$
2	$(0, 1, 0)$
3	$(0, 0, 1)$

A.2.2   Continuous Probability Distributions

- A *continuous variable* X can take values in a continuous set, $X = \mathbb{R}$ (the real line), or $X = [0, 1]$ (a bounded interval).
- Continuous variable X has an associated *probability density function*: $p(X)$, where $p : X \to \mathbb{R}^+$ (a positive real number).
- In a continuous set, there are infinitely many values that the random value can take. Therefore, the *absolute* probability of any particular value is 0.
- We can only define absolute probability on an interval, $p(a < X \leq b) = \int_a^b p(x)$. (For example, $p(2 < X \leq 3) = 0.2$, the probability that X will fall between 2 and 3 is equal to 20%.)
- The interpretation of an individual value of the density, like $p(X = 3) = 4$, is only a *relative* probability. The higher the probability $p(X = x)$, the higher the relative chance that we would observe x.
- Probability distributions always sum to 1: $\int x \in X p(x) = 1$ (note that this time we integrate instead of sum).

A.2.2.1   Parameters

We need to represent a continuous distribution with a parameterized function, which for every possible value in the sample space predicts a relative probability. Moreover, we need to obey the sum to one constraint. Therefore, there are many *parameterized continuous probability densities*. An example is the normal distribution. A continuous density is a function $p : X \to \mathbb{R}^+$ that depends on some parameters. Scaling the parameters allows variation in the location where we put probability mass.

Example: A variable X that can take values on the real line with distribution

$$p(x; \mu, \sigma) = \frac{1}{\sigma\sqrt{2\pi}} \exp\left(-\frac{(x-\mu)^2}{2\sigma^2}\right).$$

Here, the mean parameter μ and standard deviation σ are the parameters. We can change them to change the shape of the distribution, while always ensuring that it still sums to one. We draw an example normal distribution in Fig. A.2, right.

The differences between discrete and continuous probability distributions are summarized in Table A.1.

Table A.1 Comparison of discrete and continuous probability distributions

	Discrete distribution	Continuous distribution
Input/sample space	Discrete set, $X = \{0, 1\}$, with size $n = |X|$	Continuous set, $X = \mathbb{R}$
Probability function	Probability mass function (pmf) $p : X \to [0, 1]$ such that $\sum_{x \in X} p(x) = 1$	Probability density function (pdf) $p : X \to \mathbb{R}$ such that $\int_{x \in X} p(x) = 1$
Possible parameterized distributions	Various, but only need simple Discrete	Various, normal, logistic, beta, etc.
Parameters	$\{p_{x=1}, \dots, p_{x=n-1}\}$	Depends on distribution, for normal: $\{\mu, \sigma\}$
Number of parameters	$n-1 = |X|-1$ (due to sum to 1 constraint)[a]	Depends on distribution, for normal: 2
Example distribution function	$p(x = 1) = 0.2$	For example, for normal $p(x|\mu, \sigma) = \frac{1}{\sigma\sqrt{2\pi}} \exp\left(-\frac{(x-\mu)^2}{2\sigma^2}\right)$
	$p(x = 2) = 0.4$ $p(x = 3) = 0.4$	
Absolute probability	$p(x = 1) = 0.2$	$p(3 \leq x < 4) = \int_3^4 p(x) = 0.3$ (on interval)[b]
Relative probability	–	$p(x = 3) = 7.4$

[a] Due to the sum to 1 constraint, we need one parameter less than the size of the sample space, since the last probability is 1 minus all the others: $p_n = 1 - \sum_{i=1}^{n-1} p_i$

[b] Note that for continuous distributions, probabilities are only defined on *intervals*. The density function $p(x)$ only gives relative probabilities, and therefore, we may have $p(x) > 1$, like $p(x = 3) = 5.6$, which is of course not possible (one should not interpret it as an absolute probability). However, $p(a \leq x < b) = \int_a^b p(x) < 1$ by definition

A.2.3   Conditional Distributions

- A conditional distribution means that the distribution of one variable depends on the value that another variable takes.
- We write $p(X|Y)$ to indicate that the value of X depends on the value of Y.

Example: For discrete random variables, we may store a conditional distribution as a table of size $|X| \times |Y|$. A variable X that can take three values ($X = \{1, 2, 3\}$) and variable Y that can take two values ($Y = \{1, 2\}$). The conditional distribution may, for example, be

	$p(X = 1|Y)$	$p(X = 2|Y)$	$p(X = 3|Y)$
$Y = 1$	0.2	0.4	0.4
$Y = 2$	0.1	0.9	0.0

Note that for each value Y, $p(X|Y)$ should still sum to 1, and it is a valid probability distribution. In the table above, each row therefore sums to 1.

Example: We can similarly store conditional distributions for continuous random variables, this time mapping the input space to the parameters of a *continuous* probability distribution. For example, for $p(Y|X)$ we can assume a Gaussian distribution $N(\mu(x), \sigma(s))$, where the mean and standard deviation depend on $x \in \mathbb{R}$. Then we can, for example, specify:

$$\mu(x) = 2x, \quad \sigma(x) = x^2$$

and therefore have

$$p(y|x) = N(2x, x^2).$$

Note that for each value of X, $p(Y|X)$ still integrates to 1, and it is a valid probability distribution.

A.2.4   Expectation

We also need the notion of an *expectation*.

A.2.4.1 Expectation of a Random Variable

The expectation of a random variable is essentially an average. For a discrete variable, it is defined as

$$\mathbb{E}_{X \sim p(X)}[f(X)] = \sum_{x \in X} [x \cdot p(x)]. \tag{A.1}$$

For a continuous variable, the summation becomes integration.

Example:
Assume a given $p(X)$ for a binary variable:

x	$p(X = x)$
0	0.8
1	0.2
.	

The expectation is

$$\mathbb{E}_{X \sim p(X)}[f(X)] = 0.8 \cdot 0 + 0.2 \cdot 1 = 0.2. \tag{A.2}$$

A.2.4.2 Expectation of a Function of a Random Variable

More often, and also in the context of reinforcement learning, we will need the expectation *of a function of the random variable*, denoted by $f(X)$. Often, this function maps to a continuous output:

- Assume a function $f : X \rightarrow \mathbb{R}$, which, for every value $x \in X$, maps to a continuous value $f(x) \in \mathbb{R}$.
- The expectation is then defined as follows:

$$\mathbb{E}_{X \sim p(X)}[f(X)] = \sum_{x \in X} [f(x) \cdot p(x)]. \tag{A.3}$$

For a continuous variable the summation again becomes integration. The formula may look complicated, but it essentially reweights each function outcome by the probability that this output occurs, see the example below.

Example:
Assume a given density $p(X)$ and function $f(x)$:

x	$p(X = x)$	$f(x)$
1	0.2	22.0
2	0.3	13.0
3	0.5	7.4.

The expectation of the function can be computed as

$$\mathbb{E}_{X \sim p(X)}[f(X)] = 22.0 \cdot 0.2 + 13.0 \cdot 0.3 + 7.4 \cdot 0.5$$
$$= 12.0.$$

The same principle applies when $p(x)$ is a continuous density, only with the summation replaced by integration.

A.2.5 *Information Theory*

Information theory studies the amount of information that is present in distributions and the way that we can compare distributions.

A.2.5.1 Information

The *information I* of an event x observed from distribution $p(X)$ is defined as

$$I(x) = -\log p(x).$$

In words, the more likely an observation is (the higher $p(x)$), the less information we get when we actually observe the event. In other words, the information of an event is the (potential) reduction of uncertainty. On the two extremes we have:

- $p(x) = 0$: $I(x) = -\log 0 = \infty.$
- $p(x) = 1$: $I(x) = -\log 1 = 0.$

A.2.5.2 Entropy

We define the entropy H of a discrete distribution $p(X)$ as

$$
\begin{aligned}
H[p] &= \mathbb{E}_{X \sim p(X)}[I(X)] \\
&= \mathbb{E}_{X \sim p(X)}[-\log p(X)] \\
&= -\sum_{x} p(x) \log p(x).
\end{aligned} \tag{A.4}
$$

If the base of the logarithm is 2, then we measure it in *bits*. When the base of the logarithm is e, then we measure the entropy in *nats*. The continuous version of the above equation is called the *continuous entropy* or *differential entropy*.

Informally, the entropy of a distribution is a measure of the amount of "uncertainty" in a distribution, i.e., a measure of its "spread." We can nicely illustrate this with a binary variable (0/1), where we plot the probability of a 1 against the entropy of the distribution (Fig. A.3). We see that on the two extremes, the entropy of the distribution is 0 (no spread at all), while the entropy is maximal for $p(x = 1) = 0.5$ (and therefore $p(x = 0) = 0.5$), which gives maximal spread to the distribution.

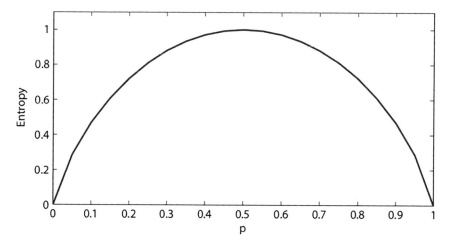

Fig. A.3 Entropy of a binary discrete variable. Horizontal axis shows the probability that the variable takes value 1, and the vertical axis shows the associated entropy of the distribution. High entropy implies high spread in the distribution, while low entropy implies little spread

Example: The entropy of the distribution in the previous example is

$$H[p] = -\sum_x p(x) \log p(x)$$

$$= -0.2 \cdot \ln 0.2 - 0.3 \cdot \ln 0.3 - 0.5 \cdot \ln 0.5 = 1.03 \text{ nats.} \qquad (A.5)$$

A.2.5.3   Cross-Entropy

The cross-entropy is defined between two distributions $p(X)$ and $q(X)$ defined over the same support (sample space). The cross-entropy is given by

$$H[p,q] = \mathbb{E}_{X \sim p(X)}[-\log q(X)]$$

$$= -\sum_x p(x) \log q(x). \qquad (A.6)$$

When we do maximum likelihood estimation in supervised learning, then we actually minimize the cross-entropy between the data distribution and the model distribution.

A.2.5.4   Kullback–Leibler Divergence

For two distributions $p(X)$ and $q(X)$ we can also define the *relative entropy*, better known as the Kullback–Leibler (KL) divergence D_{KL}:

$$D_{\text{KL}}[p||q] = \mathbb{E}_{X \sim p(X)}\left[-\log \frac{q(X)}{p(X)}\right]$$

$$= -\sum_x p(x) \log \frac{q(x)}{p(x)}. \qquad (A.7)$$

The Kullback–Leibler divergence is a measure of the distance between two distributions. The more two distributions depart from each other, the higher the KL-divergence will be. Note that the KL-divergence is not symmetrical, $D_{\text{KL}}[p||q] \neq D_{\text{KL}}[q||p]$ in general.

Finally, we can also rewrite the KL-divergence as an entropy and cross-entropy, relating the previously introduced quantities:

$$D_{\text{KL}}[p||q] = \mathbb{E}_{X \sim p(X)}\left[-\log \frac{q(X)}{p(X)}\right]$$

$$= \sum_x p(x) \log p(x) - \sum_x p(x) \log q(x)$$

$$= H[p] + H[p, q]. \tag{A.8}$$

Entropy, cross-entropy, and KL-divergence are common in many machine learning domains, especially to construct loss functions.

A.3   Derivative of an Expectation

A key problem in gradient-based optimization, which appears in parts of machine learning, is getting the gradient of an expectation. We will here discuss one well-known method:[1] the *REINFORCE* estimator (in reinforcement learning), which is in other fields also known as the *score function estimator*, *likelihood ratio method*, and *automated variational inference*.

Assume that we are interested in the gradient of an expectation, where the parameters appear in the distribution of the expectation:[2]

$$\nabla_\theta \mathbb{E}_{x \sim p_\theta(x)}[f(x)]. \tag{A.9}$$

We cannot sample the above quantity because we have to somehow move the gradient inside the expectation (and then we can sample the expectation to evaluate it). To achieve this, we will use a simple rule regarding the gradient of the log of some function $g(x)$:

$$\nabla_x \log g(x) = \frac{\nabla_x g(x)}{g(x)}. \tag{A.10}$$

This results from simple application of the chain-rule.

---

[1] Other methods to differentiate through an expectation are through the *reparameterization trick*, as for example used in variational autoencoders, but we will not further treat this topic here.

[2] If the parameters only appear in the function $f(x)$ and not in $p(x)$, then we can simply push the gradient through the expectation.

We will now expand Eq. A.9, where we midway apply the above log-derivative trick.

$$\nabla_\theta \mathbb{E}_{x \sim p_\theta(x)}[f(x)] = \nabla_\theta \sum_x f(x) \cdot p_\theta(x) \qquad \text{definition of expectation}$$

$$= \sum_x f(x) \cdot \nabla_\theta p_\theta(x) \qquad \text{push gradient through sum}$$

$$= \sum_x f(x) \cdot p_\theta(x) \cdot \frac{\nabla_\theta p_\theta(x)}{p_\theta(x)} \qquad \text{multiply/divide by } p_\theta(x)$$

$$= \sum_x f(x) \cdot p_\theta(x) \cdot \nabla_\theta \log p_\theta(x) \qquad \text{log-der. rule (Eq. A.10)}$$

$$= \mathbb{E}_{x \sim p_\theta(x)}[f(x) \cdot \nabla_\theta \log p_\theta(x)] \qquad \text{rewrite into expectation.}$$

What the above derivation essential does is *pushing the derivative inside of the sum.* This equally applies when we change the sum into an integral. Therefore, for any $p_\theta(x)$, we have

$$\nabla_\theta \mathbb{E}_{x \sim p_\theta(x)}[f(x)] = \mathbb{E}_{x \sim p_\theta(x)}[f(x) \cdot \nabla_\theta \log p_\theta(x)]. \qquad (A.11)$$

This is known as the log-derivative trick, score function estimator, or REINFORCE trick. Although the formula may look complicated, the interpretation is actually simple. We explain this idea in Fig. A.4. In Sect. 4.2.1 we apply this idea to reinforcement learning.

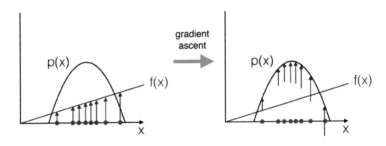

Fig. A.4 Graphical illustration of REINFORCE estimator. Left: example distribution $p_\theta(x)$ and function $f(x)$. When we evaluate the expectation of Eq. A.11, we take m samples, indicated by the blue dots (in this case $m = 8$). The magnitude of $f(x)$ is shown with the red vertical arrows. Right: when we apply the gradient update, each sample pushes up the density at that location, but the magnitude of the push is multiplied by $f(x)$. Therefore, the higher $f(x)$, the harder we push. Since a density needs to integrate to 1, we will increase the density where we push hardest (in the example on the rightmost sample). The distribution will therefore shift to the right on this update

A.4   Bellman Equations

Bellman noted that the value function can be written in recursive form, because the value is also defined at the next states. In his work on dynamic programming [1], he derived recursive equations for V and Q.

The Bellman equations for state values and state–action values are

$$V^\pi(s) = \mathbb{E}_{a \sim \pi(\cdot|s)} \mathbb{E}_{s' \sim T_a(s)} \big[r_a(s, s') + \gamma \cdot V^\pi(s') \big]$$

$$Q^\pi(s, a) = \mathbb{E}_{s' \sim T_a(s)} \big[r_a(s, s') + \gamma \cdot \mathbb{E}_{a' \sim \pi(\cdot|s')} [Q^\pi(s', a')] \big].$$

Depending on whether the state and action space are discrete or continuous, respectively, we write out these equations differently. For a discrete state space and discrete action space, we write the expectations as summations:

$$V(s) = \sum_{a \in A} \pi(a|s) \Big[\sum_{s' \in S} T_a(s, s') \big[r_a(s, s') + \gamma \cdot V(s') \big] \Big]$$

$$Q(s, a) = \sum_{s' \in S} T_a(s, s') \big[r_a(s, s') + \gamma \cdot \sum_{a \in A} \pi(a|s)[Q(s', a')] \big].$$

For continuous state and action spaces, the summations over policy and transition are replaced by integration:

$$V(s) = \int_a \pi(a|s) \Big[\int_{s'} T_a(s, s') \big[r_a(s, s') + \gamma \cdot V(s') \big] \, ds' \Big] da.$$

The same principle applies to the Bellman equation for state–action values:

$$Q(s, a) = \int_{s'} T_a(s, s') \big[r_a(s, s') + \gamma \cdot \int_{a'} [\pi(a'|s') \cdot Q(s', a')] \, da' \big] ds'.$$

We may also have a continuous state space (such as visual input) with a discrete action space (such as pressing buttons in a game):

$$Q(s, a) = \int_{s'} T_a(s, s') \big[r_a(s, s') + \gamma \cdot \sum_{a'} [\pi(a'|s') \cdot Q(s', a')] \big] ds'.$$

References

1. Richard Bellman. *Dynamic Programming*. Courier Corporation, 1957, 2013. 351
2. Andrew A Jawlik. *Statistics from A to Z: Confusing concepts clarified*. John Wiley & Sons, 2016. 341
3. Thomas Moerland. Continuous Markov decision process and policy search. Lecture notes for the course reinforcement learning, Leiden University, 2021. 337

Appendix B
Deep Supervised Learning

This appendix provides a chapter-length overview of essentials of machine learning and deep learning. Deep reinforcement learning uses much of the machinery of deep supervised learning, and a good understanding of deep supervised learning is essential. This appendix should provide you with the basics, in case your knowledge of basic machine learning and deep learning is rusty. In doubt? Try to answer the questions on page 385.

We will start with machine learning basics. We will discuss training and testing, accuracy, the confusion matrix, generalization, and overfitting.

Next, we will provide an overview of deep learning. We will look into neural networks, error functions, training by gradient descent, end-to-end feature learning, and the curse of dimensionality. For neural networks we will discuss convolutional networks, recurrent networks, LSTM, and measures against overfitting.

Finally, on the practical side, we will discuss TensorFlow, Keras, and PyTorch. We will start with methods for solving large and complex problems.

B.1 Machine Learning

The goal of machine learning is *generalization*: to create an accurate, predictive, model of the world. Such an accurate model is said to generalize well to the world.[1] In machine learning we operationalize this goal with two datasets, a training set and a test set.

---

[1] Generalization is closely related to the concepts of overfitting, regularization, and smoothness, as we will see in Sect. B.1.3.1.

© The Author(s), under exclusive license to Springer Nature Singapore Pte Ltd. 2022 353
A. Plaat, *Deep Reinforcement Learning*,
https://doi.org/10.1007/978-981-19-0638-1

The field of machine learning aims to fit a function to approximate an input–output relation. This can be, for example, a regression function or a classification function. The most basic form of machine learning is when a dataset D of input–output pairs is given. We call this form *supervised learning*, since the learning process is *supervised* by the output values.

In machine learning we often deal with large problem domains. We are interested in a function that works not just on the particular values on which it was trained, but also in the rest of the problem domain from which the data items were taken.

In this section we will first see how such a learning process for generalization works; for notation and examples we follow from [63]. Next, we will discuss the specific problems of large domains. Finally, we will discuss the phenomenon of overfitting and how it relates to the bias–variance trade-off.

B.1.1   Training Set and Test Set

A machine learning algorithm must learn a function $\hat{f}(x) \rightarrow y$ on a training set from input values x (such as images) to approximate corresponding output values y (such as image labels). The goal of the machine learning algorithm is to learn this function $\hat{f}$, such that it performs well on the training data and generalizes well to the test data. Let us see how we can measure how well the machine learning methods generalize.

The notions that are used to assess the quality of the approximation are as follows. Let us assume that our problem is a classification task. The elements that our function correctly predicts are called the true positives (TP). Elements that are correctly identified as not belonging to a class are the true negatives (TN). Elements that are mis-identified as belonging to a class are the false positives (FP), and elements that belong to the class but that the predictor misclassifies are the false negatives (FN).

Confusion Matrix: The number of true positives divided by the total number of positives (true and false, TP/(TP + FP)) is called the *precision* of the classifier. The number of true positives divided by the size of the class (TP/(TP + FN)) is the *recall*, or the number of relevant elements that the classifier could find. The term *accuracy* is defined as the total number of true positives and negatives divided by the total number of predictions (TP + TN)/(TP + TN + FP + FN): how well correct elements are predicted [11]. These numbers are often shown in the form of a *confusion matrix* (Table B.1). The numbers in the cells represent the number of true positives, etc., of the experiment.

Table B.1 Confusion matrix

		Predicted class	
		P	N
Actual	P	TP	FN
Class	N	FP	TN

		Predicted class	
		Cat	Dog
Actual	Cat	122	8
Class	Dog	3	9

In machine learning, we are often interested in the accuracy of a method on the training set, and whether the accuracy on the test set is the same as on the training set (generalization).

In regression and classification problems the term *error value* is used to indicate the difference between the true output value y and the predicted output value $\hat{f}(x)$. Measuring how well a function generalizes is typically performed using a method called k-fold cross-validation. This works as follows. When a dataset of input–output examples (x, y) is present, this set is split into a large training set and a smaller holdout test set, typically 80/20 or 90/10. The approximator is trained on the training set until a certain level of accuracy is achieved. For example, the approximator may be a neural network whose parameters θ are iteratively adjusted by gradient descent so that the error value on the training set is reduced to a suitably low value. The approximator function is said to *generalize* well, when the accuracy of the approximator is about the same on the test set as it is on the training set. (Sometimes a third dataset is used, the validation set, to allow stopping before overfitting occurs, see Sect. B.2.7.)

Since the test set and the training set contain examples that are drawn from the same original dataset, approximators can be expected to be able to generalize well. The testing is said to be *in-distribution* when training and test sets are from the same distribution. Out-of-distribution generalization to different problems (or transfer learning) is more difficult, see Sect. 9.2.1.

B.1.2   Curse of Dimensionality

The *state space* of a problem is the space of all possible different states (combinations of values of variables). State spaces grow exponentially with the number of dimensions (variables); high-dimensional problems have large state spaces, and modern machine learning algorithms have to be able to learn functions in such large state spaces.

> **Example**: A classic problem in AI is image classification: predicting what type of object is pictured in an image. Imagine that we have low-resolution

(continued)

gray scale images of 100×100 pixels, where each pixel takes a discrete value between 0 and 255 (a byte). Then, the input space $X \in \{0, 1, \ldots, 255\}^{100 \cdot 100}$, the machine learning problem has dimensionality $100 \cdot 100 = 10000$, and the state space has size 256^{10000}.

When the input space X is high dimensional, we can never store the entire state space (all possible pixels with all possible values) as a table. The effect of an exponential need for observation data as the dimensionality grows has been called the *curse of dimensionality* by Richard Bellman [8]. The curse of dimensionality states that the cardinality (the number of unique points) of a space scales exponentially in the dimensionality of the problem. In a formula, we have that

$$|X| \sim \exp(\mathrm{Dim}(X)).$$

Due to the curse of dimensionality, the size of a table to represent a function increases quickly when the size of the input space increases.

Example: Imagine we have a discrete input space $X = \{0, 1\}^D$ that maps to a real number, $Y = \mathbb{R}$, and we want to store this function fully as a table. We will show the required size of the table and the required memory when we use 32-bit (4 byte) floating point numbers:

$\mathrm{Dim}(X)$	$|X|$	Memory
1	2	8 Byte
5	$2^5 = 32$	128 Byte
10	$2^{10} = 1024$	4 KB
20	$2^{20} \approx 10^6$	4 MB
50	$2^{50} \approx 10^{15}$	4.5 million TB
100	$2^{100} \approx 10^{30}$	$5 \cdot 10^{21}$ TB
265	$2^{265} \approx 2 \cdot 10^{80}$	-

The table shows how quickly exponential growth develops. At a discrete space of size 20, it appears that we are doing alright, storing 4 Megabyte of information. However, at a size of 50, we suddenly need to store 4.5 million Terabyte. We can hardly imagine the numbers that follow. At an input dimensionality of size 265, our table would have grown to size $2 \cdot 10^{80}$, which is close to the estimated number of atoms in the universe.

Since the size of the state space grows exponentially, but the number of observations typically does not, most state spaces in large machine learning problems are sparsely populated with observations. An important challenge for machine learning algorithms is to fit good predictive models on sparse data. To reliably estimate a function, each variable needs a certain number of observations. The number of samples that are needed to maintain statistical significance increases exponentially as the number of dimensions grows. A large amount of training data is required to ensure that there are several samples for each combination of values;[2] however, datasets rarely grow exponentially.

B.1.3 Overfitting and the Bias–Variance Trade-Off

Basic statistics tells us that, according to the law of large numbers, the more observations we have of an experiment, the more reliable the estimate of their value is (that is, the average will be close to the expected value) [11]. This has important implications for the study of large problems where we would like to have confidence in the estimated values of our parameters.

In small toy problems the number of variables of the model is also small. Single-variable linear regression problems model the function as a straight line $y = a \cdot x + b$, with only one independent variable x and two parameters a and b. Typically, when regression is performed with a small number of independent variables, then a relatively large number of observations are available per variable, giving confidence in the estimated parameter values. A rule of thumb is that there should be 5 or more training examples for each variable [86].

In machine learning dimensionality typically means the number of variables of a model. In statistics, dimensionality can be a relative concept: the ratio of the number of variables compared to the number of observations. In practice, the size of the observation dataset is limited, and then absolute and relative dimensionalities do not differ much. In this book we follow the machine learning definition of dimensions meaning variables.

Modeling high-dimensional problems well typically requires a model with many parameters, the so-called *high-capacity* models. Let us look deeper into the consequences of working with high-dimensional problems.

---

[2] Unless we introduce some bias into the problem, by assuming smoothness, implying that there are dependencies between variables, and that the "true" number of independent variables is smaller than the number of pixels.

Fig. B.1 Curve fitting: does
the curvy red line or the
straight dashed blue line best
generalize the information in
the data points?

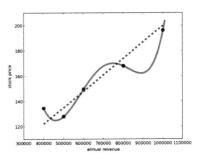

Table B.2 Observations, variables, relative dimensionality, overfitting

Environment observations		Model variables	Relative dimensionality	Chance of
n	$>$	d	Low	Underfitting, high bias
n	$<$	d	High	Overfitting, high variance

To see how to best fit our model, let us consider machine learning as a curve fitting problem, see Fig. B.1. In many problems, the observations are measurements of an underlying natural process. The observations therefore contain some measurement noise. The goal is (1) that the approximating curve fits the (noisy) observations as accurately as possible, but, (2) in order to generalize well, to aim to fit the signal, not the noise.

How complex should the approximator be to faithfully capture the essence of a natural, noisy, process? You can think of this question as: how many parameters θ the network should have, or, if the approximator is a polynomial, what the degree d of the polynomial should be? The complexity of the approximator is also called the capacity of the model, for the amount of information that it can contain (see Table B.2). When the capacity of the approximator d is lower than the number of observations n, then the model is likely to *underfit*: the curve is too simple and cannot reach all observations well, the error is large, and the accuracy is low.[3]

Conversely, when the number of coefficients d is as high or higher than the number of observations n, then many machine learning procedures will be able to find a good fit. The error on the training set will be zero, and training accuracy reaches 100%. Will this high-capacity approximator generalize well to unseen states? Most likely it will not, since the training and test observations are from a real-world process and observations contain noise from the training set. The training noise will be modeled perfectly, and the trained function will have low accuracy on the test set and other datasets. A high-capacity model ($d > n$) that trains well but tests poorly is said to *overfit* on the training data.

Underfitting and overfitting are related to the so-called bias–variance trade-off, see Fig. B.2. High-capacity models fit "peaky" high-variance curves that can fit

---

[3] Overfitting can be reduced in the loss function and training procedure, see Sect. B.2.7.

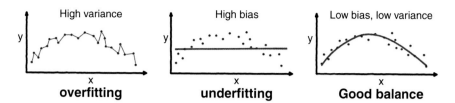

Fig. B.2 Bias–variance trade-off; few-parameter model: high bias; many-parameter model: high variance

signal and noise of the observations and tend to overfit. Low-capacity models fit "straighter" low-variance curves that have higher bias and tend to underfit the observations. Models with a capacity of $d \approx n$ can have both good bias and variance [11].

Preventing underfitting and overfitting is a matter of matching the capacity of our model so that d matches n. This is a delicate trade-off, since reducing the capacity also reduces expressive power of the models. To reduce overfitting, we can use regularization, and many regularization methods have been devised. Regularization has the effect of dynamically adjusting capacity to the number of observations.

B.1.3.1   Regularization—the World Is Smooth

Generalization is closely related to regularization. Most functions in the real world—the functions that we wish to learn in machine learning—are smooth: near similar input leads to near similar output. Few real-world functions are jagged, or locally random (when they are, they are often contrived examples).

This has led to the introduction of regularization methods, to restrict or smooth the behavior of models. Regularization methods allow us to use high-capacity models, learn the complex function, and then later introduce a smoothing procedure, to reduce too much of the randomness and jaggedness. Regularization may for example restrict the weights of variables by moving the values closer to zero, but many other methods exist, as we will see. Different techniques for regularization have been developed for high-capacity deep neural models, and we discuss them in Sect. B.2.7. The goal is to filter out random noise, but at the same time to allow meaningful trends to be recognized, to smooth the function without restricting complex shapes [29].

Before we delve into regularization methods, let us have a look in more detail at how to implement parameterized functions, for which we will use neural networks and deep learning.

B.2   Deep Learning

The architecture of artificial neural networks is inspired by the architecture of biological neural networks, such as the human brain. Neural networks consist of neural core cells that are connected by nerve cells [4]. Figure B.3 shows a drawing of a biological neuron, with a nucleus, an axon, and dendrites [89].

Figure B.4 shows a simple fully connected artificial neural network, with an input layer of two neurons, an output layer of two neurons, and a single hidden layer of five neurons. A neural network with a single hidden layer is called shallow. When the network has more hidden layers, it is called deep (Fig. B.5).

> In this section, we provide an overview of artificial neural networks and their training algorithms. We provide enough detail to understand the deep reinforcement learning concepts in this book. Space does provide a limit to how deep we can go. Please refer to specialized deep learning literature for more information, such as [29].
> We provide a conceptual overview, which should be enough to successfully use existing high-quality deep learning packages, such as TensorFlow (https://www.tensorflow.org.) [1] or PyTorch [69] (https://pytorch.org.)

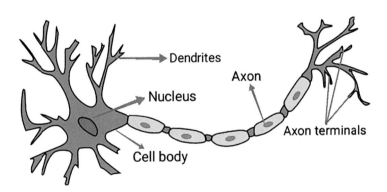

Fig. B.3 A single biological neuron [89]

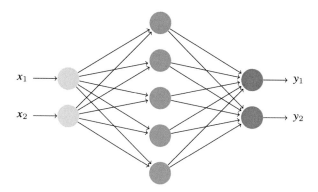

Fig. B.4 A fully connected shallow neural network with 9 neurons

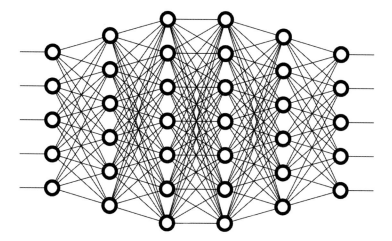

Fig. B.5 Four fully connected hidden layers [76]

B.2.1 Weights, Neurons

Neural networks consist of neurons and connections, typically organized in layers. The neurons process their input signals as a weighted combination, producing an output signal. This calculation is called the activation function, or squashing function, since it is nonlinear. Popular activation functions are the rectified linear unit (ReLU: partly linear, partly zero), the hyperbolic tangent, and the sigmoid or logistic function $\frac{1}{1+e^{-a}}$ for neuron activation a. The neurons are connected by weights. At each neuron j the incoming weights ij are summed $\sum$ and then processed by the activation function σ. The output o of neuron j is therefore:

$$o_j = \sigma \left(\sum_i o_i w_{ij} \right)$$

for weight ij of predecessor neuron o_i. The outputs of this layer of neurons are fed to the inputs for the weights of the next layer.

B.2.2   Backpropagation

The neural network as a whole is a parameterized function $f_\theta(x) \rightarrow \hat{y}$ that converts input into an output approximation. The behavior depends on the parameters θ, also known as the network weights. The parameters are adjusted such that the required input–output relation is achieved (*training* the network). This is done by minimizing the error (or loss) function that calculates the difference between the network output $\hat{y}$ and the training target y.

The training process consists of training *epochs*, individual passes in which the network weights are optimized toward the target, using a method called gradient descent (since the goal is to minimize the error function). An epoch is one complete pass over the training data. Epochs are usually done in batches. Since training is an iterative optimization process, it is typical to train for multiple epochs. Listing B.1 shows simplified pseudocode for the gradient descent training algorithm (based on [29]). When training starts, the weights of the network are initialized to small random numbers. Each epoch consists of a forward pass (recognition, usage) and a backward pass (training, adjustment). The forward pass is just the regular recognition operation for which the network is designed. The input layer is exposed to the input (the image), which is then propagated through the network to the output layers, using the weights and activation functions. The output layer provides the answer, by having a high value at the neuron corresponding to the right label (such

```
1  def train_sl(data, net, alpha=0.001):     # train classifier
2      for epoch in range(max_epochs):        # an epoch is one pass
3          sum_sq = 0                         # reset to zero for each pass
4          for (image, label) in data:
5              output = net.forward_pass(image) # predict
6              sum_sq += (output - label)**2  # compute error
7              grad = net.gradient(sum_sq)    # derivative of error
8              net.backward_pass(grad, alpha) # adjust weights
9      return net
```

Listing B.1 Network training pseudocode for supervised learning

as Cat or Dog, or the correct number), so that an error can be calculated to be used to adjust the weights in the backward pass.

The listing shows a basic version of gradient descent that calculates the gradient over all examples in the dataset and then updates the weights. Batch versions of gradient descent update the weights after smaller subsets and are typically quicker.

B.2.2.1 Loss Function

At the output layer the propagated value $\hat{y}$ is compared with the other part of the example pair, the label y. The difference with the label is calculated, yielding the *error*. The error function is also known as the loss function $\mathcal{L}$. Two common error functions are the mean squared error $\frac{1}{n}\sum_i^n (y_i - \hat{y}_i)^2$ (for regression) and the cross-entropy error $-\sum_i^M y_i \log \hat{y}_i$ (for classification of M classes). The backward pass uses the difference between the forward recognition outcome and the true label to adjust the weights, so that the error becomes smaller. This method uses the gradient of the error function over the weights and is called gradient descent. The parameters are adjusted as follows:

$$\theta_{t+1} = \theta_t - \alpha \nabla_{\theta_t} \mathcal{L}_D(f_{\theta_t})$$

where θ are the network parameters, t is the optimization time step, α is the learning rate, ∇_{θ_t} are the current gradient of the loss function of data $\mathcal{L}_D$, and f_θ is the parameterized objective function.

The training process can be stopped when the error has been reduced below a certain threshold for a single example, or when the loss on an entire validation set has dropped sufficiently. More elaborate stopping criteria can be used in relation to overfitting (see Sect. B.2.7).

Most neural nets are trained using a stochastic version of gradient descent, or SGD [80]. SGD samples a minibatch of size smaller than the total dataset and thereby computes a noisy estimate of the true gradient. This is faster per update step and does not affect the direction of the gradient too much. See Goodfellow et al. [29] for details.

B.2.3 End-to-End Feature Learning

Let us now look in more detail at how neural networks can be used to implement end-to-end feature learning.

We can approximate a function through the discovery of common features in states. Let us, again, concentrate on image recognition. Traditionally, feature discovery was a manual process. Image specialists would painstakingly pour over images to identify common features in a dataset, such as lines, squares, circles, and angles, by hand. They would write small preprocessing algorithms to recognize the features that were then used with classical machine learning methods such as decision trees, support vector machines, or principal component analysis to construct recognizers to classify an image. This hand-crafted method is a labor-intensive and error prone process, and researchers have worked to find algorithms for the full image recognition process, end-to-end. For this to work, also the features must be learned.

For example, if the function approximator consists of the sum of n features, then with hand-crafted features, only the coefficients c_i of the features in the function

$$h(s) = c_1 \times f_1(s) + c_2 \times f_2(s) + c_3 \times f_3(s) + \ldots + c_n \times f_n(s)$$

are learned. In end-to-end learning, the coefficients c_i and the features $f_i(s)$ are learned.

Deep end-to-end learning has achieved great success in image recognition, speech recognition, and natural language processing [22, 32, 53, 91]. End-to-end learning is the learning of a classifier directly from high-dimensional, raw, un-pre-processed, pixel data, all the way to the classification layer, as opposed to learning pre-processed data from intermediate (lower-dimensional) hand-crafted features.

We will now see in more detail how neural networks can perform automated feature discovery. The hierarchy of network layers together can recognize a hierarchy of low-to-high level concepts [54, 57]. For example, in face recognition (Fig. B.6) the first hidden layer may encode edges; the second layer then composes and encodes simple structures of edges; the third layer may encode higher-level concepts such as noses or eyes; and the fourth layer may work at the abstraction level of a face. Deep feature learning finds what to abstract at which level on its own [9] and can come up with classes of intermediate concepts, which work, but look counterintuitive upon inspection by humans.

Toward the end of the 1990s, the work on neural networks moved into deep learning, a term coined by Dechter in [20]. LeCun et al. [56] published an influential paper on deep convolutional nets. The paper introduced the architecture LeNet-5, a seven-layer convolutional neural net trained to classify handwritten MNIST digits from 32×32 pixel images. Listing B.2 shows a modern rendering of LeNet in Keras. The code straightforwardly lists the layer definitions.

End-to-end learning is computationally quite demanding. After the turn of the century, methods, datasets, and compute power had improved to such an extent that full raw, un-pre-processed pictures could be learned, without the intermediate step of hand-crafting features. End-to-end learning proved very powerful, achieving higher accuracy in image recognition than previous methods, and even higher than human test subjects [53]. In natural language processing, deep transformer models such as BERT and GPT-2 and 3 have reached equally impressive results [13, 22].

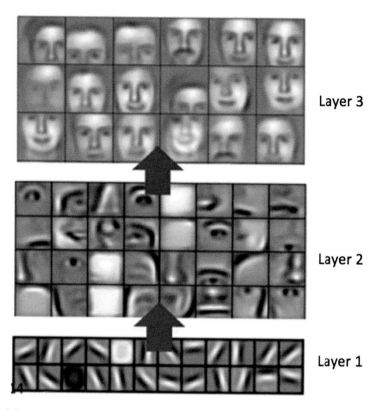

Layer 3

Layer 2

Layer 1

Fig. B.6 Layers of features of increasing complexity [57]

B.2.3.1 Function Approximation

Let us have a look at the different kinds of functions that we wish to approximate in machine learning. The most basic function establishing an input–output relation is regression, which outputs a continuous number. Another important function is classification, which outputs a discrete number. Regression and classification are often learned through supervision, with a dataset of examples (observations) and labels.

In reinforcement learning, three functions are typically approximated: the value function $V(s)$, which relates states to their expected cumulative future rewards, the action value function $Q(s, a)$ that relates actions to their values, and the policy function $\pi(s)$ that relates states to an action (or $\pi(a|s)$ to an action distribution).[4] In reinforcement learning, the functions V, Q, π are learned through reinforcement by the environment. Table B.3 summarizes these functions.

---

[4] In Chap. 5, on model-based learning, we also approximate the transition function $T_a(\cdot)$ and the reward function $R_a(\cdot)$.

```
def lenet_model(img_shape=(28, 28, 1), n_classes=10, l2_reg=0.,
        weights=None):

    # Initialize model
    lenet = Sequential()

    # 2 sets of CRP (Convolution, RELU, Pooling)
    lenet.add(Conv2D(20, (5, 5), padding="same",
            input_shape=img_shape, kernel_regularizer=l2(
                l2_reg)))
    lenet.add(Activation("relu"))
    lenet.add(MaxPooling2D(pool_size=(2, 2), strides=(2, 2)))

    lenet.add(Conv2D(50, (5, 5), padding="same",
            kernel_regularizer=l2(l2_reg)))
    lenet.add(Activation("relu"))
    lenet.add(MaxPooling2D(pool_size=(2, 2), strides=(2, 2)))

    # Fully connected layers (w/ RELU)
    lenet.add(Flatten())
    lenet.add(Dense(500, kernel_regularizer=l2(l2_reg)))
    lenet.add(Activation("relu"))

    # Softmax (for classification)
    lenet.add(Dense(n_classes, kernel_regularizer=l2(l2_reg))
            )
    lenet.add(Activation("softmax"))

    if weights is not None:
            lenet.load_weights(weights)

    # Return the constructed network
    return lenet
```

Listing B.2 LeNet-5 code in Keras [56, 90]

Table B.3 Functions that are frequently approximated

	Function	Input	Output
Dataset	Regression	Continuous (number)	Continuous (number)
	Classification	Discrete (image)	Discrete (class label)
Environment	Value V	State	Continuous (number)
	Action value Q	State × action	Continuous (number)
	Policy π	State	Action(-distribution)

B.2.4   Convolutional Networks

The first neural networks consisted of fully connected layers (Fig. B.5). In image recognition, the input layer of a neural network is typically connected directly to the input image. Higher-resolution images therefore need a higher number of input neurons. If all layers would have more neurons, then the width of the network grows quickly. Unfortunately, growing a fully connected network (see Fig. B.4) by increasing its width (the number of neurons per layer) will increase the number of parameters quadratically.

The naive solution of high-resolution problem learning is to increase the capacity of the model m. However, because of the problem of overfitting, as m grows, so must the number of examples, n.

The solution lies in using a sparse interconnection structure instead of a fully connected network. Convolutional neural nets (CNNs) take their inspiration from biology. The visual cortex in animals and humans is not fully connected, but locally connected [43, 44, 62]. Convolutions efficiently exploit prior knowledge about the structure of the data: patterns reoccur at different locations in the data (translation invariance), and therefore we can share parameters by moving a convolutional window over the image.

A CNN consists of convolutional operators or filters. A typical convolution operator has a small *receptive field* (it only connects to a limited number of neurons, say 5 × 5), whereas a fully connected neuron connects to all neurons in the layer below. Convolutional filters detect the presence of local patterns. The next layer thus acts as a feature *map*. A CNN layer can be seen as a set of learnable filters, invariant for local transformations [29].

Filters can be used to identify features. Features are basic elements such as edges, straight lines, round lines, curves, and colors. To work as a curve detector, for example, the filter should have a pixel structure with high values indicating a shape of a curve. By then multiplying and adding these filter values with the pixel values, we can detect whether the shape is present. The sum of the multiplications in the input image will be large if there is a shape that resembles the curve in the filter.

This filter can only detect a certain shape of curve. Other filters can detect other shapes. Larger activation maps can recognize more elements in the input image. Adding more filters increases the size of the network, which effectively enlarges the activation map. The filters in the first network layer process ("convolve") the input image and fire (have high values) when a specific feature that it is built to detect is in the input image. Training a convolutional net is training a filter that consists of layers of subfilters.

By going through the convolutional layers of the network, increasingly complex features can be represented in the activation maps. Once they are trained, they can be used for as many recognition tasks as needed. A recognition task consists of a single quick forward pass through the network.

Let us spend some more time on understanding these filters.

B.2.4.1   Shared Weights

In CNNs the filter parameters are shared in a layer. Each layer thus defines a filter operation. A filter is defined by few parameters but is applied to many pixels of the image; each filter is replicated across the entire visual field. These replicated units share the same parameterization (weight vector and bias) and form a feature map. This means that all the neurons in a given convolutional layer respond to the same feature within their specific response field. Replicating units in this way allows for features to be detected regardless of their position in the visual field, thus constituting the property of translation invariance.

This weight sharing is also important to prevent an increase in the number of weights in deep and wide nets and to prevent overfitting, as we shall see later.

Real-world images consist of repetitions of many smaller elements. Due to this so-called *translation invariance*, the same patterns reappear throughout an image. CNNs can take advantage of this. The weights of the links are shared, resulting in a large reduction in the number of weights that have to be trained. Mathematically, CNNs put constraints on what the weight values can be. This is a significant advantage of CNNs, since the computational requirements of training the weights of fully connected layers are prohibitive. In addition, statistical strength is gained, since the effective data per weight increases.

Deep CNNs work well in image recognition tasks, for visual filtering operations in spatial dependencies, and for feature recognition (edges, shapes) [55].[5]

B.2.4.2   CNN Architecture

Convolutions recognize features—the deeper the network, the more complex the features. A typical CNN architecture consists of a number of stacked convolutional layers. In the final layers, fully connected layers are used to then classify the inputs.

In the convolutional layers, by connecting only locally, the number of weights is dramatically reduced in comparison with a fully connected net. The ability of a single neuron to recognize different features, however, is less than that of a fully connected neuron.

By stacking many such locally connected layers on top of each other we can achieve the desired nonlinear filters whose joint effect becomes increasingly global.[6] The neurons become responsive to a larger region of pixel space, so that the network first creates representations of small parts of the input, and from

---

[5] Interestingly, this paper was already published in 1989. The deep learning revolution happened twenty years later, when publicly available datasets, more efficient algorithms, and more compute power in the form of GPUs were available.

[6] Nonlinearity is essential. If all neurons performed linearly, then there would be no need for layers. Linear recognition functions cannot discriminate between cats and dogs.

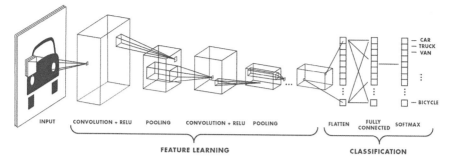

Fig. B.7 Convolutional network example architecture [76]

these representations, create larger areas. The network can recognize and represent increasingly complex concepts without an explosion of weights.

A typical CNN architecture consists of an architecture of multiple layers of convolution, max pooling, and ReLU layers, topped off by a fully connected layer (Fig. B.7).[7]

B.2.4.3   Max Pooling

A further method for reducing the number of weights is weight pooling. Pooling is a kind of nonlinear downsampling (expressing the information in lower resolution with fewer bits). Typically, a 2×2 block is sampled down to a scalar value (Fig. B.8). Pooling reduces the dimension of the network. The most frequently used form is max pooling. It is an important component for object detection [19] and is an integral part of most CNN architectures. Max pooling also allows small translations, such as shifting the object by a few pixels, or scaling, such as putting the object closer to the camera.

B.2.5   Recurrent Networks

Image recognition has had a large impact on network architectures, leading to innovations such as convolutional nets for spatial data.

Speech recognition and time series analysis have also caused new architectures to be created, for sequential data. Such sequences can be modeled by recurrent neural nets (RNN) [10, 25]. Some of the better known RNNs are Hopfield networks [42] and long short-term memory (LSTM) [41].

---

[7] Often with a softmax function. The softmax function normalizes an input vector of real numbers to a probability distribution $[0, 1]$; $p_\theta(y|x) = \text{softmax}(f_\theta(x)) = \frac{e^{f_\theta(x)}}{\sum_k e^{f_{\theta,k}(x)}}$.

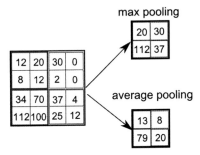

Fig. B.8 Max and average 2 × 2 pooling [29]

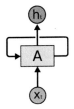

Fig. B.9 RNN x_t is an input vector, h_t is the output/prediction, and A is the RNN [68]

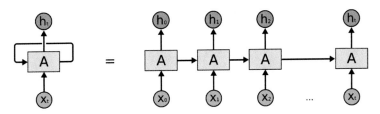

Fig. B.10 RNN unrolled in time [68]

Figure B.9 shows a basic recurrent neural network. An RNN neuron is the same as a normal neuron, with input, output, and activation function. However, RNN neurons have an extra pair of looping input–output connections. Through this structure, the values of the parameters in an RNN can evolve. In effect, RNNs have a variable-like state.

To understand how RNNs work, it helps to unroll the network, as has been done in Fig. B.10. The recurrent neuron loops have been drawn as a straight line to show the network in a deeply layered style, with connections between the layers. In reality the layers are time steps in the processing of the recurrent connections. In a sense, an RNN is a deeply layered neural net folded into a single layer of recurrent neurons.

Where deep convolutional networks are successful in image classification, RNNs are used for tasks with a sequential nature, such as captioning challenges. In a captioning task the network is shown a picture and then has to come up with a textual description that makes sense [88].

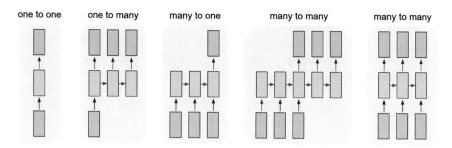

Fig. B.11 RNN configurations [48]

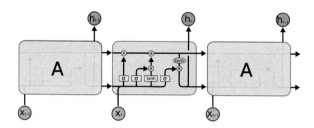

Fig. B.12 LSTM [48]

The main innovation of recurrent nets is that they allow us to work with sequences of vectors. Figure B.11 shows different combinations of sequences that we will discuss now, following an accessible and well-illustrated blog on the different RNN configurations written by Karpathy [48]. There can be sequences in the input, in the output, or in both. The figure shows different rectangles. Each rectangle is a vector. Arrows represent computations, such as matrix multiply. Input vectors are in red, output vectors are in blue, and green vectors hold the state. From left to right we see:

1. *One to one*, the standard network without RNN. This network maps a fixed-sized input to a fixed-sized output, such as an image classification task (picture in/class out).
2. *One to many* adds a sequence in the output. This can be an image captioning task that takes an image and outputs a sentence of words.
3. *Many to one* is the opposite, with a sequence in the input. Think, for example, of sentiment analysis (a sentence is classified for words with negative or positive emotional meaning).
4. *Many to many* has both a sequence for input and a sequence for output. This can be the case in machine translation, where a sentence in English is read and then a sentence in Français is produced.
5. *Many to many* is a related but different situation, with synchronized input and output sequences. This can be the case in video classification where each frame of the video should be labeled.

B.2.5.1   Long Short-Term Memory

Time series prediction is more complex than conventional regression or classification. It adds the complexity of a sequence dependence among the input variables.

LSTM (long short-term memory) is a more powerful type of neuron designed to handle sequences. Figure B.12 shows the LSTM module, allowing comparison with a simple RNN. LSTMs are designed for sequential problems, such as time series, and planning. LSTMs were introduced by Hochreiter and Schmidhuber [41].

RNN training suffers from the vanishing gradient problem. For short-term sequences this problem may be controllable by the same methods as for deep CNN [29]. For long-term remembering LSTMs are better suited. LSTMs are frequently used to solve diverse problems [28, 31, 33, 78], and we encounter them at many places throughout this book.

B.2.6   More Network Architectures

Deep learning is a highly active field of research, in which many advanced network architectures have been developed. We will describe some of the better known architectures.

B.2.6.1   Residual Networks

An important innovation on CNNs is the residual network architecture, or *ResNets*. This idea was introduced in the 2015 ImageNet challenge, which He et al. [35] won with a very low error rate of 3.57%. This error rate is actually lower than what most humans achieve: 5–10%. ResNet has no fewer than 152 layers.

Residual nets introduce *skip links*. Skip links are connections skipping one or more layers, allowing the training to go directly to other layers, reducing the effective depth of the network (Fig. B.13). Skip links create a mixture of a shallow and a deep network, preventing the accuracy degradation and vanishing gradients of deep networks [29].

B.2.6.2   Generative Adversarial Networks

Normally neural networks are used in forward mode, to discriminate input images into classes, going from high dimensional to low dimensional. Networks can also be run backward, to *generate* an image that goes with a certain class, going from low dimensional to high dimensional.[8] Going from small to large implies many

---

[8] Just like the decoding phase of autoencoders.

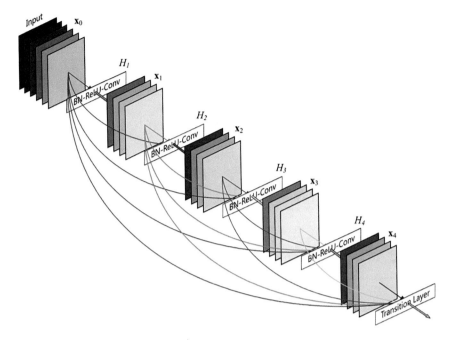

Fig. B.13 Residual net with skip links [35]

possibilities for the image to be instantiated. Extra input is needed to fill in the degrees of freedom.

Running the recognizers backward, in generative mode, has created an active research area called deep generative modeling. An important type of generative model that has made quite an impact is the generative adversarial network, or GAN [30].

Deep networks are susceptible to adversarial attacks. A well-known problem of the image recognition process is that it is brittle. It was found that if an image is slightly perturbed, and imperceptibly to the human eye, deep networks can easily be fooled to characterize an image as the wrong category [85]. This brittleness is known as the one-pixel problem: changing a single unimportant pixel in an image could cause classifiers to switch from classifying an image from cat to dog [82, 85].

GANs are used to generate input images that are slightly different from the original input. GANs generate adversarial examples whose purpose is to fool the discriminator (recognizer). The first network, the generator, generates an image. The second network, the discriminator, tries to recognize the image. The goal for the generator is to mislead the discriminator, in order to improve the robustness of the discriminator. In this way, GANs can be used to make image recognition more robust. The one-pixel problem has spawned an active area of research to understand this problem and to make deep networks more robust.

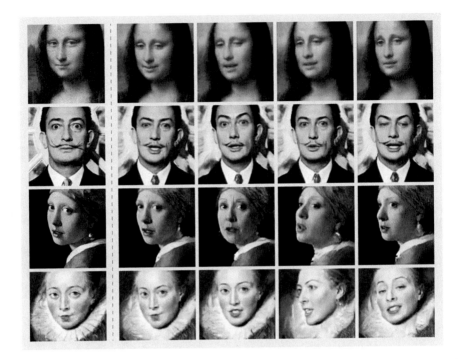

Fig. B.14   Deep fakes [92]

Another use of generative networks is to generate artificial photo-realistic images known as *deep fake* images [92] and *deep dreaming* [49], see Figs. B.14 and B.15.[9] GANs have significantly increased our theoretical understanding of deep learning.

B.2.6.3   Autoencoders

Autoencoders and variational autoencoders (VAE) are used in deep learning for dimensionality reduction, in unsupervised learning [50–52]. An autoencoder network has a butterfly-like architecture, with the same number of neurons in the input and the output layers, but a decreasing layer size as we go to the center (Fig. B.16). The input (contracting) side is said to perform a discriminative action, such as image classification, and the other (expanding) side is generative [30]. When an image is fed to both the input and the output of the autoencoder, results in the center layers are exposed to a compression/decompression process, resulting in the same image, only smoothed. The discriminative/generative process performed by autoencoders

---

[9] Deep Dream Generator at https://deepdreamgenerator.com.

Fig. B.15 Deep dream [49]

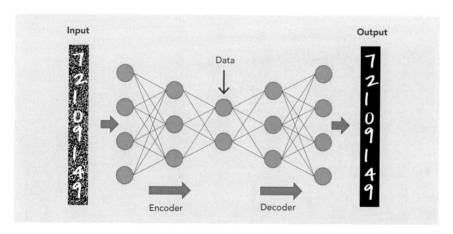

Fig. B.16 Autoencoder, finding the "essence" of the data in the middle [64]

reduces the dimensionality and is generally thought of as going to the "essence" of a problem, de-noising it [39].

Autoencoding illustrates a deep relation between supervised and unsupervised learning. The architecture of an autoencoder consists of an encoder and a decoder. The encoder is a regular discriminative network, as is common in supervised learning. The decoder is a generative network, creating high-dimensional output from low- dimensional input. Together, the encoder–decoder butterfly performs dimensionality reduction, or compression, a form of unsupervised learning[34].

Autoencoders and generative networks are an active area of research.

B.2.6.4   Attention Mechanism

Another important architecture is the attention mechanism. Sequence to sequence learning occurs in many applications of machine learning, for example, in machine translation. The dimensionality of the output of basic RNNs is the same as the input.

The *attention* architecture allows a flexible mapping between input and output dimensionalities [83]. It does so by augmenting the RNNs with an extra network that focuses attention over the sequence of encoder RNN states [3], see Fig. B.17. The extra network focuses on perception and memory access. It has been shown to achieve state-of -the-art results in machine translation and natural language tasks [3], see Fig. B.18. The attention mechanism is especially useful for time series forecasting and translation.

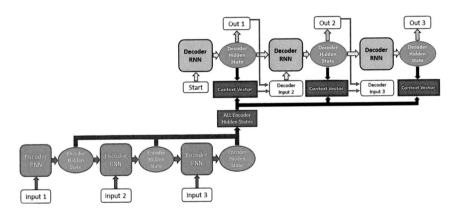

Fig. B.17 Attention architecture [3, 60]

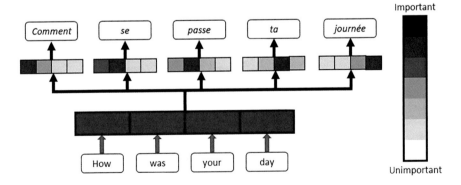

Fig. B.18 Example for attention mechanism [60]

B.2.6.5   Transformers

Finally, we discuss the transformer architecture. The transformer architecture is introduced by Vaswani et al. [87]. Transformers are based on the concept of self-attention: they use attention encoder–decoder models but weigh the influence of different parts of the data. They are central to the highly successful BERT [22] and GPT-3 [13, 74] natural language models but have also been successful in imaginative text-to-image creation.

Transformer-based foundation models play an important role in multi-modal (text/image) learning [73, 75], such as the one that we saw in Fig. 10.1. Treating deep reinforcement learning as a sequence problem, transformers are also being applied to reinforcement learning, with some early success [15, 46]. Transformers are an active area of research. For a detailed explanation of how they work, see, for example, [2, 12, 87].

B.2.7   Overfitting

Now that we have discussed network architectures, it is time to discuss an important problem of deep neural networks: overfitting, and how we can reduce it. Overfitting in neural networks can be reduced in a number of ways. Some of the methods are aimed at restoring the balance between the number of network parameters and the number of training examples, others on data augmentation and capacity reduction. Another approach is to look at the training process. Let us list the most popular approaches [29].

- *Data Augmentation* Overfitting occurs when there are more parameters in the network than examples to train on. The training dataset is increased through manipulations such as rotations, reflections, noise, rescaling, etc. A disadvantage of this method is that the computational cost of training increases.
- *Capacity Reduction* Another solution to overfitting lies in the realization that overfitting is a result of the network having too large a capacity; the network has too many parameters. A cheap way of preventing this situation is to reduce the capacity of the network, by reducing the width and depth of the network.
- *Dropout* A popular method to reduce overfitting is to introduce dropout layers into the networks. Dropout reduces the effective capacity of the network by stochastically dropping a certain percentage of neurons from the backpropagation process [40, 81]. Dropout is an effective and computationally efficient method to reduce overfitting [29].
- *L1 and L2 Regularization* Regularization involves adding an extra term to the loss function that forces the network to not be too complex. The term penalizes the model for using too high weight values. This limits flexibility but also encourages building solutions based on multiple features. Two popular versions of this method are L1 and L2 regularizations [29, 66].

- *Early Stopping* Early stopping is based on the observation that overfitting can be regarded as a consequence of the so-called overtraining (training that progresses beyond the signal, into the noise). By terminating the training process earlier, for example by using a higher stopping threshold for the error function, we can prevent overfitting from occurring [14, 71, 72]. A convenient and popular way is to add a third set to the training set/test set duo that then becomes a training set, a test set, and a holdout validation set. After each training epoch, the network is evaluated against the holdout validation set, to see if under- or overfitting occurs, and if we should stop training. In this way, overfitting can be prevented dynamically during training [11, 29, 72].
- *Batch Normalization* Another method is batch normalization. Batch normalization periodically normalizes the input to the layers [45]. This has many benefits, including a reduction of overfitting.

Overfitting and regularization are important topics of research. In fact, a basic question is why large neural networks perform so well at all. The network capacity is in the millions to billions of parameters, much larger than the number of observations, yet networks perform well. There appears to be a regime, beyond where performance suffers due to overfitting, where performance increases as we continue to increase the capacity of our networks [5, 7]. Belkin et al. have performed studies on interpolation in SGD. Their studies suggest an implicit regularization regime of many parameters beyond overfitting, where SGD generalizes well to test data, explaining in part the good results in practice of deep learning. Nakkiran et al. report similar experimental results, termed the *double descent* phenomenon [47, 65]. Research into the nature of overfitting is active [5–7, 61, 93].

B.3   Datasets and Software

We have discussed in-depth background concepts in machine learning and important aspects of the theory of neural networks. It is time to turn our attention to practical matters. Here we encounter a rich field of data, environments, software, and blogs on how to use deep learning in practice.

The field of deep reinforcement learning is an open field. Researchers release their algorithms and code allowing replication of results. Datasets and trained networks are shared. The barrier to entry is low: high-quality software is available on GitHub for you to download and start doing research in the field. We point to code bases and open software suites at GitHub throughout the book; Appendix C has pointers to software environments and open-source code frameworks.

The most popular deep learning packages are PyTorch [69] and TensorFlow [1], and its top-level language Keras [17]. Some machine learning and mathematical packages also offer deep learning tools, such as scikit-learn, MATLAB, and R. In this section we will start with some easy classification and behavior examples, using the Keras library. We will use Python as our programming language. Python has become the language of choice for machine learning packages: not just for PyTorch and TensorFlow, but also for numpy, scipy, scikit-learn, matplotlib, and many other mature and high-quality machine learning libraries. If Python is not present on your computer, or if you want to download a newer version, please go to https://www. python.org to install it. Note that due to some unfortunate version issues of the Stable Baselines and TensorFlow, you may have to install different versions of Python to get these examples to work, and to use virtual environments to manage your software versions.

Since deep learning is very computationally intensive, these packages typically support GPU parallelism, which can speed up your training tenfold or more, if you have the right GPU card in your system. Cloud providers of computing, such as AWS, Google, Azure, and others, also typically provide modern GPU hardware to support machine learning, often with student discounts.

B.3.1 MNIST and ImageNet

One of the most important elements for the success in image recognition was the availability of good datasets.

In the early days of deep learning, the field benefited greatly from efforts in handwriting recognition. This application was of great value to the postal service, where accurate recognition of handwritten zip codes or postal codes allowed great improvements to efficient sorting and delivery of the mail. A standard test set for handwriting recognition was MNIST (for Modified National Institute of Standards and Technology) [56]. Standard MNIST images are low-resolution 32×32 pixel images of single handwritten digits (Fig. B.19). Of course, researchers wanted

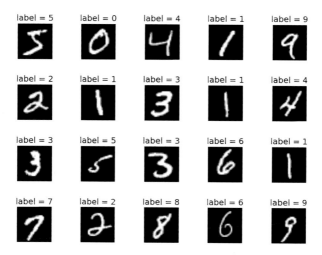

Fig. B.19 Some MNIST images [56]

to process more complex scenes than single digits and higher-resolution images. To achieve higher accuracy, and to process more complex scenes, networks (and datasets) needed to grow in size and complexity.

B.3.1.1   ImageNet

A major dataset in deep learning is ImageNet [21, 24]. It is a collection of more than 14 million URLs of images that have been hand annotated with the objects that are in the picture. It contains more than 20,000 categories. A typical category contains several hundred training images.

The importance of ImageNet for the progress in deep learning is large. The availability of an accepted standard set of labeled images allowed learning algorithms to be tested and improved, and new algorithms to be created. ImageNet was conceived by Fei-Fei Li et al. in 2006, and in later years, she developed it further with her group. Since 2010, an annual software contest has been organized, the ImageNet Large-Scale Visual Recognition Challenge (ILSVRC) [21]. Since 2012, ILSVRC has been won by deep networks, starting the deep learning boom. The network architecture that won this challenge in that year has become known as *AlexNet*, after one of its authors [53].

The 2012 ImageNet database as used by AlexNet has 14 million labeled images. The network featured a highly optimized 2D two-GPU implementation of 5 convolutional layers and 3 fully connected layers. The filters in the convolutional layers are 11×11 in size. The neurons use a ReLU activation function. In AlexNet images were scaled to 256×256 RGB pixels. The size of the network was large, with 60 million parameters. This causes considerable overfitting. AlexNet used data augmentation and dropouts to reduce the impact of overfitting.

Krizhevsky et al. won the 2012 ImageNet competition with an error rate of 15%, significantly better than the number two, who achieved 26%. Although there were earlier reports of CNNs that were successful in applications such as bioinformatics and Chinese handwriting recognition, it was this win of the 2012 ImageNet competition for which AlexNet has become well known.

B.3.2   GPU Implementations

The deep learning breakthrough around 2012 was caused by the co-occurrence of three major developments: (1) algorithmic advances that solved key problems in deep learning, (2) the availability of large datasets of labeled training data, and (3) the availability of computational power in the form of graphical processing units, GPUs.

The most expensive operations in image processing and neural network training are operations on matrices. Matrix operations are some of the most well-studied problems in computer science. Their algorithmic structure is well understood, and

Fig. B.20 SIMD: connection machine 1 and GPU

for basic linear algebra operations, high-performance parallel implementations for CPU exist, such as the BLAS [16, 23].

GPUs were originally designed for smooth graphics performance in video games. Graphical processing requires fast linear algebra computations such as matrix multiply. These are precisely the kind of operations that are at the core of deep learning training algorithms. Modern GPUs consist of thousands of small arithmetic units that are capable of performing linear algebra matrix operations very fast in parallel. This kind of data parallel processing is based on SIMD computing, for single-instruction multiple-data [26, 36]. SIMD data parallelism goes back to designs from 1960s and 1970s vector supercomputers such as the Connection Machine series from Thinking Machines [37, 38, 58]. Figure B.20 shows a picture of the historic CM-1 and of a modern GPU.

Modern GPUs consist of thousands of processing units optimized to process linear algebra matrix operations in parallel [59, 77, 79], offering matrix performance that is orders of magnitude faster than CPUs [18, 67, 84].

B.3.3 Hands On: Classification Example

It is high time to try out some of the materials in practice. Let us see if we can do some image recognition ourselves.

B.3.3.1 Installing TensorFlow and Keras

We will first install TensorFlow. It is possible to run TensorFlow in the cloud, in Colab, or in a Docker container. Links to ready to run Colab environments are on the TensorFlow website. We will, however, assume a traditional local installation on your own computer. All major operating systems are supported: Linux/Ubuntu, macOS, Windows.

The programming model of TensorFlow is complex and not very user friendly. Fortunately, an easy to use language has been built on top of TensorFlow: Keras. The Keras language is easy to use, well-documented, and many examples exist to get you started. When you install TensorFlow, Keras is installed automatically as well.

To install TensorFlow and Keras, go to the TensorFlow page on https://www. tensorflow.org. It is recommended to make a virtual environment to isolate the package installation from the rest of your system. This is achieved by typing

```
python3 -m venv --system-site-packages ./venv
```

(or equivalent) to create the virtual environment. Using the virtual environment requires activation:[10]

```
source ./venv/bin/activate
```

You will most likely run into version issues when installing packages. Note that for deep reinforcement learning we will be making extensive use of reinforcement learning agent algorithms from the so-called Stable Baselines.[11] The Stable Baselines work with version 1 of TensorFlow and with PyTorch, but, as of this writing, not with version 2 of TensorFlow. TensorFlow version 1.14 has been tested to work.[12]

Installing is easy with Python's pip package manager: just type

```
pip install tensorflow==1.14
```

---

[10] More installation guidance can be found on the TensorFlow page at https://www.tensorflow.org/install/pip.

[11] https://github.com/hill-a/stable-baselines.

[12] You may be surprised that so many version numbers were mentioned. Unfortunately, not all versions are compatible; some care is necessary to get the software to work: Python 3.7.9, TensorFlow 1.14.0, Stable Baselines 2, pyglet 1.5.11 worked at the time of writing. This slightly embarrassing situation is because the field of deep reinforcement learning is driven by a community of researchers, who collaborate to make code bases work. When new insights trigger a rewrite that loses backward compatibility, as happened with TensorFlow 2.0, frantic rewriting of dependent software occurs. The field is still a new field, and some instability of software packages will remain with us for the foreseeable future.

```
1  from tensorflow.keras.models import Sequential
2  from tensorflow.keras.layers import Dense
3
4  model = Sequential()
5
6  model.add(Dense(units=64, activation='relu'))
7  model.add(Dense(units=10, activation='softmax'))
8
9  model.compile(loss='categorical_crossentropy',
10               optimizer='sgd',
11               metrics=['accuracy'])
12
13 # x_train and y_train are Numpy arrays --just like in the Scikit-
       Learn API.
14 model.fit(x_train, y_train, epochs=5, batch_size=32)
15
16 # evaluation on test data is simple
17 loss_and_metrics = model.evaluate(x_test, y_test, batch_size=128)
18
19 # as is predicting output
20 classes = model.predict(x_test, batch_size=128)
```

Listing B.3 Sequential Model in Keras

or, for GPU support,

```
pip install tensorflow-gpu==1.14
```

This should now download and install TensorFlow and Keras. We will check if everything is working by executing the MNIST training example with the default training dataset.

B.3.3.2 Keras MNIST Example

Keras is built on top of TensorFlow and is installed with TensorFlow. Basic Keras mirrors the familiar Scikit-learn interface [70].

Each Keras program specifies a model that is to be learned. The model is the neural network that consists of weights and layers (of neurons). You will specify the architecture of the model in Keras and then fit the model on the training data. When the model is trained, you can evaluate the loss function on test data, or perform predictions of the outcome based on some test input example.

Keras has two main programming paradigms: sequential and functional. Listing B.3 shows the most basic Sequential Keras model, from the Keras documentation, with a two-layer model, a ReLU layer, and a softmax layer, using simple SGD for backpropagation. The sequential model in Keras has an object-oriented syntax.

```
1  # Get the data as Numpy arrays
2  (x_train, y_train), (x_test, y_test) = keras.datasets.mnist.
       load_data()
3
4  # Build a simple model
5  inputs = keras.Input(shape=(28, 28))
6  x = layers.experimental.preprocessing.Rescaling(1.0 / 255)(inputs
       )
7  x = layers.Flatten()(x)
8  x = layers.Dense(128, activation="relu")(x)
9  x = layers.Dense(128, activation="relu")(x)
10 outputs = layers.Dense(10, activation="softmax")(x)
11 model = keras.Model(inputs, outputs)
12 model.summary()
13
14 # Compile the model
15 model.compile(optimizer="adam", loss="
       sparse_categorical_crossentropy")
16
17 # Train the model for 1 epoch from Numpy data
18 batch_size = 64
19 print("Fit on NumPy data")
20 history = model.fit(x_train, y_train, batch_size=batch_size,
       epochs=1)
21
22 # Train the model for 1 epoch using a dataset
23 dataset = tf.data.Dataset.from_tensor_slices((x_train, y_train)).
       batch(batch_size)
24 print("Fit on Dataset")
25 history = model.fit(dataset, epochs=1)
```
Listing B.4 Functional MNIST Model in Keras

The Keras documentation is at https://keras.io/getting_started/intro_to_keras_for_researchers/. It is quite accessible, and you are encouraged to learn Keras by working through the online tutorials.

A slightly more useful example is fitting a model on MNIST, see Listing B.4. This example uses a more flexible Keras syntax, the functional API, in which transformations are chained on top of the previous layers. This example of Keras code loads MNIST images in a training set and a test set and creates a model of dense ReLU layers using the functional API. It then creates a Keras model of these layers and prints a summary of the model. Then the model is trained from numpy data, with the fit function, for a single epoch, and also from a dataset, and the loss history is printed.

The code of the example shows how close Keras is to the way in which we think and reason about neural networks. The examples only show the briefest of glimpses to what is possible in Keras. Keras has options for performance monitoring, for checkpointing of long training runs, and for interfacing with TensorBoard, to visualize the training process. TensorBoard is an indispensible tool, allowing you to debug your intuition of what should be going on in your network, and what is going on.

Deep reinforcement learning is still very much a field with more degrees of freedom in experimentation than established practices, and being able to plot the progress of training processes is essential for a better understanding of how the model behaves. The more you explore Keras, the better you will be able to progress in deep reinforcement learning [27].

Exercises

Below are some questions to check your understanding of deep learning. Each question is a closed question where a simple, single sentence answer is expected.

Questions

1. Datasets are often split into two sub-datasets in machine learning. Which are those two, and why are they split?
2. What is generalization?
3. Sometimes a third sub-dataset is used. What is it called and what is it used for?
4. If we consider observations and model parameters, when is a problem high dimensional and when is it low dimensional?
5. How do we measure the capacity of a machine learning model?
6. What is a danger with low-capacity models?
7. What is a danger with high-capacity models?
8. What is the effect of overfitting on generalization?
9. What is the difference between supervised learning and reinforcement learning?
10. What is the difference between a shallow network and a deep network?
11. Supervised learning has a model and a dataset, reinforcement learning has which two central concepts?
12. What phases does a learning epoch have? What happens in each phase?
13. Name three factors that were essential for the deep learning breakthrough, and why.
14. What is end-to-end learning? Do you know an alternative? What are the advantages of each?
15. What is underfitting, and what causes it? What is overfitting, and what causes it? How can you see if you have overfitting?
16. Name three ways to prevent overfitting.
17. Which three types of layers does a neural network have?
18. How many hidden layers does a shallow neural network have?

19. Describe how adjusting weights works in a neural network. Hint: Think of examples, labels, forward phase, a backward phase, error functions, and gradients.
20. What is the difference between a fully connected network and a convolutional neural network?
21. What is max pooling?
22. Why are shared weights advantageous?
23. What is feature learning?
24. What is representation learning?
25. What is deep learning?
26. What is an advantage of convolutional neural networks of fully connected neural networks?
27. Name two well-known image recognition datasets.

Exercises

Let us now start with some exercises. If you have not done so already, install PyTorch[13] or TensorFlow and Keras (see Sect. 2.2.4.2 or go to the TensorFlow page).[14] Be sure to check the right versions of Python, TensorFlow, and the Stable Baselines to make sure they work well together. The exercises below are meant to be done in Keras.

1. *Generalization* Install Keras. Go to the Keras MNIST example. Perform a classification task. Note how many epochs the training takes, and in testing, how well it generalizes. Perform the classification on a smaller training set, how does learning rate change, how does generalization change. Vary other elements: try a different optimizer than adam, try a different learning rate, try a different (deeper) architecture, and try wider hidden layers. Does it learn faster? Does it generalize better?
2. *Overfitting* Use Keras again, but this time on ImageNet. Now try different over-fitting solutions. Does the training speed change? Does generalization change? Now try the holdout validation set. Do training and generalization change?
3. *Confidence* How many runs did you do in the previous exercises, just a single run to see how long training took and how well generalization worked? Try to run it again. Do you get the same results? How large is the difference? Can you change the random seeds of Keras or TensorFlow? Can you calculate the confidence interval, how much does the confidence improve when you do 10 randomized runs? How about 100 runs? Make graphs with error bars.

---

[13] https://pytorch.org.
[14] https://www.tensorflow.org.

4. *GPU* It might be that you have access to a GPU machine that is capable of running PyTorch or TensorFlow in parallel to speed up the training. Install the GPU version and check that it recognizes the GPU and is indeed using it.
5. *Parallelism* It might be that you have access to a multicore CPU machine. When you are running multiple runs in order to improve confidence, then an easy way to speed up your experiment is to spawn multiple jobs at the shell, assigning the output to different log files, and write a script to combine results and draw graphs. Write the scripts necessary to achieve this, test them, and do a large-confidence experiment.

References

1. Martín Abadi, Paul Barham, Jianmin Chen, Zhifeng Chen, Andy Davis, Jeffrey Dean, Matthieu Devin, Sanjay Ghemawat, Geoffrey Irving, Michael Isard, Manjunath Kudlur, Josh Levenberg, Rajat Monga, Sherry Moore, Derek Gordon Murray, Benoit Steiner, Paul A. Tucker, Vijay Vasudevan, Pete Warden, Martin Wicke, Yuan Yu, and Xiaoqiang Zheng. TensorFlow: A system for large-scale machine learning. In *12th USENIX Symposium on Operating Systems Design and Implementation (OSDI 16)*, pages 265–283, 2016. 360, 378
2. Jay Alammer. The illustrated transformer. https://jalammar.github.io/illustrated-transformer/. 377
3. Dzmitry Bahdanau, Kyunghyun Cho, and Yoshua Bengio. Neural machine translation by jointly learning to align and translate. *arXiv preprint arXiv:1409.0473*, 2014. 376, 376, 376
4. Mark F Bear, Barry W Connors, and Michael A Paradiso. *Neuroscience*, volume 2. Lippincott Williams & Wilkins, 2007. 360
5. Mikhail Belkin, Daniel Hsu, Siyuan Ma, and Soumik Mandal. Reconciling modern machine learning and the bias-variance trade-off. *arXiv preprint arXiv:1812.11118*, 2018. 378, 378
6. Mikhail Belkin, Daniel Hsu, and Ji Xu. Two models of double descent for weak features. *arXiv preprint arXiv:1903.07571*, 2019.
7. Mikhail Belkin, Daniel J Hsu, and Partha Mitra. Overfitting or perfect fitting? Risk bounds for classification and regression rules that interpolate. In *Advances in Neural Information Processing Systems*, pages 2300–2311, 2018. 378, 378
8. Richard Bellman. *Dynamic Programming*. Courier Corporation, 1957, 2013. 356
9. Yoshua Bengio, Aaron Courville, and Pascal Vincent. Representation learning: A review and new perspectives. *IEEE Transactions on Pattern Analysis and Machine Intelligence*, 35(8):1798–1828, 2013. 364
10. R Bertolami, H Bunke, S Fernandez, A Graves, M Liwicki, and J Schmidhuber. A novel connectionist system for improved unconstrained handwriting recognition. *IEEE Transactions on Pattern Analysis and Machine Intelligence*, 31(5), 2009. 369
11. Christopher M Bishop. *Pattern Recognition and Machine Learning*. Information science and statistics. Springer Verlag, Heidelberg, 2006. 354, 357, 359, 378
12. Peter Bloem. Transformers http://peterbloem.nl/blog/transformers. 377
13. Tom B. Brown, Benjamin Mann, Nick Ryder, Melanie Subbiah, Jared Kaplan, Prafulla Dhariwal, Arvind Neelakantan, Pranav Shyam, Girish Sastry, Amanda Askell, Sandhini Agarwal, Ariel Herbert-Voss, Gretchen Krueger, Tom Henighan, Rewon Child, Aditya Ramesh, Daniel M. Ziegler, Jeffrey Wu, Clemens Winter, Christopher Hesse, Mark Chen, Eric Sigler, Mateusz Litwin, Scott Gray, Benjamin Chess, Jack Clark, Christopher Berner, Sam McCandlish, Alec Radford, Ilya Sutskever, and Dario Amodei. Language models are few-shot learners. In *Advances in Neural Information Processing Systems*, 2020. 364, 377

14. Rich Caruana, Steve Lawrence, and C Lee Giles. Overfitting in neural nets: Backpropagation, conjugate gradient, and early stopping. In *Advances in Neural Information Processing Systems*, pages 402–408, 2001. 378
15. Lili Chen, Kevin Lu, Aravind Rajeswaran, Kimin Lee, Aditya Grover, Michael Laskin, Pieter Abbeel, Aravind Srinivas, and Igor Mordatch. Decision transformer: Reinforcement learning via sequence modeling. *arXiv preprint arXiv:2106.01345*, 2021. 377
16. Jaeyoung Choi, Jack J Dongarra, and David W Walker. PB-BLAS: a set of parallel block basic linear algebra subprograms. *Concurrency: Practice and Experience*, 8(7):517–535, 1996. 381
17. François Chollet. *Deep learning with Python*. Manning Publications Co., 2017. 378
18. Dan Cireşan, Ueli Meier, Luca Maria Gambardella, and Jürgen Schmidhuber. Deep, big, simple neural nets for handwritten digit recognition. *Neural Computation*, 22(12):3207–3220, 2010. 381
19. Dan Cireşan, Ueli Meier, and Jürgen Schmidhuber. Multi-column deep neural networks for image classification. In *2012 IEEE Conference on Computer Vision and Pattern Recognition, Providence, RI, US*, pages 3642–3649, 2012. 369
20. Rina Dechter. *Learning while searching in constraint-satisfaction problems*. AAAI, 1986. 364
21. Jia Deng, Wei Dong, Richard Socher, Li-Jia Li, Kai Li, and Li Fei-Fei. ImageNet: A large-scale hierarchical image database. In *2009 IEEE Conference on Computer Vision and Pattern Recognition*, pages 248–255. IEEE, 2009. 380, 380
22. Jacob Devlin, Ming-Wei Chang, Kenton Lee, and Kristina Toutanova. BERT: pre-training of deep bidirectional transformers for language understanding. In *Proceedings of the 2019 Conference of the North American Chapter of the Association for Computational Linguistics: Human Language Technologies, NAACL-HLT 2019, Minneapolis*, 2018. 364, 364, 377
23. Jack J Dongarra, Jeremy Du Croz, Sven Hammarling, and Richard J Hanson. An extended set of FORTRAN basic linear algebra subprograms. *ACM Transactions on Mathematical Software*, 14(1):1–17, 1988. 381
24. Li Fei-Fei, Jia Deng, and Kai Li. ImageNet: Constructing a large-scale image database. *Journal of Vision*, 9(8):1037–1037, 2009. 380
25. Santiago Fernández, Alex Graves, and Jürgen Schmidhuber. An application of recurrent neural networks to discriminative keyword spotting. In *International Conference on Artificial Neural Networks*, pages 220–229. Springer, 2007. 369
26. Michael J Flynn. Some computer organizations and their effectiveness. *IEEE Transactions on Computers*, 100(9):948–960, 1972. 381
27. Aurélien Géron. *Hands-on machine learning with Scikit-Learn and TensorFlow: concepts, tools, and techniques to build intelligent systems*. O'Reilly Media, Inc., 2019. 385
28. Felix A Gers, Jürgen Schmidhuber, and Fred Cummins. Learning to forget: Continual prediction with LSTM. In *Ninth International Conference on Artificial Neural Networks ICANN 99*. IET, 1999. 372
29. Ian Goodfellow, Yoshua Bengio, and Aaron Courville. *Deep Learning*. MIT Press, Cambridge, 2016. 359, 360, 362, 363, 367, 370, 372, 372, 377, 377, 377, 378
30. Ian Goodfellow, Jean Pouget-Abadie, Mehdi Mirza, Bing Xu, David Warde-Farley, Sherjil Ozair, Aaron Courville, and Yoshua Bengio. Generative adversarial nets. In *Advances in Neural Information Processing Systems*, pages 2672–2680, 2014. 373, 374
31. Alex Graves, Santiago Fernández, and Jürgen Schmidhuber. Bidirectional LSTM networks for improved phoneme classification and recognition. In *International Conference on Artificial Neural Networks*, pages 799–804. Springer, 2005. 372
32. Alex Graves, Abdel-rahman Mohamed, and Geoffrey Hinton. Speech recognition with deep recurrent neural networks. In *2013 IEEE International Conference on Acoustics, Speech and Signal Processing*, pages 6645–6649. IEEE, 2013. 364
33. Klaus Greff, Rupesh K Srivastava, Jan Koutník, Bas R Steunebrink, and Jürgen Schmidhuber. LSTM: A Search Space Odyssey. *IEEE Transactions on Neural Networks and Learning Systems*, 28(10):2222–2232, 2017. 372
34. Peter D Grünwald. *The minimum description length principle*. MIT press, 2007. 375

35. Kaiming He, Xiangyu Zhang, Shaoqing Ren, and Jian Sun. Deep residual learning for image recognition. In *Proceedings of the IEEE Conference on Computer Vision and Pattern Recognition*, pages 770–778, 2016. 372, 373

36. John L Hennessy and David A Patterson. *Computer Architecture: a Quantitative Approach.* Elsevier, 2017. 381

37. W Daniel Hillis. New computer architectures and their relationship to physics or why computer science is no good. *International Journal of Theoretical Physics*, 21(3–4):255–262, 1982. 381

38. W Daniel Hillis and Lewis W Tucker. The CM-5 connection machine: A scalable supercomputer. *Communications of the ACM*, 36(11):30–41, 1993. 381

39. Geoffrey E Hinton and Ruslan R Salakhutdinov. Reducing the dimensionality of data with neural networks. *Science*, 313(5786):504–507, 2006. 375

40. Geoffrey E Hinton, Nitish Srivastava, Alex Krizhevsky, Ilya Sutskever, and Ruslan R Salakhutdinov. Improving neural networks by preventing co-adaptation of feature detectors. *arXiv preprint arXiv:1207.0580*, 2012. 377

41. Sepp Hochreiter and Jürgen Schmidhuber. Long short-term memory. *Neural Computation*, 9(8):1735–1780, 1997. 369, 372

42. John J Hopfield. Neural networks and physical systems with emergent collective computational abilities. *Proceedings of the National Academy of Sciences*, 79(8):2554–2558, 1982. 369

43. David H Hubel and Torsten N Wiesel. Shape and arrangement of columns in cat's striate cortex. *The Journal of Physiology*, 165(3):559–568, 1963. 367

44. David H Hubel and Torsten N Wiesel. Receptive fields and functional architecture of monkey striate cortex. *The Journal of Physiology*, 195(1):215–243, 1968. 367

45. Sergey Ioffe. Batch renormalization: Towards reducing minibatch dependence in batch-normalized models. In *Advances in Neural Information Processing Systems*, pages 1945–1953, 2017. 378

46. Michael Janner, Qiyang Li, and Sergey Levine. Reinforcement learning as one big sequence modeling problem. *arXiv preprint arXiv:2106.02039*, 2021. 377

47. Dimitris Kalimeris, Gal Kaplun, Preetum Nakkiran, Benjamin L. Edelman, Tristan Yang, Boaz Barak, and Haofeng Zhang. SGD on neural networks learns functions of increasing complexity. In *Advances in Neural Information Processing Systems*, pages 3491–3501, 2019. 378

48. Andrej Karpathy. The unreasonable effectiveness of recurrent neural networks. http://karpathy.github.io/2015/05/21/rnn-effectiveness/. Andrej Karpathy Blog, 2015. 371, 371, 371

49. Tero Karras, Timo Aila, Samuli Laine, and Jaakko Lehtinen. Progressive growing of GANs for improved quality, stability, and variation. In *International Conference on Learning Representations*, 2018. 374, 375

50. Diederik P Kingma and Max Welling. Auto-encoding variational Bayes. In *International Conference on Learning Representations*, 2014. 374

51. Diederik P Kingma and Max Welling. An introduction to variational autoencoders. *Found. Trends Mach. Learn.*, 12(4):307–392, 2019.

52. Mark A Kramer. Nonlinear principal component analysis using autoassociative neural networks. *AIChE journal*, 37(2):233–243, 1991. 374

53. Alex Krizhevsky, Ilya Sutskever, and Geoffrey E Hinton. ImageNet classification with deep convolutional neural networks. In *Advances in Neural Information Processing Systems*, pages 1097–1105, 2012. 364, 364, 380

54. Yann LeCun, Yoshua Bengio, and Geoffrey Hinton. Deep learning. *Nature*, 521(7553):436, 2015. 364

55. Yann LeCun, Bernhard Boser, John S Denker, Donnie Henderson, Richard E Howard, Wayne Hubbard, and Lawrence D Jackel. Backpropagation applied to handwritten zip code recognition. *Neural Computation*, 1(4):541–551, 1989. 368

56. Yann LeCun, Léon Bottou, Yoshua Bengio, and Patrick Haffner. Gradient-based learning applied to document recognition. *Proceedings of the IEEE*, 86(11):2278–2324, 1998. 364, 366, 379, 379

57. Honglak Lee, Roger Grosse, Rajesh Ranganath, and Andrew Y Ng. Convolutional deep belief networks for scalable unsupervised learning of hierarchical representations. In *Proceedings of*

the 26th Annual International Conference on Machine Learning, pages 609–616. ACM, 2009. 364, 365

58. Charles E. Leiserson, Zahi S. Abuhamdeh, David C. Douglas, Carl R. Feynman, Mahesh N. Ganmukhi, Jeffrey V. Hill, W. Daniel Hillis, Bradley C. Kuszmaul, Margaret A. St. Pierre, David S. Wells, Monica C. Wong, Shaw-Wen Yang, and Robert C. Zak. The network architecture of the connection machine CM-5. In *Proceedings of the fourth annual ACM Symposium on Parallel Algorithms and Architectures*, pages 272–285, 1992. 381

59. Hui Liu, Song Yu, Zhangxin Chen, Ben Hsieh, and Lei Shao. Sparse matrix-vector multiplication on NVIDIA GPU. *International Journal of Numerical Analysis & Modeling, Series B*, 3(2):185–191, 2012. 381

60. Gabriel Loye. The attention mechanism. https://blog.floydhub.com/attention-mechanism/. 376, 376

61. Siyuan Ma, Raef Bassily, and Mikhail Belkin. The power of interpolation: Understanding the effectiveness of SGD in modern over-parametrized learning. In *International Conference on Machine Learning*, pages 3331–3340, 2018. 378

62. Masakazu Matsugu, Katsuhiko Mori, Yusuke Mitari, and Yuji Kaneda. Subject independent facial expression recognition with robust face detection using a convolutional neural network. *Neural Networks*, 16(5–6):555–559, 2003. 367

63. Thomas Moerland. Continuous Markov decision process and policy search. Lecture notes for the course reinforcement learning, Leiden University, 2021. 354

64. Hussain Mujtaba. Introduction to autoencoders. https://www.mygreatlearning.com/blog/autoencoder/, 2020. 375

65. Preetum Nakkiran, Gal Kaplun, Yamini Bansal, Tristan Yang, Boaz Barak, and Ilya Sutskever. Deep double descent: Where bigger models and more data. In *8th International Conference on Learning Representations, ICLR 2020, Addis Ababa, Ethiopia*, 2020. 378

66. Andrew Y Ng. Feature selection, L1 vs. L2 regularization, and rotational invariance. In *Proceedings of the Twenty-first International Conference on Machine Learning*, page 78. ACM, 2004. 377

67. Kyoung-Su Oh and Keechul Jung. GPU implementation of neural networks. *Pattern Recognition*, 37(6):1311–1314, 2004. 381

68. Chris Olah. Understanding LSTM networks. http://colah.github.io/posts/2015-08-Understanding-LSTMs/, 2015. 370, 370

69. Adam Paszke, Sam Gross, Francisco Massa, Adam Lerer, James Bradbury, Gregory Chanan, Trevor Killeen, Zeming Lin, Natalia Gimelshein, Luca Antiga, Alban Desmaison, Andreas Köpf, Edward Z. Yang, Zachary DeVito, Martin Raison, Alykhan Tejani, Sasank Chilamkurthy, Benoit Steiner, Lu Fang, Junjie Bai, and Soumith Chintala. PyTorch: An imperative style, high-performance deep learning library. In *Advances in Neural Information Processing Systems*, pages 8024–8035, 2019. 360, 378

70. Fabian Pedregosa, Gaël Varoquaux, Alexandre Gramfort, Vincent Michel, Bertrand Thirion, Olivier Grisel, Mathieu Blondel, Peter Prettenhofer, Ron Weiss, Vincent Dubourg, Jake VanderPlas, Alexandre Passos, David Cournapeau, Matthieu Brucher, Matthieu Perrot, and Edouard Duchesnay. Scikit-learn: Machine learning in Python. *Journal of Machine Learning Research*, 12(Oct):2825–2830, 2011. 383

71. Lutz Prechelt. Automatic early stopping using cross validation: quantifying the criteria. *Neural Networks*, 11(4):761–767, 1998. 378

72. Lutz Prechelt. Early stopping-but when? In *Neural Networks: Tricks of the trade*, pages 55–69. Springer, 1998. 378, 378

73. Alec Radford, Jong Wook Kim, Chris Hallacy, Aditya Ramesh, Gabriel Goh, Sandhini Agarwal, Girish Sastry, Amanda Askell, Pamela Mishkin, Jack Clark, Gretchen Krueger, and Ilya Sutskever. Learning transferable visual models from natural language supervision. In *International Conference on Machine Learning*, 2021. 377

74. Alec Radford, Karthik Narasimhan, Tim Salimans, and Ilya Sutskever. Improving language understanding by generative pre-training. https://openai.com/blog/language-unsupervised/, 2018. 377

75. Aditya Ramesh, Mikhail Pavlov, Gabriel Goh, Scott Gray, Chelsea Voss, Alec Radford, Mark Chen, and Ilya Sutskever. Zero-shot text-to-image generation. In *International Conference on Machine Learning*, 2021. 377

76. Sumit Saha. A comprehensive guide to convolutional neural networks—the ELI5 way. https:// towardsdatascience.com/a-comprehensive-guide-to-convolutional-neural-networks-the-eli5-way-3bd2b1164a53. Towards Data Science, 2018. 361, 369

77. Jason Sanders and Edward Kandrot. *CUDA by example: an introduction to general-purpose GPU programming*. Addison-Wesley Professional, 2010. 381

78. Jürgen Schmidhuber, F Gers, and Douglas Eck. Learning nonregular languages: A comparison of simple recurrent networks and LSTM. *Neural Computation*, 14(9):2039–2041, 2002. 372

79. Fengguang Song and Jack Dongarra. Scaling up matrix computations on shared-memory manycore systems with 1000 CPU cores. In *Proceedings of the 28th ACM International Conference on Supercomputing*, pages 333–342. ACM, 2014. 381

80. Mei Song, A Montanari, and P Nguyen. A mean field view of the landscape of two-layers neural networks. In *Proceedings of the National Academy of Sciences*, volume 115, pages E7665–E7671, 2018. 363

81. Nitish Srivastava, Geoffrey Hinton, Alex Krizhevsky, Ilya Sutskever, and Ruslan Salakhutdinov. Dropout: a simple way to prevent neural networks from overfitting. *The Journal of Machine Learning Research*, 15(1):1929–1958, 2014. 377

82. Jiawei Su, Danilo Vasconcellos Vargas, and Kouichi Sakurai. One pixel attack for fooling deep neural networks. *IEEE Transactions on Evolutionary Computation*, 23(5):828–841, 2019. 373

83. Ilya Sutskever, Oriol Vinyals, and Quoc V Le. Sequence to sequence learning with neural networks. In *Advances in Neural Information Processing Systems*, pages 3104–3112, 2014. 376

84. Vivienne Sze, Yu-Hsin Chen, Tien-Ju Yang, and Joel S Emer. Efficient processing of deep neural networks: A tutorial and survey. *Proceedings of the IEEE*, 105(12):2295–2329, 2017. 381

85. Christian Szegedy, Wojciech Zaremba, Ilya Sutskever, Joan Bruna, Dumitru Erhan, Ian Goodfellow, and Rob Fergus. Intriguing properties of neural networks. In *International Conference on Learning Representations*, 2013. 373, 373

86. Sergios Theodoridis and Konstantinos Koutroumbas. *Pattern recognition*. Academic Press, 1999. 357

87. Ashish Vaswani, Noam Shazeer, Niki Parmar, Jakob Uszkoreit, Llion Jones, Aidan N Gomez, Lukasz Kaiser, and Illia Polosukhin. Attention is all you need. In *Advances in Neural Information Processing Systems*, pages 5998–6008, 2017. 377, 377

88. Oriol Vinyals, Alexander Toshev, Samy Bengio, and Dumitru Erhan. Show and tell: A neural image caption generator. In *Proceedings of the IEEE Conference on Computer Vision and Pattern Recognition*, pages 3156–3164, 2015. 370

89. Loc Vu-Quoc. Neuron and myelinated axon. https://commons.wikimedia.org/w/index.php?curid=72816083, 2018. 360, 360

90. Eddie Weill. LeNet in Keras on GitHub. https://github.com/eweill/keras-deepcv/tree/master/models/classification. 366

91. Wayne Xiong, Lingfeng Wu, Fil Alleva, Jasha Droppo, Xuedong Huang, and Andreas Stolcke. The Microsoft 2017 Conversational Speech Recognition System. In *IEEE International Conference on Acoustics, Speech and Signal Processing (ICASSP)*, pages 5934–5938. IEEE, 2018. 364

92. Xin Yang, Yuezun Li, and Siwei Lyu. Exposing deep fakes using inconsistent head poses. In *ICASSP 2019-2019 IEEE International Conference on Acoustics, Speech and Signal Processing*, pages 8261–8265. IEEE, 2019. 374, 374

93. Amy Zhang, Nicolas Ballas, and Joelle Pineau. A dissection of overfitting and generalization in continuous reinforcement learning. *arXiv preprint arXiv:1806.07937*, 2018. 378

Appendix C
Deep Reinforcement Learning Suites

Deep reinforcement learning is a highly active field of research. One reason for the progress is the availability of high- quality algorithms and code: high-quality environments, algorithms, and deep learning suites are all being made available by researchers along with their research papers. This appendix provides pointers to these codes.

C.1 Environments

Progress has benefited greatly from the availability of high-quality environments, on which the algorithms can be tested. We provide pointers to some of the environments (Table C.1).

C.2 Agent Algorithms

For value-based and policy-based methods most mainstream algorithms have been collected and are freely available. For two-agent, multi-agent, hierarchical, and meta learning, the agent algorithms are also on GitHub, but not always in the same place as the basic algorithms. Table C.2 provides pointers.

C.3 Deep Learning Suites

The two most well-known deep learning suites are TensorFlow and PyTorch. Base-TensorFlow has a complicated programming model. Keras has been developed as

© The Author(s), under exclusive license to Springer Nature Singapore Pte Ltd. 2022   393
A. Plaat, *Deep Reinforcement Learning*,
https://doi.org/10.1007/978-981-19-0638-1

Table C.1 Reinforcement learning environments

Environment	Type	URL	Ref
Gym	ALE, MuJoCo	https://gym.openai.com	[4]
ALE	Atari	https://github.com/mgbellemare/Arcade-Learning-Environment	[3]
MuJoCo	Simulated Robot	http://www.mujoco.org	[31]
DeepMind Lab	3D navigation	https://github.com/deepmind/lab	[2]
Control Suite	Physics tasks	https://github.com/deepmind/dm_control	[28]
Behavior Suite	Core RL	https://github.com/deepmind/bsuite	[26]
StarCraft	Python interface	https://github.com/deepmind/pysc2	[34]
Omniglot	Images	https://github.com/brendenlake/omniglot	[21]
Mini ImageNet	Images	https://github.com/yaoyao-liu/mini-imagenet-tools	[33]
ProcGen	Procedurally Gen.	https://openai.com/blog/procgen-benchmark/	[8]
OpenSpiel	Board Games	https://github.com/deepmind/open_spiel	[22]
RLlib	Scalable RL	https://github.com/ray-project/ray	[23]
Meta Dataset	Dataset of datas.	https://github.com/google-research/meta-dataset	[32]
Meta World	Meta RL	https://meta-world.github.io	[36]
Alchemy	Meta RL	https://github.com/deepmind/dm_alchemy	[35]
Garage	Reproducible RL	https://github.com/rlworkgroup/garage	[13]
Football	Multi-agent	https://github.com/google-research/football	[20]
Emergence	Hide and Seek	https://github.com/openai/multi-agent-emergence-environments	[1]
Unplugged	Offline RL	https://github.com/deepmind/deepmind-research/tree/master/rl_unplugged	[14]
Unity	3D	https://github.com/Unity-Technologies/ml-agents	[19]
PolyGames	Board games	https://github.com/teytaud/Polygames	[7]
Dopamine	RL framework	https://github.com/google/dopamine	[6]

an easy to use layer on top of TensorFlow. When you use TensorFlow, start with Keras. Or use PyTorch.

TensorFlow and Keras are at https://www.tensorflow.org.

PyTorch is at https://pytorch.org.

Table C.2 Agent algorithms

Repo	URL	Algorithms	Ref
Spinning Up	https://spinningup.openai.com	Tutorial on DDPG, PPO, etc	OpenAI
Baselines	https://github.com/openai/baselines	DQN, PPO, etc	[10]
Stable Basel.	https://stable-baselines.readthedocs.io/en/master/	Refactored baselines	[17]
PlaNet	https://planetrl.github.io	Latent Model	[15]
Dreamer	https://github.com/danijar/dreamerv2	Latent Model	[16]
VPN	https://github.com/junhyukoh/value-prediction-network	Value Prediction Network	[25]
MuZero	https://github.com/kaesve/muzero	Reimplementation of MuZero	[9]
MCTS	Pip install mcts	MCTS	[5]
AlphaZ. Gen	https://github.com/suragnair/alpha-zero-general	AZ in Python	[29]
ELF	https://github.com/facebookresearch/ELF	Framework for Game research	[30]
PolyGames	https://github.com/teytaud/Polygames	Zero Learning	[7]
CFR	https://github.com/bakanaouji/cpp-cfr	CFR	[37]
DeepCFR	https://github.com/EricSteinberger/Deep-CFR	Deep CFR	[27]
MADDPG	https://github.com/openai/maddpg	Multi-agent DDPG	[24]
PBT	https://github.com/voiler/PopulationBasedTraining	Population- Based Training	[18]
Go-Explore	https://github.com/uber-research/go-explore	Go Explore	[11]
MAML	https://github.com/cbfinn/maml	MAML	[12]

References

1. Bowen Baker, Ingmar Kanitscheider, Todor Markov, Yi Wu, Glenn Powell, Bob McGrew, and Igor Mordatch. Emergent tool use from multi-agent autocurricula. *arXiv preprint arXiv:1909.07528*, 2019. 394

2. Charles Beattie, Joel Z. Leibo, Denis Teplyashin, Tom Ward, Marcus Wainwright, Heinrich Küttler, Andrew Lefrancq, Simon Green, Víctor Valdés, Amir Sadik, Julian Schrittwieser, Keith Anderson, Sarah York, Max Cant, Adam Cain, Adrian Bolton, Stephen Gaffney, Helen King, Demis Hassabis, Shane Legg, and Stig Petersen. DeepMind Lab. *arXiv preprint arXiv:1612.03801*, 2016. 394

3. Marc G Bellemare, Yavar Naddaf, Joel Veness, and Michael Bowling. The Arcade Learning Environment: An evaluation platform for general agents. *Journal of Artificial Intelligence Research*, 47:253–279, 2013. 394

4. Greg Brockman, Vicki Cheung, Ludwig Pettersson, Jonas Schneider, John Schulman, Jie Tang, and Wojciech Zaremba. OpenAI Gym. *arXiv preprint arXiv:1606.01540*, 2016. 394

5. Cameron B Browne, Edward Powley, Daniel Whitehouse, Simon M Lucas, Peter I Cowling, Philipp Rohlfshagen, Stephen Tavener, Diego Perez, Spyridon Samothrakis, and Simon Colton. A survey of Monte Carlo Tree Search methods. *IEEE Transactions on Computational Intelligence and AI in Games*, 4(1):1–43, 2012. 395

6. Pablo Samuel Castro, Subhodeep Moitra, Carles Gelada, Saurabh Kumar, and Marc G Bellemare. Dopamine: A research framework for deep reinforcement learning. *arXiv preprint arXiv:1812.06110*, 2018. 394

7. Tristan Cazenave, Yen-Chi Chen, Guan-Wei Chen, Shi-Yu Chen, Xian-Dong Chiu, Julien Dehos, Maria Elsa, Qucheng Gong, Hengyuan Hu, Vasil Khalidov, Cheng-Ling Li, Hsin-I Lin, Yu-Jin Lin, Xavier Martinet, Vegard Mella, Jérémy Rapin, Baptiste Rozière, Gabriel Synnaeve, Fabien Teytaud, Olivier Teytaud, Shi-Cheng Ye, Yi-Jun Ye, Shi-Jim Yen, and Sergey Zagoruyko. Polygames: Improved zero learning. *arXiv preprint arXiv:2001.09832*, 2020. 394, 395

8. Karl Cobbe, Chris Hesse, Jacob Hilton, and John Schulman. Leveraging procedural generation to benchmark reinforcement learning. In *International Conference on Machine Learning*, pages 2048–2056. PMLR, 2020. 394

9. Joery A. de Vries, Ken S. Voskuil, Thomas M. Moerland, and Aske Plaat. Visualizing MuZero models. *arXiv preprint arXiv:2102.12924*, 2021. 395

10. Prafulla Dhariwal, Christopher Hesse, Oleg Klimov, Alex Nichol, Matthias Plappert, Alec Radford, John Schulman, Szymon Sidor, Yuhuai Wu, and Peter Zhokhov. OpenAI baselines. https://github.com/openai/baselines, 2017. 395

11. Adrien Ecoffet, Joost Huizinga, Joel Lehman, Kenneth O Stanley, and Jeff Clune. First return, then explore. *Nature*, 590(7847):580–586, 2021. 395

12. Chelsea Finn, Pieter Abbeel, and Sergey Levine. Model-Agnostic Meta-Learning for fast adaptation of deep networks. In *International Conference on Machine Learning*, pages 1126–1135. PMLR, 2017. 395

13. The garage contributors. Garage: A toolkit for reproducible reinforcement learning research. https://github.com/rlworkgroup/garage, 2019. 394

14. Caglar Gulcehre, Ziyu Wang, Alexander Novikov, Tom Le Paine, Sergio Gómez Colmenarejo, Konrad Zolna, Rishabh Agarwal, Josh Merel, Daniel Mankowitz, Cosmin Paduraru, et al. RL unplugged: Benchmarks for offline reinforcement learning. *arXiv preprint arXiv:2006.13888*, 2020. 394

15. Danijar Hafner, Timothy Lillicrap, Ian Fischer, Ruben Villegas, David Ha, Honglak Lee, and James Davidson. Learning latent dynamics for planning from pixels. In *International Conference on Machine Learning*, pages 2555–2565, 2019. 395

16. Danijar Hafner, Timothy Lillicrap, Mohammad Norouzi, and Jimmy Ba. Mastering Atari with discrete world models. In *International Conference on Learning Representations*, 2021. 395

17. Ashley Hill, Antonin Raffin, Maximilian Ernestus, Adam Gleave, Anssi Kanervisto, Rene Traore, Prafulla Dhariwal, Christopher Hesse, Oleg Klimov, Alex Nichol, Matthias Plappert, Alec Radford, John Schulman, Szymon Sidor, and Yuhuai Wu. Stable baselines. https://github.com/hill-a/stable-baselines, 2018. 395

18. Max Jaderberg, Valentin Dalibard, Simon Osindero, Wojciech M. Czarnecki, Jeff Donahue, Ali Razavi, Oriol Vinyals, Tim Green, Iain Dunning, Karen Simonyan, Chrisantha Fernando, and Koray Kavukcuoglu. Population based training of neural networks. *arXiv preprint arXiv:1711.09846*, 2017. 395

19. Arthur Juliani, Vincent-Pierre Berges, Ervin Teng, Andrew Cohen, Jonathan Harper, Chris Elion, Chris Goy, Yuan Gao, Hunter Henry, Marwan Mattar, and Danny Lange. Unity: A general platform for intelligent agents. *arXiv preprint arXiv:1809.02627*, 2018. 394

20. Karol Kurach, Anton Raichuk, Piotr Stańczyk, Michał Zając, Olivier Bachem, Lasse Espeholt, Carlos Riquelme, Damien Vincent, Marcin Michalski, Olivier Bousquet, and Sylvain Gelly. Google research football: A novel reinforcement learning environment. In *Proceedings of the AAAI Conference on Artificial Intelligence*, volume 34, pages 4501–4510, 2020. 394

21. Brenden Lake, Ruslan Salakhutdinov, Jason Gross, and Joshua Tenenbaum. One shot learning of simple visual concepts. In *Proceedings of the Annual Meeting of the Cognitive Science Society*, volume 33, 2011. 394

22. Marc Lanctot, Edward Lockhart, Jean-Baptiste Lespiau, Vinícius Flores Zambaldi, Satyaki Upadhyay, Julien Pérolat, Sriram Srinivasan, Finbarr Timbers, Karl Tuyls, Shayegan Omidshafiei, Daniel Hennes, Dustin Morrill, Paul Muller, Timo Ewalds, Ryan Faulkner, János Kramár, Bart De Vylder, Brennan Saeta, James Bradbury, David Ding, Sebastian Borgeaud, Matthew Lai, Julian Schrittwieser, Thomas W. Anthony, Edward Hughes, Ivo Danihelka, and Jonah Ryan-Davis. OpenSpiel: A framework for reinforcement learning in games. *arXiv preprint arXiv:1908.09453*, 2019. 394

23. Eric Liang, Richard Liaw, Philipp Moritz, Robert Nishihara, Roy Fox, Ken Goldberg, Joseph E Gonzalez, Michael I Jordan, and Ion Stoica. RLlib: abstractions for distributed reinforcement learning. In *International Conference on Machine Learning*, pages 3059–3068, 2018. 394

24. Ryan Lowe, Yi Wu, Aviv Tamar, Jean Harb, Pieter Abbeel, and Igor Mordatch. Multi-agent Actor-Critic for mixed cooperative-competitive environments. In *Advances in Neural Information Processing Systems*, pages 6379–6390, 2017. 395

25. Junhyuk Oh, Satinder Singh, and Honglak Lee. Value prediction network. In *Advances in Neural Information Processing Systems*, pages 6118–6128, 2017. 395

26. Ian Osband, Yotam Doron, Matteo Hessel, John Aslanides, Eren Sezener, Andre Saraiva, Katrina McKinney, Tor Lattimore, Csaba Szepesvári, Satinder Singh, Benjamin Van Roy, Richard S. Sutton, David Silver, and Hado van Hasselt. Behaviour suite for reinforcement learning. In *8th International Conference on Learning Representations, ICLR 2020, Addis Ababa, Ethiopia*, April 2020. 394

27. Eric Steinberger. Single deep counterfactual regret minimization. *arXiv preprint arXiv:1901.07621*, 2019. 395

28. Yuval Tassa, Yotam Doron, Alistair Muldal, Tom Erez, Yazhe Li, Diego de Las Casas, David Budden, Abbas Abdolmaleki, Josh Merel, Andrew Lefrancq, Timothy Lillicrap, and Martin Riedmiller. DeepMind control suite. *arXiv preprint arXiv:1801.00690*, 2018. 394

29. Shantanu Thakoor, Surag Nair, and Megha Jhunjhunwala. Learning to play Othello without human knowledge. Stanford University CS238 Final Project Report, 2017. 395

30. Yuandong Tian, Qucheng Gong, Wenling Shang, Yuxin Wu, and C Lawrence Zitnick. ELF: An extensive, lightweight and flexible research platform for real-time strategy games. In *Advances in Neural Information Processing Systems*, pages 2659–2669, 2017. 395

31. Emanuel Todorov, Tom Erez, and Yuval Tassa. MuJoCo: A physics engine for model-based control. In *IEEE/RSJ International Conference on Intelligent Robots and Systems (IROS)*, pages 5026–5033, 2012. 394

32. Eleni Triantafillou, Tyler Zhu, Vincent Dumoulin, Pascal Lamblin, Utku Evci, Kelvin Xu, Ross Goroshin, Carles Gelada, Kevin Swersky, Pierre-Antoine Manzagol, and Hugo Larochelle. Meta-dataset: A dataset of datasets for learning to learn from few examples. In *International Conference on Learning Representations*, 2020. 394

33. Oriol Vinyals, Charles Blundell, Timothy Lillicrap, Daan Wierstra, et al. Matching networks for one shot learning. In *Advances in Neural Information Processing Systems*, pages 3630–3638, 2016. 394

34. Oriol Vinyals, Timo Ewalds, Sergey Bartunov, Petko Georgiev, Alexander Sasha Vezhnevets, Michelle Yeo, Alireza Makhzani, Heinrich Küttler, John P. Agapiou, Julian Schrittwieser, John Quan, Stephen Gaffney, Stig Petersen, Karen Simonyan, Tom Schaul, Hado van Hasselt, David Silver, Timothy P. Lillicrap, Kevin Calderone, Paul Keet, Anthony Brunasso, David Lawrence, Anders Ekermo, Jacob Repp, and Rodney Tsing. StarCraft II: A new challenge for reinforcement learning. *arXiv:1708.04782*, 2017. 394

35. Jane X. Wang, Michael King, Nicolas Porcel, Zeb Kurth-Nelson, Tina Zhu, Charlie Deck, Peter Choy, Mary Cassin, Malcolm Reynolds, H. Francis Song, Gavin Buttimore, David P. Reichert, Neil C. Rabinowitz, Loic Matthey, Demis Hassabis, Alexander Lerchner, and Matthew Botvinick. Alchemy: A structured task distribution for meta-reinforcement learning. *arXiv:2102.02926*, 2021. 394

36. Tianhe Yu, Deirdre Quillen, Zhanpeng He, Ryan Julian, Karol Hausman, Chelsea Finn, and Sergey Levine. Meta-world: A benchmark and evaluation for multi-task and meta reinforcement learning. In *Conference on Robot Learning*, pages 1094–1100. PMLR, 2020. 394
37. Martin Zinkevich, Michael Johanson, Michael Bowling, and Carmelo Piccione. Regret minimization in games with incomplete information. In *Advances in Neural Information Processing Systems*, pages 1729–1736, 2008. 395

Glossary

A action in a state. 34
A advantage function in actor critic. 116
C_p exploration/exploitation constant in MCTS; high is more exploration. 194
D dataset. 358
I_ω initiation set for option ω. 271
Q state-action value. 42
R reward. 37
S state. 33
T transition. 35
V state value. 41
Ω set of options ω (hierarchical reinforcement learning). 272
α learning rate. 51
$\beta_\omega(s)$ termination condition for option ω at state s. 271
ϵ-**greedy**   exploration/exploitation rule that selects an ϵ fraction of random exploration actions. 53
γ discount rate, to reduce the importance of future rewards. 37
$\mathcal{L}$ loss function. 367
$\mathcal{T}_i$ base-learning task $\mathcal{T}_i = (D_{i,train}, \mathcal{L}_i)$, part of a meta-learning task. 303
ω option (hierarchical reinforcement learning). 271
ω hyperparameters (meta-learning). 303
ϕ parameters for the value network in actor critic (as opposed to θ, the policy parameters). 114
π policy. 37
τ trajectory, trace, episode, sequence. 39
θ parameters, weights in the neural network. 78, 114, 359

A2C   advantage actor critic. 117
A3C   asynchronous advantage actor critic. 117

accuracy   the total number of true positives and negatives divided by the total number of predictions. 359

ACO   ant colony optimization. 244

ALE   atari learning environment. 75

BERT   bidirectional encoder representations from transformers. 311

bootstrapping   old estimates of a value are refined with new updates. 50

CFR   counterfactual regret minimization. 234

CPU   central processing unit. 385

D4PG   distributional distributed deep deterministic policy gradient. 128

DDDQN   dueling double deep Q-network. 91

DDPG   deep deterministic policy gradient. 123

DDQN   double deep Q-network. 89

deep learning   training a deep neural network to approximate a function, used for high-dimensional problems. 2

deep reinforcement learning   approximating value, policy, and transition functions with a deep neural network. 3

deep supervised learning   approximating a function with a deep neural network; often for regression of image classification. 357

DQN   deep Q-network. 84

entropy   measure of the amount of uncertainty in a distribution. 351

exploitation   selecting actions as suggested by the current best policy $\pi(s)$. 53

exploration   selecting other actions than those that the policy $\pi(s)$ suggests. 52

few-shot learning   task with which meta learning is often evaluated, to see how well the meta-learner can learn with only a few training examples. 301

finetuning   training the pre-trained network on the new dataset. 296

function approximation   approximation of a mathematical function, a main goal of machine learning, often performed by deep learning . 13

GAN   generative adversarial network. 377

GPT-3   generative pretrained transformer 3. 311

GPU   graphical processing unit. 383

hyperparameters   determine the behavior of a learning algorithm; base-learning learns parameters θ, meta-learning learns hyperparameters ω. 315

LSTM   long short-term memory. 374

machine learning   learning a function or model from data. 13
MADDPG   multi-agent DDPG. 239
MAML   model-agnostic meta-learning. 305
Markov decision process   stochastic decision process that has the Markov (no-memory) property: the next state depends only on the current state and the action. 32
meta-learning   learning to learn hyperparameters; use a sequence of related tasks to learn a new task quicker. 296
MuJoCo   multi-joint dynamics with contact. 106

optimization   find an optimal element in a space; used in many aspects in machine learning. 8
overfitting   high-capacity models can overtrain, where they model the signal and the noise, instead of just the signal. 363

parameters   the parameters θ (weights of a neural network) connect the neurons, together they determine the functional relation between input and output. 359
PBT   population-based training. 245
PETS   probabilistic ensemble with trajectory sampling. 147
PEX   prioritized experience replay. 91
PILCO   probabilistic inference for learning control. 146
PPO   proximal policy optimization. 120
pretraining   parameter transfer of the old task to the new task. 296

REINFORCE   REward Increment = Non-negative Factor × Offset Reinforcement × Characteristic Eligibility. 112
reinforcement learning   agent learns a policy for a sequential decision problem from environment feedback on its actions. 3

SAC   soft actor critic. 121
SARSA   state action reward state action. 55
sequential decision problem   problem consisting of a sequence of decisions. 4
supervised learning   training a predictive model on a labeled dataset. 3

TD   temporal difference. 50
TPU   tensor processing unit. 204

transfer learning   using part of a network (pretraining) to speed up learning (fine-tuning) on a new dataset. 296

TRPO   trust region policy optimization. 119

unsupervised learning   clustering elements in an unlabeled dataset based on an inherent metric. 16

VI   value iteration. 44
VIN   value iteration network. 153
VPN   value prediction network. 148

XAI   explainable artificial intelligence. 334

zero-shot learning   an example has to be recognized as belonging to a class without ever having been trained on an example of this class. 309
ZSL   zero-shot learning. 309

Index

Printed in the United States
by Baker & Taylor Publisher Services